T0327533

MASS SPECTROMETRY IN GRAPE AND WINE CHEMISTRY

WILEY-INTERSCIENCE SERIES IN MASS SPECTROMETRY

Series Editors

Dominic M. Desiderio
Departments of Neurology and Biochemistry
University of Tennessee Health Science Center

Nico M. M. Nibbering
Vrije Universiteit Amsterdam, The Netherlands

A complete list of the titles in this series appears at the end of this volume.

MASS SPECTROMETRY IN GRAPE AND WINE CHEMISTRY

RICCARDO FLAMINI

CRA, Centro di Ricerca per la Viticoltura, Conegliano (TV), Italy

PIETRO TRALDI

CNR, Istituto di Scienze e Tecnologie Molecolari, Padova, Italy

A JOHN WILEY & SONS, INC., PUBLICATION

Published by John Wiley & Sons, Inc., Hoboken, New Jersey
Published simultaneously in Canada

For general information on our other products and services or for technical support, please
contact our Customer Care Department within the United States at (800) 762-2974, outside the
United States at (317) 572-3993 or fax (317) 572-4002.

Wiley also publishes its books in a variety of electronic formats. Some content that appears in
print may not be available in electronic formats. For more information about Wiley products,
visit our web site at www.wiley.com.

Library of Congress Cataloging-in-Publication Data:
Flamini, Riccardo, 1968–
   Mass spectrometry in grape and wine chemistry / Riccardo Flamini, Pietro Traldi.
      p. cm.
   Includes bibliographical references and index.
   ISBN 978-0-470-39247-8 (cloth)
   1. Wine and wine making–Analysis. 2. Wine and wine making–Chemistry. 3. Mass
spectrometry. I. Traldi, Pietro. II. Title.
   TP548.5.A5T73 2010
   663'.200284–dc22

                                                                              2009019923

10 9 8 7 6 5 4 3 2 1

CONTENTS

PREFACE

Science is based on the transfer of knowledge on specific subjects. Only by comparison of results and experiences can some fixed points be defined. These points represent the foundation of further investigations. This finding is particularly true when the knowledge is found in different research areas: here the researcher interests operate in a collaborative effort, which leads to a feedback process between the two groups. Often it seems that the verb *to collaborate* has a different meaning from that given by Webster's dictionary; that is, *to work jointly with others especially in an intellectual endeavour*. This definition implies a transfer of knowledge between the collaborating groups in two directions, but what generally happens is that one direction is highly privileged. The right balance between the two arrows is due to the coscience of each partner of the efforts, difficulties, and views of the other: once this has been reached the collaboration becomes more complete and completes the professional relationship in a friendly manner.

This book was written by two friends, who are different in age, experiences, and knowledge, who started to collaborate many years ago on the application of mass spectrometry to the field of grape and wine chemistry. The availability of new mass spectrometric approaches and the desire to test their capabilities in the analysis of complex natural matrices, such as grape and wine, led the authors to undertake a series of research, projects, which give a more detailed view of the chemistry involved in these natural substrates.

In the last decades, the increased consumption of table grapes and wines has been encouraged by the amply demonstrated beneficial effects of these substances on human illness, such as cardiovascular diseases, brain degeneration, and certain carcinogenic diseases. Improving the quality of grapes is achieved by selecting the best clones and varieties, the use of more appropriate growing techniques, and taking into account the environmental effects on the vineyard. The quality of wines is increased by optimizing the wine-making processes, such as extraction of grape compounds, alcoholic fermentation, malolactic fermentation, and barrel– and bottle aging.

The legislation of the European Community (EC) and of single countries is devoted to protecting consumer health and internal markets from the sometimes harmful effects that may be caused by low-quality products. Legal limits are defined and quality certificates are often required (for pesticides, toxins, etc.).

In this framework, knowledge of the chemical composition of grapes and wines is essential. Mass spectrometry (MS) is proving to be the most powerful tool with which to achieve this result: This book presents the match between the high structural identification power of MS techniques and the variegated chemistry of grape and wine.

The volume is divided into two parts: Part I (Chapters 1–3) gives a general view of the mass spectrometric methods usually employed in the field of interest; Part II (Chapters 4–10) is divided into seven chapters by subject and describes the grape and wine chemistry, as well as both the traditional and more recent applications of MS.

This book was perceived as both an up-to-date source for students beginning work in the field of oenological (and in general of foods) analytical chemistry, and as support to Research and Quality Control Laboratories.

RICCARDO FLAMINI and PIETRO TRALDI

Padua, Italy

ACKNOWLEDGMENTS

We want to thank all the contributors whose work made this volume possible: Laura Molin and Luca Raveane for their kind support in preparing the first part of the book, Mirko De Rosso, Annarita Panighel, and Antonio Dalla Vedova for their support in organizing the second section, and Professor Rocco Di Stefano of Palermo University and Professor Paolo Cabras of Cagliari University for the kindly provided data.

Lastly, we want to thank John Wiley & Sons, Inc. for giving us the opportunity to prepare this book.

INTRODUCTION

Nowadays, mass spectrometry (MS) strongly interacts with most chemical research areas, from studies of gas-phase reactivity of ions of interest to biomedical investigations. This finding is the result of the many efforts from different research groups around the world, working to develop instrumental arrangements suitable for specific analytical and fundamental studies.

Half a century ago mass spectrometers were considered (and really they were!) very complex and expensive instruments, requiring well-experienced personnel for their management. Then, "mass spectrometry labs" were present at the departmental level and were used mainly by research groups operating in the field of organic chemistry. Now the situation is completely different: The (relatively) low prices and ease of instrumental management has moved mass spectrometers from dedicated labs to the utilizer environment, resulting in a capillary diffusion of medium high-performances instruments.

Surely, this is the result on one hand to the development of ionization methods alternative to electron ionization, able to generate ions from highly polar, high-mass molecules and are easily coupled with chromatographic systems. This aspect has been well recognized by the entire scientific community with the assignment in 2002 of the Nobel

Mass Spectrometry in Grape and Wine Chemistry, by Riccardo Flamini
and Pietro Traldi
Copyright © 2010 John Wiley & Sons, Inc.

Prize for Chemistry to John B. Fenn and Koichi Tanaka for *"their development of soft ionization methods for mass spectrometric analyses of biological macromolecules"*.

On the other hand, the development of compact mass analyzers, which are easy to use and fully controlled by data systems, led to mass spectrometers no longer covering an area of some square meters (as the early magnetic sector-based ones), but bench top machines, whose dimensions are sometimes smaller than those of the chromatographic devices with which they are coupled. Most of these instruments are based on the interaction of ions with quadrupolar electrical fields and were developed by the W. Paul (Nobel Prize for Physics, 1989) group at Bonn University.

These developments [together with the availability of high-performance instruments, e.g., Fourier transform–mass spectrometry (FT–MS) and Orbitrap] make possible the application of MS in many different fields. The problem is to individualize the best instrumental choices and the related parameterization to obtain the analytically more valid results, which allows to propose new, highly specific analytical methods.

As complex as the analytical substrate of interest might be, specificity plays a fundamental role. This is the case for grape and wine, highly complex natural substrates, for which the use of different mass spectrometric techniques allowed to obtain a clear (but still not complete!) view of the chemical pathways present in them.

Viticulture and oenology play an important role in the economy of many countries, and considerable efforts are devoted to improve the quality of products and to match the broadest demands of the market. Many industrial processes are finalized to obtain products with peculiar characteristics: the inoculum of selected yeast permits a regular alcoholic fermentation with minimum secondary processes by other microorganisms, which favor formation of positive sensory compounds and limit the negative ones; extraction of grape components is enhanced by maceration of grape skins in controlled conditions during fermentation and addition of specific enzymes; malolactic fermentation to improve organoleptic characteristics and to add biological stability to the wine; barrel- and bottle-aging refines the final product (Flamini, 2003). European Community (EC) laws, as well as those of a single country, are devoted to protecting consumer health, rather than the market, from the introduction of low-quality products. This goal is achieved by accurate foods controls. Consequently, quality certificates are often required, for exporting wine and enological products. Of particular concern are the presence of pesticides, heavy metals, ethyl carbamate, and toxins, for which legal limits are often defined. To prevent frauds and to confirm product identity, accordance between the real-product characteristics and the producer declarations (e.g., variety, geographic origin, quality, vintage), has to be verified. Researchers and control organism activities are devoted to developing new analytical methods. These methods are applied to verify the product origin (Ogrinc et al., 2001), to detect illegal additions and adulteration (sugar-beet, cane sugar or ethanol addition, watering) (Guillou et al., 2001), to protect the consumer health by determination of contaminants (Szpunar et al., 1998; MacDonald et al., 1999; Wong and Halverson, 1999).

On the other hand, to expand the worldwide market considerable efforts of the main wine producing countries are devoted to improve the image of products. Consequently, the product characteristics and origin have to be well defined. Research in viticulture and oenology tries to enhance the typical characteristics of grape varieties by selection of best clones, and to identify the more suitable parameters for product characterization (Di Stefano, 1996; Flamini et al., 2001). For the variety characterization, several parameters of plant and grape, such as deoxyribonucleic acid (DNA), amphelography, isoenzymes, and chemical compounds of grape, are studied (Costacurta et al., 2001). To define characteristics and identify products, secondary metabolites of grape and wine (compounds mainly linked to a specific variety, but not indispensable for the plant survivor, also if environmental and climatic variables can influence their contents in the fruit) are studied (Di Stefano, 1996). These compounds are included in the chemical classes of terpenes and terpenols, methoxy-pyrazines, volatile sulfur compounds, benzenoids, nor-isoprenoids, and polyphenols (e.g., flavanols, flavonols, anthocyanins, procyanidins, and tannins). Volatile compounds and polyphenols are transferred from the grape to the wine in winemaking conferring fragrance, taste, and color to the products.

The first structural studies by gas chromatography–mass spectrometry–electron impact (GC/MS–EI) of grape and wine compounds were performed in the early 1980s. A number of new volatile wine compounds formed by yeasts during alcoholic fermentation, and aroma compounds from grapes, were identified (Rapp and Knipser, 1979; Rapp et al., 1980; 1983; 1984; 1986; Williams et al., 1980; 1981; 1982; Shoseyov et al., 1990; Versini et al., 1991; Strauss et al., 1986; 1987a; 1987b; Winterhalter et al., 1990; Winterhalter, 1991; Humpf et al., 1991). It was confirmed that grape varieties with an evident floral aroma were classified as "aromatic varieties" (e.g., Muscats, Malvasie, Riesling, Müller-Thurgau, and Gewürztraminer) and are characterized by their high monoterpenol contents. These characteristics increase during the final stages of ripening (Di Stefano, 1996), and during fermentation. Wine aging chemical transformations involving these compounds lead to formation of new monoterpenols (Williams et al., 1980; Di Stefano, 1989; Di Stefano et al., 1992). It was found also that several norisoprenoid compounds are important in the aroma formation of grapes and wines (Strauss et al., 1986; 1987a; 1987b; Winterhalter et al., 1990; Winterhalter, 1991; Humpf et al., 1991).

In the 1990s, studies of the Sauvignon grapes and wines revealed that several sulfur compounds and methoxypyrazines (grassy note) are typical aroma compounds of these varieties (Harris et al., 1987; Lacey et al., 1991; Allen et al., 1994; 1995; Tominaga et al., 1996; Bouchilloux et al., 1998).

Mass spectrometry is also applied in the control of pesticides and other contaminants (e.g., 2,4,6-trichloroanisole), detection of compounds formed by yeast and bacteria, determination of illegal additions to the wine. Liquid chromatography/mass spectroscopy (LC/MS) methods for determination of toxins in the wine (e.g., ochratoxin A) have been proposed (Zöllner et al., 2000; Flamini and Panighel, 2006; Flamini et al., 2007).

Currently, LC/MS and multiple mass spectrometry (MS/MS) have been used to study the grape polyphenols (anthocyanins, flavonols, tannins and proanthocyanidins, hydroxycinnamic, and hydroxycinnamoyltartaric acids), which allow to structurally characterize and understand the mechanisms involved in stabilizing the color in wines (Flamini, 2003).

To be able to estimate the potential of the grape and how it may be transferred to the wine, a good knowledge of enological chemistry is essential. In this framework, the MS played, and, by the new technologies introduced in the recent years, plays a fundamental role.

REFERENCES

Allen, M.S., Lacey, M.J., and Boyd, S.J. (1994). Determination of methoxy-pyrazines in red wines by stable isotope diluition gas chromatography-mass spectrometry, *J. Agric. Food. Chem.*, **42**(8), 1734–1738.

Allen, M.S., Lacey, M.J., and Boyd, S.J. (1995). Methoxypyrazines in red wines: occurrence of 2-methoxy-3-(1-methylethyl) pyrazine, *J. Agric. Food. Chem.*, **43**(3), 769–772.

Bouchilloux, P., Darriet, P., and Dubourdieu, D. (1998). Identification of a very odoriferous thiol, 2 methyl-3-furanthiol, in wines, *Vitis*, **37**(4), 177–180.

Costacurta, A., Calò, A., Crespan, M., Milani, M., Carraro, R., Aggio, L., Flamini, R., and Ajmone-Marsan, P. (2001). Morphological, aromatic and molecular characteristics of Moscato vine varieties and research on phylogenetic relations, *Bull. O.I.V.*, **841–842**, 133–150.

Di Stefano, R., Maggiorotto, G., and Gianotti, S. (1992). Trasformazioni di nerolo e geraniolo indotte dai lieviti, *Riv. Vitic. Enol.*, **1**, 43–49.

Di Stefano, R. (1989). Evoluzione dei composti terpenici liberi e glucosidici e degli actinidioli durante la conservazione dei mosti e dei vini in funzione del pH, *Riv. Vitic. Enol.*, **2**, 11–23.

Di Stefano, R. (1996). Metodi chimici nella caratterizzazione varietale, *Riv. Vitic. Enol.*, **1**, 51–56.

Flamini, R., Dalla Vedova, A., and Calò, A. (2001). Study on the monoterpene contents of 23 accessions of Muscat grape: correlation between aroma profile and variety, *Riv. Vitic. Enol.*, **2**(3), 35–49.

Flamini, R. (2003). Mass spectrometry in grape and wine chemistry. Part I: Polyphenols, *Mass Spec. Rev.*, **22**(4), 218–250.

Flamini, R. and Panighel, A. (2006). Mass spectrometry in grape and wine chemistry. Part II: The Consumer Protection, *Mass Spectrosc. Rev.*, **25**(5), 741–774.

Flamini, R., Dalla Vedova, A., De Rosso, M., and Panighel, A. (2007). A new sensitive and selective method for analysis of ochratoxin A in grape and wine by direct liquid chromatography/surface activated chemical ionization-tandem mass spectrometry, *Rapid Commun. Mass Spectrom.*, **21**, 3737–3742.

Guillou, C., Jamin, E., Martin, G.J., Reniero, F., Wittkowski, R., and Wood, R. (2001). Isotopic analyses of wine and of products derived from grape, *Bull. O.I.V.*, **839–840**, 27–36.

Harris, R.L.N., Lacey, M.J., Brown, W.V., and Allen, M.S. (1987). Determination of 2-methoxy-3-alkylpyrazines in wine by gas chromatography/mass spectrometry, *Vitis*, **26**, 201–207.

Humpf, H.U., Winterhalter, P., and Schreier, P. (1991). 3,4-Dihydroxy-7,8-dihydro-β-ionone β-D-glucopyranoside: natural precursor of 2,2,6,8-tetramethyl-7,11-dioxatricyclo[6.2.1.0^{1,6}]undec-4-ene (Riesling acetal) and 1,1,6-trimethyl-1,2-dihydronaphthalene in Red currant (*Ribes Rubrum* L) leaves, *J. Agric. Food. Chem.*, **39**, 1833–1835.

Lacey, M.J., Allen, M.S., Harris, R.L.N., and Brown, W.V. (1991). Methoxypyrazines in Sauvignon blanc grapes and wines, *Am. J. Enol. Vitic.*, **42**(2), 103–108.

MacDonald, S., Wilson, P., Barnes, K., Damant, A., Massey, R., Mortby, E., and Shepherd, M.J. (1999). Ochratoxin A in dried vine fruit: method development and survey, *Food Addit. Contam.*, **16**(6), 253–260.

Ogrinc, N., Košir, I.J., Kocjančič, M., and Kidrič, J. (2001). Determination of authenticy, regional origin, and vintage of Slovenian wines using a combination of IRMS and SNIF–NMR analyses, *J. Agric. Food. Chem.*, **49**(3), 1432–1440.

Rapp, A., Knipser, W., and Engel, L. (1980). Identification of 3,7-dimethyl-octa-1,7-dien-3,6-diol in grape and wine aroma of Muscat varieties, *Vitis*, **19**, 226–229.

Rapp, A. and Knipser, W. (1979). 3,7-Dimethyl-octa-1,5-dien-3,7-diol- a new terpenoid component of grape and wine aroma, *Vitis*, **18**, 229–233.

Rapp, A., Mandery, H., and Niebergall, H. (1986). New monoterpenediols in grape must and wine and in cultures of *Botrytis cinerea*, *Vitis*, **25**, 79–84.

Rapp, A., Mandery, H., and Ullemeyer, H. (1983). 3,7-Dimethyl-1,7-octandiol- a new terpene compound of the grape and wine volatiles, *Vitis*, **22**, 225–230.

Rapp, A., Mandery, H., and Ullemeyer, H. (1984). New monoterpenoic alcohols in grape must and wine and their significance for the biogenesis of some cyclic monoterpene ethers, *Vitis*, **23**, 84–92.

Shoseyov, O., Bravdo, B.A., Siegel, D., Goldman, A., Cohen, S., and Ikan, R. (1990). *Iso*-geraniol (3,7-dimethyl-3,6-octadien-1-ol): A novel monoterpene in *Vitis vinifera* L. cv. Muscat Roy, *Vitis*, **29**, 159–163.

Strauss, C.R., Dimitriadis, E., Wilson, B., and Williams, P.J. (1986). Studies on the hydrolysis of two megastigma-3,6,9-triols rationalizing the origins of some volatile C_{13} norisoprenoids of *Vitis vinifera* grapes, *J. Agric. Food. Chem.*, **34**, 145–149.

Strauss, C.R., Gooley, P.R., Wilson, B., and Williams, P.J. (1987a). Application of droplet countercurrent chromatography to the analysis of conjugated forms of terpenoids, phenols, and other constituents of grape juice, *J. Agric. Food. Chem.*, **35**, 519–524.

Strauss, C.R., Wilson, B., and Williams, P.J. (1987b) 3-Oxo-α-ionol, vomifoliol and roseoside in *Vitis vinifera* fruit, *Phytochem.*, **26**(7), 1995–1997.

Szpunar, J., Pellerin, P., Makarov, A., Doco, T., Williams, P., Medina, B., and Łobiński, R. (1998). Speciation analysis for biomolecular complexes of lead in wine by size-exclusion high-performance liquid chromatography-inductively coupled plasma mass spectrometry, *J. Anal. At. Spectrom.*, **13**, 749–754.

Tominaga, T., Darriet, P., and Dubourdieu, D. (1996). Identification of 3-mercaptohexyl acetate in Sauvignon wine, a powerful aromatic compound exhibiting box-tree odor, *Vitis*, **35**(4), 207–210.

Versini, G., Rapp, A., Reniero, F., and Mandery, H. (1991). Structural identification and presence of some p-menth-1-enediols in grape products, *Vitis*, **30**, 143–149.

Williams, P.J., Strauss, C.R., Wilson, B., and Massy-Westropp, R.A. (1982). Use of C_{18} reversed-phase liquid chromatography for the isolation of monoterpene glycosides and nor-isoprenoid precursors from grape juice and wines, *J. Chromatogr.*, **235**, 471–480.

Williams, P.J., Strauss, C.R., and Wilson, B. (1980). Hydroxylated linalool derivatives as precursors of volatile monoterpenes of Muscat grapes, *J. Agric. Food. Chem.*, **28**, 766–771.

Williams, P.J., Strauss, C.R., and Wilson, B. (1981). Classification of the monoterpenoid composition of Muscat grapes, *Am. J. Enol. Vitic.*, **32**(3), 230–235.

Winterhalter, P., Sefton, M.A., and Williams, P.J. (1990). Two-dimensional GC-DCCC analysis of the glycoconjugates of monoterpenes, norisoprenoids, and shikimate-derived metabolites from Riesling wine, *J. Agric. Food. Chem.*, **38**, 1041–1048.

Winterhalter, P. (1991). 1,1,6-Trimethyl-1,2-dihydronaphthalene (TDN) formation in wine. 1. Studies on the hydrolysis of 2,6,10,10-tetramethyl-1-oxaspiro[4.5]dec-6-ene-2,8-diol rationalizing the origin of TDN and related C_{13} norisoprenoids in Riesling wine, *J. Agric. Food. Chem.*, **39**, 1825–1829.

Wong, J.W. and Halverson, C.A. (1999). Multiresidue analysis of pesticides in wines using C-18 solid-phase extraction and gas chromatography-mass spectrometry, *Am. J. Enol. Vitic.*, **50**(4), 435–442.

Zöllner, P., Leitner, A., Luboki, D., Cabrera, K., and Lindner, W. (2000). Application of a Chromolith SpeedROD RP-18e HPLC column: Determination of ochratoxin A in different wines by high-performance liquid chromatography-tandem mass spectrometry, *Chromatographia*, **52** (11/12), 818–820.

PART I

MASS SPECTROMETRY

1

IONIZATION METHODS

Electron ionization (EI) is surely the ionization method most widely employed (Mark and Dunn, 1985). This method was proposed and used from the early days of mass spectrometry (MS) applications in the chemical world and is still of wide interest. This interest is due to the presence of libraries of EI mass spectra, which allows easy identification of unknown previously studied analytes. The EI method suffers from two limitations: It is based on the gas-phase interactions between the neutral molecules of the analyte and an electron beam of mean energy 70 eV. This interaction leads to the deposition of internal energy in the molecules of the analyte, which is reflected in the production of odd-electron molecular ($[M]^{+\bullet}$) and fragment ions. These ions are highly diagnostic from a structural point of view.

Then, the first limitation of EI is related to sample vaporization, usually obtained by heating the sample under vacuum conditions (10^{-5}–10^{-6} Torr) present in the ion source. Unfortunately, for many classes of compounds the intermolecular bonds (usually through hydrogen bridges) are stronger than the intramolecular ones and the result of heating is the pyrolysis of the analyte. The EI spectrum so obtained is not that of analyte, but that of its pyrolysis products. As examples of

Mass Spectrometry in Grape and Wine Chemistry, by Riccardo Flamini and Pietro Traldi
Copyright © 2010 John Wiley & Sons, Inc.

this behavior one can consider saccarides, peptides, and generally all highly polar compounds.

The second limitation of EI is related to internal energy deposition. For many classes of compounds it is too high, leading to extensive fragmentation of the molecule and to the absence of a molecular ion, generally considered the most important information received from a mass spectrometric measurement.

However, EI can be, and is, successfully employed in the analysis of volatile compounds and is mainly employed linked to gas chromatographic methods (GC/MS). This approach has been extensively used in the field of grape and wine chemistry, allowing to obtain valid results on low molecular weight, low-medium polarity compounds, as described in Part II.

To overcome the second limitation described above, in the 1960s a new ionization method was proposed, based not on a physical interaction, but on gas-phase reactions of the analyte with acid or basic ions present in excess inside an ion source, and operating at a pressure in the order of 10^{-1}–10^{-2} Torr. This method is usually called chemical ionization (CI) (Harrison, 1983).

Generally, analyte protonation reactions are more widely employed. The occurrence of such reactions is related to the proton affinity (PA) of M and that of the reactant gas. The internal energy of the obtained species are related to the difference between these proton affinities. Thus, as an example, consider an experiment performed on an organic molecule M with a PA value of 180 kcal/mol (PA_M), it can be protonated by reaction with CH_5^+ ($PA_{CH_4} = 127$ kcal/mol) H_3O^+ ($PA_{H_2O} = 165$ kcal/mol), but not with NH_4^+ ($PA_{NH_3} = 205$ kcal/mol). This example shows an important point about CI: It can be effectively employed to select species of interest in complex matrices. In other words, by a suitable selection of a reacting ion $[AH]^+$ one could produce $[MH]^+$ species of molecules with a PA higher than that of A. Furthermore, the extension of fragmentation can be modified in terms of the difference of $[PA_M - PA_A]$.

From an operative point of view, CI is simply obtained by introducing the neutral reactant species inside an EI ion source in a "close" configuration, by which quite high reactant pressure can be obtained. If the operative conditions are properly set, the formation of the abundant $[AH]^+$ species (or, in the case of negative ions, B^- species) is observed in high yield. Of course, particular attention must be paid to the case of quantitative analysis that carefully reproduces these experimental conditions, because they reflect substantially on the values of the limit of detection (LOD).

The CI, as well as EI, requires the presence of samples in the vapor phase and consequently it cannot be applied to nonvolatile analytes. Efforts have been made from the 1960s to develop ionization methods overcoming this aspect. Among them, field desorption (FD) (Beckey, 1975) and fast-atom bombardment (FAB) (Barber et al., 1982) resulted in highly effective methods and opened new applications for MS. More recently, new techniques have become available and are currently employed for nonvolatile samples: atmospheric pressure chemical ionization (APCI) (Bruins, 1991), electrospray ionization (ESI) (Yamashita and Fenn, 1984a), atmospheric pressure photoionization (APPI) (Robb et al., 2000), and matrix-assisted laser desorption–ionization (MALDI) (Karas et al., 1991) now represent the most used techniques for the analysis of high molecular weight, high-polarity samples.

Considering the wide, positive impact that these techniques had with the grape and wine chemistry in past years (as will be described in detail in Part II) we focus now on the in-depth description of these new methods, in order to give the reader a useful background for critical evaluation of results obtained with their use.

1.1   ELECTROSPRAY IONIZATION

Electrospray is based on droplet production in the presence of strong electrical fields. The first electrospray experiments were performed by Jean-Antoine Nollet (physicist and Abbé), who in 1750 observed that water flowing from a small hole of an electrified metal container aerosolized when placed near the electrical ground. At that time, physics, chemistry, physiology, and medicine were very often seen as a unique science and some experiments were performed at the physiological level. Abbé Nollet observed that "a person, electrified by connection to a high-voltage generator (hopefully, well insulated from the ground!—authors' note) would not bleed normally if he was to cut himself; blood sprays from the wound" (ORNL Review, 1995).

About one century later, Lord Kelvin studied the charging between water dripping from two different liquid nozzles, which leads to electrospray phenomena at the nozzles themselves (Smith, 2000). In the last century, a series of systematic studies on electrospray were carried out by Zeleny (Zeleny, 1917) and Taylor (Taylor, 1964a and b) allowing a detailed description of the phenomenon. In the middle of the century, electrospray started to be used on the industrial scale, in the application of paints and coatings to metal surfaces. The fine spray results in very smooth even films, with the paint actually attracted to the metal. Miniaturized versions of electrospray are even finding their way into the next generation of microsatellites: The electrostatic plume makes an efficient, although very low power, ion propulsion engine.

Electrospray became of analytical interest in 1968, when Dole and co-workers produced gas-phase, high molecular weight polystyrene ions by electrospraying a benzene–acetone solution of the polymer (Dole et al., 1968). Quite strangely, these results did not lead to further applications until 1984, when the studies of Yamashita and Fenn (Yamashita and Fenn, 1984b) brought electrospray to the analytical world and from which electrospray applications have grown fantastically.

This technique can be considered the ionization method that the entire scientific community was waiting for. This method is an effective and valid approach for the direct study of analytes present in solution, without the need of analyte vaporization and, consequently, for an easy coupling with LC methods. Furthermore, in the same time period methods able to give information on large biomolecules was growing. In this framework, the behavior in electrospray conditions of proteins and peptides (as well as oligonucleotides), reflecting on the production of multiply charged ions, makes this ionization method essential in biomedical studies and in proteome investigations. For this last reason, the Nobel Price in 2002 was assigned to Fenn, with the official sentence "for the development of soft desorption ionization methods for mass spectrometric analyses of biological macromolecules".

This chapter aims to offer a concise description of chemical–physical phenomena that are at the base of the ESI.

1.1.1   The Taylor Cone

The instrumental setup for ESI experiments is schematized in Fig. 1.1. The solution is injected into a stainless steel capillary (10^{-4} m o.d.). A voltage on the order of kilovolts is applied between this capillary and a counterelectrode, which is placed a few-tenths of a millimeter away from it. In general, as the liquid begins to exit from the needle, it charges up and assumes a conical shape, referred to as the Taylor cone, in honor of Taylor who first described the phenomenon in 1964. The liquid assumes this shape because when charged up, a conic shape can hold more charge than a sphere. The formation of this cone-shaped structure can be justified by the presence of charged species inside the solution that experiment with the effect of the electrostatic field existing between the capillary and the counterelectrode. What is the origin of this charged species, in the absence of ionic solute? It emphasises that even in the absence of ionic analytes, protic solvents produce ionic

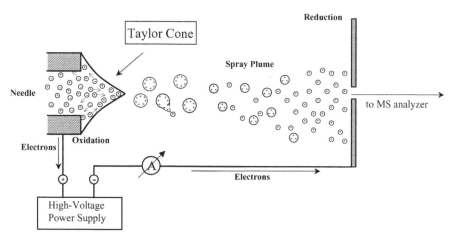

Figure 1.1. Schematic of an electrospray source showing the production of charged droplets from the Taylor cone.

species, due to their dissociation. Thus, for example, taking into account K_w at $20\,^{\circ}C$ is $10^{-14.16}$, the H_3O^+ concentration at $20\,^{\circ}C$ is in the order of $8.3 \times 10^{-8}\,M$. Analogously, $K_a(CH_3OH) = 10^{-15.5}$. Consequently, the solvents usually employed for electrospray experiments already produce ions in solution, which can be considered responsible for the cone formation. Of course, the presence of dopant analytes (e.g., acids), as well as traces of inorganic salts, strongly enhance this phenomenon.

If the applied electrical field is high enough, the formation of charged droplets from the cone apex is observed which, due to their charge, further migrate through the atmosphere to the counterelectrode. Experimental data have shown that the droplet formation is strongly influenced by

- Solvent chemical–physical characteristics (viscosity, surface tension, pK_a).
- Concentration and chemical nature of ionic analytes.
- Concentration and chemical nature of inorganic salts.
- Voltage applied between capillary and counterelectrode.

In the case of positive-ion analysis, the capillary is usually placed at a positive voltage while the counterelectrode is placed at a negative voltage (this is the case shown in Fig. 1.1). The reverse is used in the case of negative-ion analysis. In both cases, a high number of positive (or negative) charges are present on the droplet surface.

The formation of the Taylor cone and the subsequent charged droplet generation can be enhanced by the use of a coaxial nitrogen gas stream. This instrumental setup is usually employed in the commercially available electrospray sources: Then the formation of charged droplets is due to either electrical and pneumatic forces.

Blades et al. (Blades et al., 1991) showed that the electrospray mechanism consists of the early separation of positive from negative electrolyte ions present in solution. This phenomenon requires a charge balance with conversion of ions to electrons occurring at the metal–liquid interface of the ESI capillary, in the case of positive-ion analysis. The processes that lead to a deep change of composition of ions in the spray solution are those occurring at the metal–liquid capillary interface and the related oxidation reactions were studied by the use of a Zn capillary tip. This experiment was carried out by using three different capillary structures. The passivated stainless steel capillary normally employed was substituted with one where the tip was made by Zn having a very low reaction potential ($Zn_{(s)} \rightarrow Zn^{2+} + 2e$, $E^0_{red} = -0.76\,V$). Actually, under these conditions abundant Zn^{2+} ions were detected in the mass spectrum simply by spraying methanol (CH_3OH) at a flow rate of $20\,\mu L/min$. This result suggests that the oxidation reaction took place at the zinc–liquid capillary interface. In order to be confident of this hypothesis, and to exclude other possible origins of Zn^{2+} production, a further experiment was carried out. This experiment consisted of placing a Zn capillary before the electrospray capillary line and keeping it electrically insulated. In this case, Zn^{2+} ions were not detected. These results provide qualitative and quantitative evidence that in the case of positive-ion instrumental setup (Fig. 1.1), electrochemical oxidation takes place at the liquid–metal interface of the electrospray capillary tip.

Now, ESI can be considered as an electrolysis cell and the ion transport takes place in the liquid, not the gas phase. The oxidation reaction yield depends on the electrical potential applied to the capillary, as well as on the electrochemical oxidation potentials from the different possible reactions. Kinetic factors can exhibit only minor effects, considering the low current involved.

The effect of oxidation reactions at the capillary tip will be the production of an excess of positive ions, together with the production of an electron current flowing through the metal (see Fig. 1.1). An excess of positive ions could be the result of two different phenomena; that is, the production of positive ions themselves or the removal of negative ions from the solution.

In the case of a negative-ion source setup (spray capillary placed at negative voltage), reduction reactions usually take place with the formation of deprotonated species.

The electrical current due to the droplets motion can be measured easily by the amperometer (A) shown in Fig. 1.1. This measurement allows to estimate, from a quantitative point of view, the total number of elementary charges leaving the capillary and which, theoretically, may correspond to gas-phase ions. The droplet current I, the droplet radii R, and charge q were originally calculated by Pfeifer and Hendricks (Pfeifer and Hendricks, 1968):

$$I = \left[\left(\frac{4\pi}{\varepsilon} \right)^3 (9\gamma)^2 \varepsilon_0^5 \right]^{1/7} (KE)^{3/7} (V_f)^{4/7} \tag{1.1}$$

$$R = \left(\frac{3\varepsilon\gamma^{1/2}V_f}{4\pi\varepsilon_0^{1/2} KE} \right)^{2/7} \tag{1.2}$$

$$q = 0.5 \left[8(\varepsilon_0\gamma R^3)^{1/2} \right] \tag{1.3}$$

where γ is the surface tension of the solvent; K is the conductivity of the infused solution; E is the electrical field; ε is the dielectric constant of the solvent; ε_0 is the dielectric constant of the vacuum; and V_f is the flow rate.

De La Mora and Locertales (De La Mora and Locertales, 1994) found, based on both theoretical calculation and experimental data, the following equations for the same quantities:

$$I = f\left(\frac{\varepsilon}{\varepsilon_0} \right)\left(\gamma K V_f \frac{\varepsilon}{\varepsilon_0} \right)^{1/2} \tag{1.4}$$

$$R \approx \left(\frac{V_f \varepsilon}{K} \right)^{1/3} \tag{1.5}$$

$$q = 0.7 \left[8\pi(\varepsilon_0\gamma R^3)^{1/2} \right] \tag{1.6}$$

where $f(\varepsilon/\varepsilon_0)$ is a function of the $\varepsilon/\varepsilon_0$ ratio.

Equations 1.1 and 1.4 at first seem to be strongly different, but they indicate an analogous dependence of I from the two most relevant experimental parameters (i.e., the flow rate and the conductivity).

Equations 1.2 and 1.5 both show a decrease of the droplets dimension by increasing the solution conductivity. These relationships are particularly relevant because in solution, when different electrolytes are present, the conductivity K may be obtained as the sum of the conductivities due to the different species and is proportional to the ion concentration:

$$K = \sum_i \lambda_{0,m,i} C_i \qquad (1.7)$$

where $\lambda_{0,m,i}$ is the molar conductivity of the electrolyte i.

The charged droplets, generated by solution spraying, decrease their radius due to solvent evaporation, but their total charge amount remains constant. The first step is to determine the energy required for solvent evaporation that is due to environmental thermal energy. In a second step, this process is enhanced through further heating obtained by the use of a heated capillary or by collisions with heated gas molecules. The maintenance of the total charge during this evaporation phase can be explained because the ion emission from the solution to the gas phase is an endothermic process.

The decrease of the droplet radius can be described by Eq. 1.8, where $\bar{v}$ is the average thermal speed of solvent molecules in the vapor phase:

$$\frac{dR}{dt} = -\frac{\alpha \bar{v}}{4\rho} \frac{p^0 M}{R_g T} \qquad (1.8)$$

where p^0 is the solvent vapor pressure at the droplet temperature, M is the solvent molecular weight; ρ is the solvent density; R_g is the gas constant; T is the droplet temperature; and α is the solvent condensation coefficient.

This relationship showed all factors that can influence the droplet dimensions, and consequently the effectiveness of ESI.

The decrease of the droplet radius with respect to time leads to an increase of surface charge density. When the radius reaches the Rayleigh stability limit (given by Eq. 1.9) the electrostatic repulsion is identical to the attraction because of surface tension. For lower radii, the charged droplet is unstable and decomposes through a process generally defined as Columbic fission (Rayleigh, 1882). This fission is not regular (i.e., the two parts originated by it do not necessarily have analogous dimensions).

$$q_{R_y} = 8\pi(\varepsilon_0 \gamma R^3)^{1/2} \qquad (1.9)$$

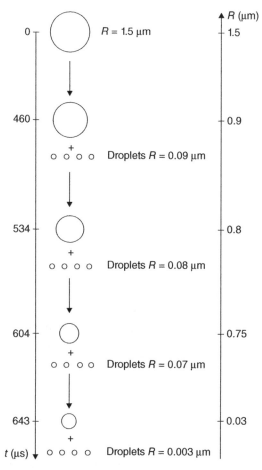

Figure 1.2. Data obtained by theoretical calculation of drop dimensions and lifetime.

Detailed studies performed by Gomez and Tang (Gomez and Tang, 1994) allowed to calculate the lifetime and the fragmentation of droplets. An example is shown in Fig. 1.2.

Until now two different mechanisms have been proposed to give a rationale for the formation of ions from small charged droplets. The first recently was discussed by Cole (Cole, 2000) and Kebarle and Peschke (Kebarle and Peschke, 2000). It describes the process as a series of scissions that lead at the end to the production of small droplets having one or more charges, but only one analyte molecule. When the last few solvent molecules evaporate, the charges are localized on the analyte substructure, which give rise to the most stable gas-phase ion. This model is usually called the charged residue mechanism (CRM) (see lower part of Fig. 1.3).

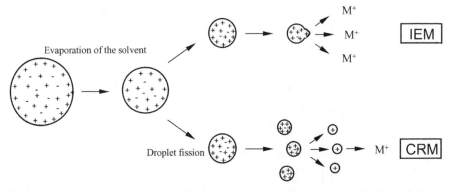

Figure 1.3. Mechanisms proposed for the formation of ions from small charged drop-lets: ion evaporation mechanism (IEM) and charge residue mechanism (CRM).

Thomson and Iribarne (Thomson and Iribarne, 1979) proposed a different mechanism, in which a direct emission of ions from the droplet is considered. It occurs only after the droplets have reached a critical radius. This process is called ionic evaporation mechanism (IEM) and is dominant with respect to Columbic fission from particles with radii $r < 10\,nm$ (see upper part of Fig. 1.3).

Both CRM and IEM are able to explain many of the behaviors observed in ESI experiments. However, a clear distinction between the two mechanisms lies in the way by which an analyte molecule is sepa-rated from the other molecules (either of analyte or solvent present in droplets). In the case of IEM, this separation takes place when a single-analyte molecule, bringing a part of the charge in excess of the droplet, is desorbed in the gas phase, thus reducing the Columbic repulsion of the droplets. In the CRM mechanism, this separation occurs through successive scissions, reducing the droplet dimensions until only one single molecule of analyte is present in them. In general, the CRM model remains valid in the process of gas-phase ion formation for high molecular weight molecules.

1.1.2 Some Further Considerations

What was just described can give us an idea of the high complexity of the ESI process: The ion formation depends on many different mechanisms occurring either in solution or during the charged droplets production and ion generation from the droplets themselves.

First, the ESI users must consider that the concentration of the analyte present in the original solution does not correspond to that present in the droplets generating the gas-phase ions. This point must be carefully considered when the original solution is far from neutrality. In this case, the pH will show sensible changes. Gatling and Turecek (Gatling and Turecek, 1994) studying the $Fe^{2+}(bpy)_3$ $Ni^{2+}(bpy)_3$ (bpy = 2,2'-bipyridine) complex dissociation under electrospray conditions, found that an apparent increase of $[H_3O^+]$ in the order of 10^3–10^4-fold with respect to the bulk solution is observed. Furthermore, due to solvent evaporation, the pH value is not homogeneous inside the droplet. A spherical microdroplet is estimated to maintain a pH 2.6–3.3 in a 5–27-nm thick surface layer without exceeding the Rayleigh limit. This limit implies that complex dissociations occur near the droplet surface of high-local acidity.

For polar compounds, the surface charge density present in the droplet can activate some decomposition reaction of the analyte. In the study of $[Pt(\eta^3\text{-allyl})XP(C_6H_5)_3]$ complexes (X = Cl or Br), the formation of any molecular species in the ESI condition was not observed (Favaro et al., 1997). This result is quite surprising, considering that the same compounds lead to molecular species either in fast-atom bombardment (FAB) or under electron ionization conditions (i.e., in experimental conditions surely "harder" than ESI). This result has been linked to the occurrence of phenomena strictly related to the ESI condition and explained by the high positive charge density present on the droplet surface. It can activate the formation, in the polar molecule under study, of an ion pair consisting of X^- and $[M–X]^+$ (X = Cl or Br) (see Fig. 1.4). The latter are the only species detectable in positive-ion ESI conditions.

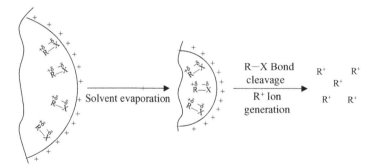

Figure 1.4. Formation of R^+ species ($[Pt(\eta^3\text{-allyl})P(C_6H_5)_3]^+$) from $[Pt(\eta^3\text{-allyl}) XP(C_6H_5)_3]$ complexes, due to the formation of an ion pair catalyzed by the high surface charge density present in the droplet.

1.1.3   Positive- and Negative-Ion Modes

As described in Section 1.1.2, the ESI source can lead to the production of positive or negative ions, depending on the potentials applied to the sprayer and the related counterelectrode. Some producers follow the original ESI source design, placing the sprayer at some kilovolts (positive for positive-ion analysis, negative for negative-ion production) and the counterelectrode (i.e., the entrance to the mass analyzer) grounded or at a few volts (see the left-hand side of Fig. 1.5 for positive-ion analysis). Some other producers use a different potential profile, placing the sprayer at ground potential and the counterelectrode at + or − kilovolts for production of negative or positive ions respectively (right-hand side of Fig. 1.5).

In ESI conditions, aside from the formation of protonated ($[M + H]^+$) and deprotonated ($[M − H]^-$) molecules arising from the oxidation–reduction reactions at the sprayer, some cationization and anionization reactions can take place, due to the presence, inside the solution, of cations and anions. As an example, the positive- and negative-ion spectra of secoisolariciresinol diglucoside (SDG), obtained by injecting the methanol solution (10^{-5} M with 0.1% of formic acid) in the ESI source at a flow rate of 15 μL/min, are reported in Fig. 1.6a and b, respectively. In the former case, the protonated molecule is detectable at m/z 687, but the most abundant peaks are present at m/z 704 and 709, due to cationization reactions with NH_4^+ and Na^+, respectively. A scarcely abundant adduct with K^+ is also detectable at m/z 725. The negative-ion spectrum (Fig. 1.6b) shows an abundant peak due to a deprotonated molecule (m/z 685), together with those due to adducts with Cl^- and $HCOO^-$ (m/z 721 and 731, respectively).

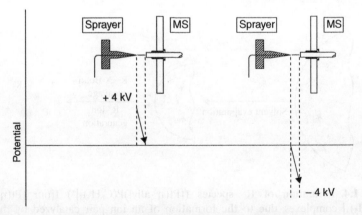

Figure 1.5. Potential profiles usually employed in ESI/MS for positive-ion analysis.

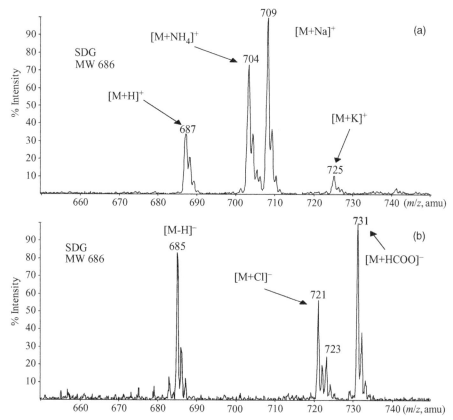

Figure 1.6. The ESI spectra of SDG showing the presence of a wide number of adducts: (a) positive and (b) negative ions.

The privileged formation of molecular species makes the ESI method highly interesting for the analysis of complex mixtures, without the need of previous chromatographic separation. By direct infusion of the mixture dissolved in a suitable solvent, it is possible to obtain a map of the molecular species present in the mixture itself. Furthermore, by operating in the positive-ion mode it is possible to see the compounds with the highest proton affinity values (i.e., the most basic ones), while in the negative-ion mode the formation of ions from the most acidic species will be privileged .

This aspect is well described by the ESI spectra reported in Fig. 1.7, which is obtained by direct infusion of a CH_3OH/H_2O (1:1) solution of an extract of Cynara scolymus. The positive-ion spectrum (Fig. 1.7a) is highly complex either for the complexity of the mixture under analysis, or for the possible ionization of the various molecular species by

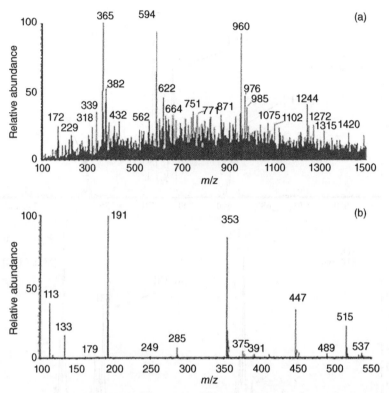

Figure 1.7. Positive (a) and negative (b) ion ESI spectra of an extract of Cynara scoly-mus, obtained by direct infusion of its (CH_3OH/H_2O, 1:1) solution.

addition of H^+, Na^+, K^+, and NH_4^+. On the contrary, the negative-ion spectrum (Fig. 1.7b) is due to few, well-defined ionic species. The chemical background observed in positive-ion mode is completely sup-pressed. The two most abundant ions at m/z 191 and 353 correspond, as proved by MS–MS experiments, to molecular anion species ($[M - H]^-$) of isoquercitrin and chlorogenic acid, respectively. In the higher m/z region, the ion at m/z 515 originates from cinarin. Ions at m/z 447 and 285, detected in low abundance, correspond to luteolin-7-O-glucoside and luteolin, respectively. It is interesting to observe that these species are completely undetectable in the positive-ion ESI spectrum (Fig. 1.7a), being completely lost in the chemical background. This result can be explained by the chemical nature of these compounds. The carbox-ylic and phenolic hydroxyl groups present in them are easily deproton-ated to give the corresponding molecular anion species, whereas they do not easily undergo protonation to give the corresponding molecular cation species, which are thermodynamically unfavorable.

1.1.4   Micro- and Nano-LC/ESI/MS

Electrospray is surely the ionization method most widely employed for the liquid chromatography (LC)-MS coupling (Cappiello, 2007). The possibility of performing ionization at atmospheric pressure [also obtained in the case of atmospheric pressure chemical ionization (APCI) and atmospheric pressure photoionization (APPI), allows the direct analysis of analyte solutions. However, some problems arise from the intrinsically different operative conditions of the two analytical methods. First, there are the high-vacuum conditions that must be present at the mass analyzer level. Second, the mass spectrometers generally exhibit a low tolerance for the nonvolatile mobile-phase components, usually employed in LC conditions to achieve high chromatographic resolution.

Summarizing, the difficulties in LC/MS coupling can be related to the following aspects:

- Sample restriction: The differences among different classes of samples in terms of molecular weight, polarity, and stability (either from the chemical or the chemical–physical point of view) requires an accurate setup of the ESI source conditions;
- Solvent restriction: The LC mobile phase is generally a solvent mixture of variable composition. This variability necessarily reflect on the formation in ESI conditions of droplets of different dimension and lifetime (i.e., under somewhat different ionization conditions). Also in this case an indepth evaluation of the ESI source parameters must be performed to achieve results as close as possible.
- Chromatographic eluate flow, which must be compatible with the sprayer operative flow.

The scheme of a LC/ESI/MS system is shown in Fig. 1.8. Depending of the LC solvent flow, the splitter S can be employed to reduce the

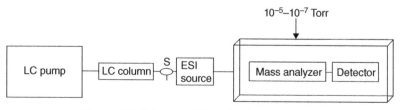

Figure 1.8. Scheme of the LC/ESI/MS system.

flow itself to values suitable for the ESI sprayer. Of course, the split ratio reflects on a decrease of sensitivity (a portion of the sample is dropped away). Analytical columns with internal diameters (i.d.) in the range 2.1–4.6 mm require the use of the splitter, while columns with i.d. ≤1 mm can be directly connected with the ESI source.

Note that, aside from the splitting problem, the i.d. reduction of LC columns leads to a sensible increase in sensitivity of the LC/ESI/MS system. In fact, as shown schematically in Fig. 1.9, the i.d. reduction leads to a higher analyte concentration, due to the volume reduction: Then, passing from a 4.6 to a 1.5 mm i.d. column, a decreased volume of one order of magnitude is obtained. This result reflects in a 10 times increase of analyte concentration and the consequent increase of the MS signal.

This aspect has led to the production of micro- and nanoelectrospray sources, where the chromatographic eluate flow is in the range $1–10^{-2}$ μL/min. A typical instrument setup for nano-ESI experiments is shown in Fig. 1.10. In this case, the supplementary gas flow for spray generation is no longer present and the spray formation is only due to the action of the electrical field. The sprayer capillary, with an internal

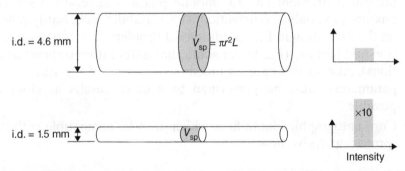

Figure 1.9. Comparison of the behavior of two LC columns of different internal diameter, operating with the same linear velocity. In the case of a low i.d. column, a higher analyte concentration is present, reflecting in a higher signal intensity.

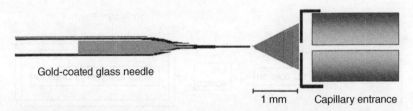

Figure 1.10. Typical instrumental configuration for nano-ESI experiments.

diameter in the range 5–20 μm, is coated with a conductive film (e.g., gold film) in order to be placed at the correct electrical potential.

Just to give an idea of the nano-ESI performances, when the electrical field is applied, a spray is generated with a flow rate in the order of 25–100 nL/min. This finding means that 1 μL of sample can be sprayed for ~40 min! This result reflects in a system with high sensitivity and requires a very low sample quantity.

The main reason for moving to low-flow LC/MS systems is to handle the analytes in the smallest possible volume. Thus, by using an enrichment column, the sample can be trapped and then eluted in the smallest possible volume, so as to reach maximum concentration levels. This approach has required the development of a nanopump. The related technology is nowadays available on the market, but the weak point of the nanopump–nanocolumn–nano-ESI system is mainly related to the connections among the three components, introducing dead volumes and the possibility of small, undetectable leaks.

Recently, a new approach was proposed to overcome these problems, based on a microfluidic chip device (Agilent, 2007) (Fig. 1.11). This finding includes the enrichment column, the separation column, and the nanospray emitter. By use of a robotic system, the chip is automatically positioned in front of the entrance orifice of the mass spectrometer.

The full LC-chip/MS system includes an autosampler and capillary LC pump for delivery of the sample to the enrichment column, which allows sample loading of larger volume samples in a short period of

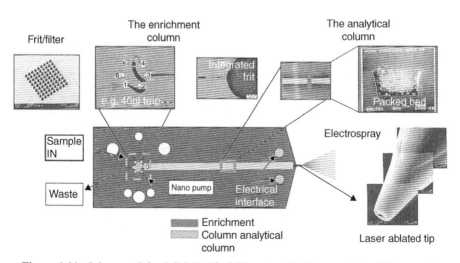

Figure 1.11. Scheme of the full LC-chip-MS system. (Agilent, technical literature).

time, by using a higher loading flow rate (e.g., 4 μL/min). The analytical column is driven by a nano-LC pump typically operating at 100–600 nL/min.

1.2 ATMOSPHERIC PRESSURE CHEMICAL IONIZATION

As described in the introduction, CI was developed in the 1960s as an alternative ionization method to EI being able to induce a low-energy deposition in the molecule of interest, reflecting on the privileged formation of charged molecular species (Harrison, 1983). Chemical ionization is based on the production in the gas phase, at a pressure in the range 10^{-1}–1 Torr, of acidic or basic species, which further react with a neutral molecule of analyte leading to $[M + H]^+$ or $[M - H]^-$ ions, respectively.

Atmospheric pressure chemical ionization (Bruins, 1991) was developed starting from the assumption that the yield of a gas-phase reaction depends not only on the partial pressure of the two reactants, but also on the total pressure of the reaction environment. For this reason, the passage from the operative pressure of 0.1–1 Torr, present inside a classical CI source, to atmospheric pressure would, in principle, lead to a relevant increase in ion production, which consequently leads to a relevant sensitivity increase. Furthermore, the presence of air at atmospheric pressure can play a positive role in promoting ionization processes.

At the beginning of research devoted to the development of an APCI method, the problem was the choice of an ionizing device. The most suitable and effective one was, and still is, a corona discharge. The importance of this ionization method lies in its possible application to the analysis of compounds of interest dissolved in suitable solvents: The solution is injected into a heated capillary (typical temperatures in the range 350–400 °C), which behaves as a vaporizer. The solution is vaporized and reaches, outside from the capillary, the atmospheric pressure region where the corona discharge takes place. Usually, vaporization is assisted by a nitrogen flow coaxial to the capillary (Figs. 1.12 and 1.13). The ionization mechanisms are typically the same as those present in CI experiments.

The needle generates a discharge current of ~2–3 μA, which ionizes air producing primary ions (mainly $N_2^{+\cdot}$, $O_2^{+\cdot}$, $H_2O^{+\cdot}$, and $NO^{+\cdot}$ in the positive mode, $O_2^{-\cdot}$, $O^{-\cdot}$, NO_2^-, NO_3^-, $O_3^{-\cdot}$, and $CO_3^{-\cdot}$ in the negative mode). Primary ions react very rapidly (within 10^{-6} s) transferring their charge to solvent molecules, in a reaction controlled by the

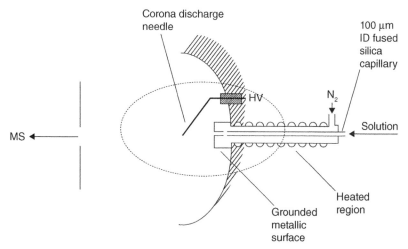

Figure 1.12. Scheme of the APCI ion source.

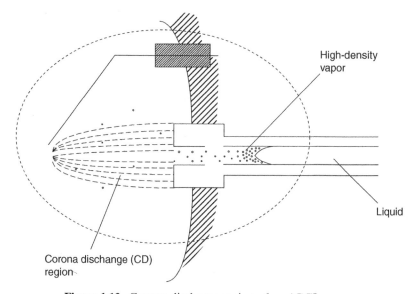

Figure 1.13. Corona discharge region of an APCI source.

recombination energy of the primary ions themselves, to produce the effective CI reactant ions. These are characterized by a longer lifetime ($\sim$0.5 $\times$ 10^{-3}s) and react with analyte molecules to produce analyte quasimolecular ions by charge- or proton-transfer reactions, according to the proton affinity of the analyte itself. The total reaction time in the source corresponds in practice to the final proton transfer ($\sim$0.5 $\times$ 10^{-3}s) as the time of the preceding solvent ionization can be

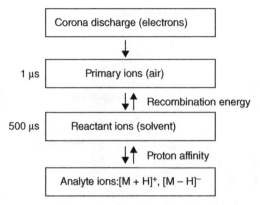

Figure 1.14. Sequence and time scale of the reactions occurring in an APCI ion source.

disregarded. The whole ionization cascade is represented in Fig. 1.14. Under these conditions, the formation of protonated ($[M + H]^+$) or deprotonated ($[M - H]^-$) molecules is generally observed operating in the positive- or negative-ion mode, respectively.

One problem that APCI exhibited at the beginning of its development was the presence of still solvated analyte molecules (i.e., the presence of clusters of analyte molecules with different numbers of solvent molecules). To obtain a declustering of these species, different approaches have been proposed, among which nonreactive collision with target gases (usually nitrogen) and thermal treatments, are those considered to be most effective and currently employed. Different instrumental configurations, based on a different angle between the vaporizer and entrance capillary (or skimmer) have been proposed; 180° (in line) and 90° (orthogonal) geometries are those most widely employed.

1.3   ATMOSPHERIC PRESSURE PHOTOIONIZATION

Photoionization has been considered, from the beginning of analytical MS, to be highly attractive; it exhibits some theoretical advantages with respect to electron ionization, but also has some severe limitations (Morrison, 1986).

In general, the energy transfer involved in ionization of atoms and molecules must be enough to excite one electron from a bond to an unquantized orbital. The ionization energy is just the lowest energy value required for the occurrence of this phenomenon.

For electron ionization, this process can be written as:

$$M + e^- \rightarrow [M^*]^- \rightarrow M^{+\cdot} + 2e^-$$

where three events occur, according to Wigner (Wigner, 1968): (1) approach of an electron to a neutral molecule, (2) formation of a collisional complex, (3) dissociation of the complex in a positive ion and two electrons. The probability of ionization is critically dependent on step (3). Wannier (Wannier, 1953) has shown that this probability depends on the number of freedom degrees n for sharing the excess energy between the electrons. By defining E_c the minimum energy, the ionization probability can be defined as:

$$P(E - E_c) = k(E - E_c)^n \qquad (1.10)$$

and, in the case of emission of two electrons from the collision complex, $n = 1$.

For photoionization, the basic reaction becomes

$$M + h\upsilon \rightarrow [M^*] \rightarrow M^{+\cdot} + e^-$$

and, by using the Wigner and Wannier arguments, in this case the probability expressed by Eq. 1.10 will have $n < 1$ of the value for the EI-induced process. This finding implies that the ionization probability as a function of photon energy will be zero until the ionization energy is reached. When the ionization energy (IE) is reached, the probability will rise immediately to the value determined by the electronic transition probability for the process. In other words, the necessary condition to obtain the photoionization of a molecule M is that $IE_M \leq h\upsilon$ (i.e., that the ionization energy of M is lower than the photon energy).

The main limitation to the extensive use of photoionization in MS was that at the light frequencies suitable to produce ionization of most organic compounds (IE ranging up to 13 eV) it is not possible to use optical windows in the path of the light beam. All the window materials are essentially opaque at this photon energy. Consequently, the light source, usually involving a gas discharge, must be mounted inside the ion source housing operating under high-vacuum conditions. A further aspect that in the past limited the common use of photoionization was surely the low sensitivity of the method. When operating under high-vacuum conditions, typical of classical ion sources, the formation of ions is some orders of magnitude lower than is observed with the same sample density in EI conditions. This finding can be related to the photon cross section.

However, note that photoionization has been used since 1976 as a detection method in GC, proving that, when the sample density is high enough, good sensitivity can be achieved, together with the specificity related to the wavelength employed.

Only a few papers appeared in the past on analytical applications of photoionization in MS. Among them, that by Chen et al. showed the analytical power of the method (Chen et al., 1983) by employing an argon resonance lamp emitting photons with energies of 11.6 and 11.8 eV with an intensity of 3×10^{12} photons s^{-1}. The interaction of a mixture of alkanes in nitrogen (at a pressure of 10^{-2} Torr) with the light beam led to good quality mass spectra, with a detection limit of ~10 ppb. Analogous results were achieved by Revel'skii et al. (Revel'skii et al., 1985).

Of course, with high-power lasers, the photoionization is no longer limited to photons whose energy exceeds that of the ionization energy, since multiphoton processes now become operative. However, the use of a "conventional" light beam interacting with high-density vapors coming from the vaporization of the sample solution has been considered of interest and the Bruins' group (Robb et al., 2000) developed and tested the first experimental apparatus, devoted to LC/MS experiments. By considering the analogies with the well-established APCI technique, this new method was called APPI.

In an APPI source, a series of different processes can be activated by photon irradiation. Calling ABC the analyte molecule, S the solvent, and G other gaseous species present in the source (N_2, O_2, and H_2O at trace level), the first step can be considered their photoexcitation:

$$ABC + h\upsilon \rightarrow ABC^*$$

$$S + h\upsilon \rightarrow S^*$$

$$G + h\upsilon \rightarrow G^*$$

At this stage, inside the source, a collection of excited and nonexcited species is present and a series of further processes can occur, as:

$$ABC^* \rightarrow ABC + h\upsilon \quad \text{Radiative decay}$$

$$ABC^* \rightarrow AB^{\cdot} + B^{\cdot} \quad \text{Photodissociation}$$

$$ABC^* \rightarrow ACB^* \quad \text{Isomerization}$$

$$\left. \begin{array}{l} ABC^* + S \rightarrow ABC + S^* \\ ABC^* + G \rightarrow ABC + G^* \end{array} \right\} \quad \text{Collisional quenchin}$$

$$ABC + S^* \rightarrow ABC^* + S$$

Only when $h\upsilon \geq IE$, can ionization take place

$$ABC^* \rightarrow ABC^{+\cdot} + e^-$$

$$S^* \rightarrow S^{+\cdot} + e^-$$

$$G^* \rightarrow G^{+\cdot} + e^-$$

Recombination processes can occur as well. Hence, inside an ion source at atmospheric pressure, a highly complex mixture of ions, neutrals at ground and excited states, radicals and electrons, is photogenerated.

To put some order in this complex environment, the photon energy can be chosen in order to avoid the ionization processes of the S and G species.

The plot of intensity versus frequency of the most commonly employed lamps is reported in Fig. 1.15. By these data, it can be deducted that the Kr lamp is most suitable for analytical purposes. It shows an energy distribution from 8 to 10 eV, and consequently it does not lead, in principle, to ionization of O_2 (IE = 12.07 eV), N_2 (IE = 15.58 eV), H_2O (IE = 12.62 eV), CH_3OH (IE = 10.86 eV), and CH_3CN (IE = 12.26 eV), typical G and S species present inside the source. The negative counterpart is that, by its use, the direct photo-ionization of organic compounds with IE > 10 eV cannot be obtained and in these cases the use of a dopant (D) has been proposed. A suitable substance, added in relatively large amounts (with an IE value ≤10 eV), reasonably leads to the production of a large number of analyte ions through charge exchange (electron transfer) and/or proton transfer (Chapman, 1993). In the former case, the related mechanism can be described simply as:

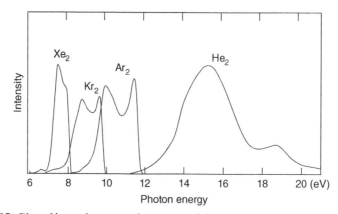

Figure 1.15. Plot of intensity versus frequency of the most commonly employed lamps.

$$D + h\upsilon \rightarrow D^{+\bullet}$$

$$D^{+\bullet} + ABC \rightarrow D + ABC^{+\bullet}$$

and the internal energy of $ABC^{+\bullet}$ can be calculated by

$$E_{int} = RE(D^{+\bullet}) - IE(ABC)$$

where RE $(D^{+\bullet})$ is the recombination energy of the dopant ion and IE (ABC) is the ionization energy of the sample molecule. Relatively intense molecular ions are expected if $RE(D^{+\bullet})$ is >IE(ABC). By considering that benzene and toluene, often used as dopants, exhibit RE values of ~9 eV (Einolf and Murson, 1972), the simple dopant mechanism described above seemed to be quite improbable.

Mainly two different instrumental configurations are employed for APPI experiments. The in-line geometry, due to Robb et al., has been derived from the standard heated nebulizer of the PE/Sciex 300 and 3000 (MDS Sciex, Concord, Ontario, Canada) series triple quadrupole mass spectrometers; in this case, the lamp is mounted perpendicular to the ion guide tube (Fig. 1.16). In the orthogonal geometry source, developed by Syagen Technology, starting from the scheme of the Agilent Technologies APCI source (Fig. 1.17), the heated nebulizer and the Kr lamp are, respectively, perpendicular and in-line with respect to the mass spectrometer ion path and there is no guide tube present.

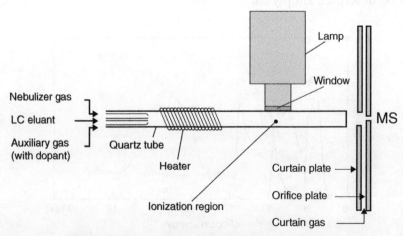

Figure 1.16. The APPI source with in-line geometry.

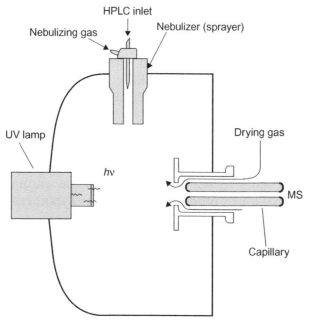

Figure 1.17. The APPI source with orthogonal geometry. (High-performance liquid chromatography = HPLC and UV = ultraviolet.)

1.4 SURFACE-ACTIVATED CHEMICAL IONIZATION

Recently, surface-activated chemical ionization (SACI) has been proposed as an effective alternative approach to APCI and ESI in the MS analysis of biologically relevant molecules (Cristoni et al., 2005). A SACI experiment is particularly simple to realize. In a conventional APCI ion source, the corona discharge needle is substituted by a metallic surface and placed at a potential of a few hundred volts. The sample solution is vaporized in the usual way by the APCI nebulizer operating at a temperature in the range 350–400 °C. Even if, in principle, no ionizing conditions are present (i.e., electrons from the corona discharge are completely suppressed and the vaporizing conditions are far from those typical for the thermospray approach), the production of ionized molecular species (i.e., $[M + H]^+$, $[M + Na]^+$, $[M + K]^+$ ions) is observed in high yields.

In an investigation of the effect of several instrumental parameters on the efficiency of SACI, most attention was paid to the evaluation of vaporization parameters to test the hypothesis that ion evaporation can play an important role in the SACI mechanism. The data so obtained partially supported this hypothesis; by increasing the flow rates of either vaporizing gas (F_g) or solution (F_s) in the range 0.6–2.5 L/min and 10–150 μL/min, respectively, a reasonably linear relationship of F_s/ F_g versus ion intensity was obtained over a narrow range, after which saturation phenomena were observed. The positive role of the surface was proved by increasing its dimensions; in fact, a linear relationship between surface area and signal intensity was found by varying the former from 1 to 4 cm^2.

After that investigation, a series of questions naturally arose that led to the need to understand the real role of the surface. Is it an active region for sample ionization, or does it simply behave as an electro-static mirror leading to better focusing into the entrance capillary orifice of ions previously produced? If the latter hypothesis is true, how are the ions generated?

A reasonable ionizing mechanism might involve collisional phe-nomena occurring either in the high-density vapor or in the dilute gas-phase region. The kinetic energy acquired by the molecules of the expanding gas could lead to effective collisional phenomena with the neutral species present inside the SACI source. The high density (atmospheric pressure) of the gas could lead to effective internal energy deposition through multicollisional phenomena, with the for-mation of ionic species, which in turn activate possible ion–molecule reactions. Alternatively, it could be hypothesized that protonated molecules are generated by collisionally induced decomposition of solvated analyte molecules through charge permutation and/or proton exchange processes.

The kinetic energies of the vaporized species were evaluated by simple calculations, performed on the basis of the experimental setup: They are in the 1–10-meV range, at least three order of magnitude less than those necessary to promote effective gas-phase collision-induced ionization and decomposition processes and, consequently, collisional phenomena cannot be held responsible for ion production.

Then, another aspect was considered, moving from the physical to a chemical phenomenon. At 20 °C, the dissociation constant for water (pK_w) is 14.1669, while it is 13.0171 at 60 °C and 12.4318 at 90 °C. Considering that

$$K_w = [H^+][OH^-]$$

it follows that the [H$^+$] concentration changes from 8.3×10^{-8} to 6.1×10^{-7} M passing from 20 to 90 °C. In other words, the [H$^+$] concen-tration shows a clear increase with respect to temperature. These values are just related to pure water: The possible (and expected!) presence of

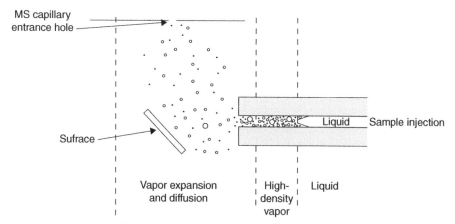

Figure 1.18. Scheme of a SACI ion source.

electrolytes even at low concentration would lead to a further increase of [H⁺] concentration. Consequently, it is more than reasonable to assume that during heating, but before solution vaporization, a decrease in pH of the solution takes place with the formation of protonated molecules from the analyte.

However, these considerations cannot explain the large increase in ion production (typically from 10^4 to 10^6 counts/s) observed when a metallic surface is mounted at 45°C with respect to the direction of vapor emission (Fig. 1.18). In the investigations of the SACI mechanism, it was shown that the best results are obtained not by holding the surface at ground potential, but by leaving it floating (i.e., insulated from ground). Interestingly, it was observed that in the latter condition the ionization efficiency increased with increasing nebulizing time; that is, the total elapsed time after nebulization was started. This result might be explained by considering that a number of ions, generated by the above described mechanisms, are deposited on the surface; by increasing the nebulizing time, this number increases to the point where a suitable potential is induced on the surface. The deposited ions then act as a protonating agent.

To investigate these ideas, a simple experiment based on the deposition on the metallic surface of a thin layer of deuterated glycerol was performed. Under this condition, when the analysis of the PHGGGWGQPHGGGWGQ peptide was performed by SACI, the signal of the [M + D⁺] ion at m/z 1573 was observed to dominate (Fig. 1.19). This evidence is good for the participation of the chemicals present on the surface in the ionization phenomena occurring in SACI.

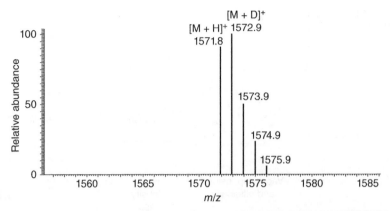

Figure 1.19. The SACI spectrum of the peptide PHGGGWGQPHGGGWGQ obtained by the pretreatment of the metallic surface with deuterated glycerol.

1.5 MATRIX-ASSISTED LASER DESORPTION–IONIZATION

Matrix-assisted laser desorption–ionization (Karas et al., 1991) is based on the interaction of a laser beam [usually generated by an ultraviolet (UV) laser, $\lambda = 337\,nm$] with a crystal of a suitable matrix containing, at a very low level, the analyte of interest (usually the analyte/matrix molar ratio is on the order of 10^{-4}). As depicted in Fig. 1.20, the laser beam–crystal interaction leads to the vaporization of a microvolume of the solid sample, with the formation of a cloud rapidly expanding in the space. The interaction gives rise to the formation of ionic species from the matrix (Ma), which exhibit an absorption band in correspondence to the laser wavelength, as $Ma^{+\cdot}$ (odd electron molecular ions), Fr_i^+ (fragment ions), MaH^+ (protonated molecules), and Ma_nH^+ (protonated matrix clusters). These species, through gas-phase, ion–molecule reactions, gives rise to analyte positive ions (usually protonated molecules). Analogously, the formation of [M–H]$^-$ anions from the matrix can lead to deprotonated molecules of analyte.

A detailed description of the MALDI mechanism is highly complex, due to the presence of many different phenomena:

1. First, the choice of the matrix is relevant to obtain effective and well-reproducible data.
2. The photon–phonon transformation, obtained when a photon interacts with a crystal and gives information on the vibrational levels of the crystal lattice, cannot be applied to the laser induced

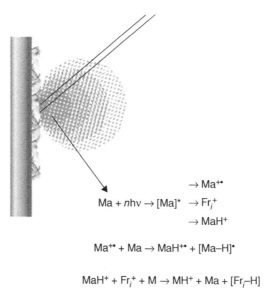

$$\text{Ma} + n\text{h}\nu \rightarrow [\text{Ma}]^* \quad \begin{array}{l} \rightarrow \text{Ma}^{+\bullet} \\ \rightarrow \text{Fr}_i^+ \\ \rightarrow \text{MaH}^+ \end{array}$$

$$\text{Ma}^{+\bullet} + \text{Ma} \rightarrow \text{MaH}^{+\bullet} + [\text{Ma}-\text{H}]^\bullet$$

$$\text{MaH}^+ + \text{Fr}_i^+ + \text{M} \rightarrow \text{MH}^+ + \text{Ma} + [\text{Fr}_i-\text{H}]$$

Figure 1.20. Interaction between laser beam and matrix (Ma) and analyte (M) solid samples, leading to the formation of protonated analyte molecules (MH$^+$).

vaporization observed in MALDI experiments, due to the inhomogeneity of the solid sample.

3. The laser irradiance (laser power/cm^2) is an important parameter: Different irradiance values lead to a vapor cloud of different density, and consequently the ion–molecule reactions can take place with highly different yields.

4. The solid-sample preparation is usually achieved by the deposition on a metallic surface of the solution of matrix and analyte with a concentration suitable to obtain the desired analyte/matrix ratio. The solution is left to dry under different conditions (simply at atmospheric pressure, reduced pressure, or under a nitrogen stream). This method is usually called the Dried Droplet Method. In all cases, what is observed is the formation of an inhomogeneous solid sample, due to the different crystallization rate of the matrix and analyte. Consequently, the 10^{-4} molar ratio is only a theoretical datum: In the solid sample, different ratios will be found in different positions and the only way to overcome this is to average a high number of spectra corresponding to laser irradiation of different points.

To overcome this negative point, a new sample preparation method recently has been proposed (Molin et al., 2008). It is based on the

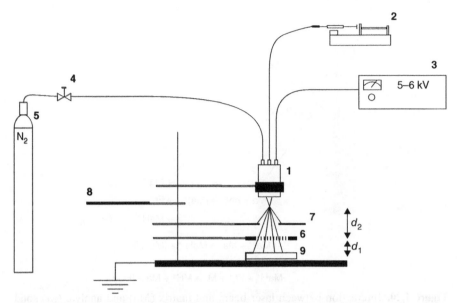

Figure 1.21. Schematic description of the sieve-based-device: **1** sprayer; **2** syringe pump; **3** DC power supply; **4** auxiliary gas flow regulator; **5** N_2 cylinder; **6** sieve mounted on a stainless steel frame; **7** screen; **8** beam interceptor; **9** sample holder.

electrospraying of the matrix–analyte solution on the MALDI sample holder surface through a sieve (38 µm, 450 mesh) (see Fig. 1.21). Under these conditions, a uniform but discrete sample deposition is obtained by the formation of microcrystals of the same dimension as the sieve holes.

With this approach:

1. Identical MALDI spectra are obtained by irradiation of different areas of the sample.
2. In the case of oligonucleotides, a clear increase of both sensitivity and resolution is achieved (see, i.e., Fig. 1.22).

The latter point can be explained in terms of the phenomena occurring by irradiation of a limited quantity of sample–matrix crystals. As the laser spot is elliptically shaped, with a maximum diameter of 100–150 µm, it follows that more than one sample-matrix crystal is irradiated and, considering the untreated sample holder (conductive glass) areas around the crystals, it may be hypothesized that the thermal energy associated with the laser beam is mostly deposited on the conductive glass surface, resulting in highly effective sample heating and

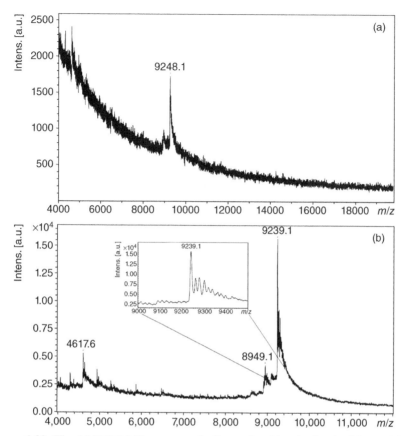

Figure 1.22. The MALDI–MS spectra of oligonucleotide obtained with traditional Dried Droplet deposition method (a) and obtained by spraying the matrix–analyte solution by a sieve-based device (b).

thus effective sample desorption. This result may explain the very high signal intensity achieved by this approach. In fact, irradiation of crystals obtained by the classical Dried Droplet method yields a signal intensity of 2000 arbitrary units (a.u.) that increases to 15,000 a.u. with SBD deposition (see Fig. 1.22).

The MALDI data originate from a series of physical phenomena and chemical interactions originating from the parameterization (matrix nature, analyte nature, matrix/analyte molar ratio, laser irradiation value, averaging of different single spectra), which must be kept under control as much as possible. However, the results obtained by MALDI are of great interest, due to its applicability in fields not covered by other ionization methods. Due to the pulsed nature of ionization phenomena (an N_2 laser operating with pulses of 10^2 ns and with a repetition rate of 5 MHz) the analyser usually employed to obtain the MALDI spectrum is the time-of-flight (TOF), which will be described in Section 2.4.

REFERENCES

Agilent, technical literature (2007).

Barber, M., Bordoli, R.S., Elliott, G.J., Sedgwick, R.D., and Tyler, A.N. (1982). Fast atom bombardment mass spectrometry, *Anal. Chem.*, **54**, 645–657 A.

Beckey, H.D. (1975). Principles of Field Ionization and Field Desorption, *Mass Spectrometry*, Pergamon, London.

Blades, A.T., Ikonomou, M.G., and Kebarle, P. (1991). Mechanism of electrospray mass spectrometry. Electrospray as an electrolysis cell, *Anal. Chem.*, **63**, 2109–2114.

Bruins, A.P. (1991). Mass Spectrometry with ion sources operating at atmospheric pressure, *Mass Spectrom. Rev.*, **10**, 53–77.

Cappiello, A. (2007). *Advances in LC–MS Instrumentation*, Elsevier, Amsterdam.

Chapman, J. R. (1993). *Pratical Org Mass Spectrom*, 2nd ed., Wiley, Chichester, pp. 74–82.

Chen, H.N., Genuit, W., Boerboom, A.J.H., and Los, J. (1983). Selective ion source for trace gas analysis, *Int. J. Mass Spectr. Ion Phys.*, **51**, 207.

Cole, R.B. (2000). Some tenets pertaining to electrospray ionization mass spectrometry, *J. Mass Spectrom.*, **35**, 763–772.

Cristoni, S., Rossi Bernardi, L., Guidugli, F., Tubaro, M., and Traldi, P. (2005). The role of different phenomena in surface-activate chemical ionization (SACI) performance, *J. Mass Spectrom.*, **40**, 1550–1557.

De La Mora, J.F. and Locertales, I.G.J. (1994). The current emitted by highly conducting Taylor cones, *Fluid Mech.*, **260**, 155–184.

Dole, M., Mack, L.L., Hines, R.L., Mobley, R.C., Ferguson, L.D., and Alice, M.B. (1968). Molecular beams of macroions, *J. Chem. Phys.*, **49**, 2240–2249.

Einolf, N. and Murson, B. (1972). High pressure change exchange mass spectrometry, *Int. J. Mass Spectr. Ion Phys.*, **9**, 141–160.

Favaro, S., Pandolfo, L., and Traldi, P. (1997). The behaviour of [Pt (η^3-allyl) XP (C_6H_5)$_3$] complexes in electrospray ionization conditions compared with those achieved by other ionization methods, *Rapid Commun. Mass Spectrom.*, **11**, 1859–1866.

Gatling, C.L. and Turecek, F. (1994). Acidity determination in droplets formed by electrospraying methanol–water solution, *Anal. Chem.*, **66**, 712–718.

Gomez, A. and Tang, K. (1994). Charge and fission of droplets in electrostatic sprays, *Phys. Fluids*, **6**, 404.

Harrison, A.G. (1983). *Chemical Ionization Mass Spectrometry*, CRC Press, Boca Raton.

Karas, M., Bahr, U., and Gießmann U. (1991). Matrix-Assisted Laser Desorpion Ionization Mass Spectrometry, *Mass Spectrom. Rev.*, **10**, 335–357.

Kebarle, P. and Peschke, M. (2000). On the mechanisms by which the charged droplets produced by electrospray lead to gas phase ions, *Anal. Chim. Acta*, **11**, 406.

March, R.E. and Todd J.F. (1995). Practical aspects of ion trap mass spectrometry VOL I,II,III CRC Press, Boca Raton.

Mark, T.D. and Dunn, G.H. (Eds.) (1985). *Electron Impact Ionization*, Springer-Verlag, Wien.

Molin, L., Cristoni, S., Crotti, S., Bernardi, L.R., Seraglia, R., and Traldi, P. (2008). Sieve-based device for MALDI sample preparation. I. Influence of sample deposition conditions in oligonucleotide analysis to achieve significant increases in both sensitivity and resolution. *J. Mass Spectrom.*, **43**, 1512–1520.

Morrison, J.C. (1986). In J. H. Futtrell (Ed.), *Gaseous Ion Chemistry and Mass Spectrometry.*, Wiley, New York, pp. 95–106.

ORNL Review. (1995). *State of the laboratory*. Vol. **29** no. 1–2 http://www.ornl.gov/info/ornlreview/rev29-12/text/environ.htm

Pfeifer, R.J. and Hendricks, C.D. (1968). Parametric studies of electrohydrodynamic spraying, *AIAA J*, **6**, 496–502.

Rayleigh, L. (1882). On the equilibrium of liquid conducting masses charged with electricity, *Philos. Mag.*, **14**, 184–186.

Revel'skii, I.A., Yashin, Y.S., Kurochkin, V.K., and Kostyanovshii, R.G. (1985). Pat. SU 1159412.

Robb, D.B., Covery, T.R., and Briuns, A.P. (2000). Atmospheric Pressure Photoionization: an Ionization Method for Liquid Chromatography-Mass Spectrometry, *Anal. Chem.*, **72**, 3653–3659.

Smith, J.N. (2000). Fundamental Studies of Droplet Evaporation and Discharge Dynamics in Electrospray Ionization, California Institute of Tecnology.

Taylor, G. (1964a). Disintegration of water drops in an electric field, *Proc. R. Soc. London, Ser. A*, **280**, 283.

Taylor, G. (1964b). Studies in electrohydrodynamics. I. The circulation produced in a drop by electrical field, *Proc. R. Soc. London, Ser. A*, **291**, 159–166.

Thomson, B.A. and Iribarne, J.V.J. (1979). Field-induced ion evaporation from liquid surfaces at atmospheric pressure, *Chem. Phys.*, **71**, 4451–4463.

Wannier, G.H. (1953). The threshold law for single ionization at atoms or ions by electrons, *Phys. Rev.*, **90**, 817–825.

Wigner, E.P. (1968). Symmetry principles in old and new physics, *Phys. Rev.*, **73**, 1002.

Yamashita, M. and Fenn, J.B. (1984a). Electrospray ion source. Another variation on the free-jet theme, *J. Phys. Chem.*, **88**, 4451–4459.

Yamashita, M. and Fenn, J.B. (1984b). Negative ion production with the electrospray ion source, *J. Phys. Chem.*, **88**, 4671–4675.

Zelely, J. (1917). Instability of electrified liquid surfaces, *Phys. Rev.*, **10**, 1.

2

MASS ANALYZERS AND ACCURATE MASS MEASUREMENTS

The devices devoted to the separation of ions with respect to their mass-to-charge (m/z) ratio are usually called mass analyzer. They are based on the knowledge of the physical laws that govern the interaction of charged particles with electrical and magnetic fields. Most have been projected and developed during the twentieth century, but in the last few years some interesting, highly promising, new approaches have been proposed.

Two main characteristics are relevant for a mass analyzer; ion transmission and mass resolution. The former can be defined as the capability of a mass analyzer to bring to the detector all the ions that have entered into it. Of course, the ion transmission will reflect on sensitivity (or, better, to the detection limit) of the instrument. Mass resolution is usually defined as the analyzers capability to separate two neighboring ions. The resolution necessary to separate two ions of mass M and $(M + \Delta M)$ is defined as:

$$R = \frac{M}{\Delta M}$$

Then, as an example, the resolution necessary to separate N_2^+ (exact mass = 28.006158) from CO^+ (exact mass = 27.994915) is

Mass Spectrometry in Grape and Wine Chemistry, by Riccardo Flamini and Pietro Traldi

$$R = \frac{M}{\Delta M} = \frac{28}{0.011241} = 2490$$

From the theoretical point of view, the resolution parameters can be described as shown in Fig. 2.1. It follows that a relevant parameter is the valley existing between the two peaks. Usually, resolution data are related to 10% valley definition.

If the peak shape is approximately Gaussian, the resolution can be obtained by a single peak. In fact, as shown by Fig. 2.1, the mass difference, ΔM, is equal to the peak width at 5% of its height and, accordingly to the gaussian definition, it is about two times the fwhm. Consequently, with this approach it is possible to estimate the resolution of a mass analyzer simply by looking at a single peak, without introduction of two isobaric species of different accurate mass.

The resolution present in different mass analyzers can be affected by different parameters and different definitions can be employed. Thus, in the case of a magnetic sector instrument the above 10% valley definition is usually employed, while in the case of a quadrupole mass filter the operating conditions are such to keep the ΔM constant through the entire mass range. Consequently, in the case of a quadrupole mass filter the resolution will be 1000 at m/z 1000 and 100 at m/z 100, while in the case of the magnetic sector the resolution will be, for example, 1000 at m/z 1000 and 10,000 at m/z 100. This parameter will be useful to evaluate and compare the performances of different mass analyzers.

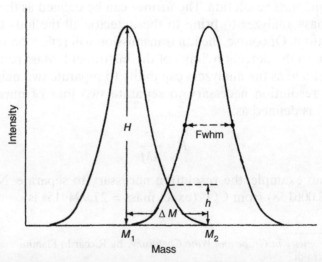

Figure 2.1. Mass resolution parameters. (Full width at half-maximum = Fwhm.)

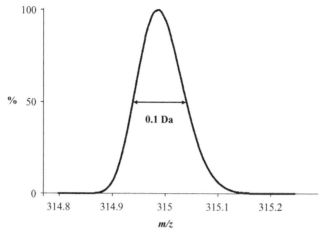

Mass:315
Peak width (50%):0.1
Resolution (FHM):315/0.1=3150
True mass:315.0000
Measured mass:315.002
Difference:0.002 or 2 ppm
Error:0.002/315=6 ppm

Figure 2.2. Full width half-maximum definition.

A more general approach that can be employed for all mass analyzers [as time-of-flight (TOF), ion cyclotron resonance (Fourier transform–mass spectroscopy, FT–MS), Q–TOF] is based upon measuring the full width half-maximum, as shown in Fig. 2.2. In the same figure, the definition of mass accuracy is also reported. This parameter reflects on the specificity of mass measurements. In fact, on one hand it allows to determine the accurate mass of a selected ion (and, consequently, its elemental composition). On the other hand it allows to operate in accurate mass mode in order to identify species of interest present in complex matrices on the basis of their elemental composition.

2.1   DOUBLE-FOCUSING MASS ANALYZERS

Until 1960s, mass spectrometers were mainly based on these analyzers, originally developed by the early Thomson, Aston, and Dempter's studies at the beginning of the last century (Beynon, 1960). Actually, they are still present on the market, but their use is mainly confined in environmental monitoring; in fact, the official Environmental Protection Agency (EPA) methods for dioxines and polychlorinated biphenyls (PCB$_S$) analysis specifically require the employment of this instrumental configuration.

An essential part of the double-focusing analyzer is a magnetic sector, an electromagnet with a well-defined geometry. Ions are ejected from an ion source by the action of an acceleration field. Their potential energy is fully transformed in kinetic energy

$$zV = \frac{1}{2}mv^2 \tag{2.1}$$

where m and z are the mass and charge of the ions, V is the acceleration potential, and v is the speed acquired by the ions after the acceleration phase.

The ions interact with the magnetic field in a specific region (as shown in Fig. 2.3) and are subjected to a force described by the Lorenz law: $F = zvB$.

They will follow a circular pathway of radius r, due to the equality of centrifugal and centripetal forces

$$\frac{mv^2}{r} = zvB \tag{2.2}$$

By combining Eqs. (2.1) and (2.2) it follows that:

$$\frac{m}{z} = \frac{B^2r^2}{2V} \tag{2.3}$$

This equation shows that for B and V const, ions of different m/z values will follow circular pathways of different radius r. Alternatively, keeping const V and r (the latter condition can be obtained easily by the use of a slit placed after the magnetic field), by varying B it is possible to focalize through the exit slit ions of different m/z value. The B scanning is the approach usually employed for mass analysis.

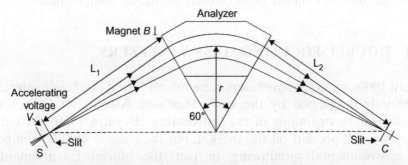

Figure 2.3. Scheme of magnetic sector analyzer showing its focusing capabilities.

The use of magnetic sectors leads not only to the separation of ions of different m/z value, but allows to compensate the dishomogeneity in direction of ion beams emerging from the ion source slit and reaching the magnetic sector. In fact, even if a series of electrostatic lenses can lead to a well-concentrated ion beam, during the pathway from the ion source to the magnet the beam will naturally spread in the direction orthogonal to its motion, because of space charge effects (i.e., to the repulsion of charged species of the same sign). Consequently, the spred ion beam will enter into the magnetic sector with different angles, as shown in Fig. 2.3. The use of magnetic sectors of suitable geometry compensate for this point. In fact, by considering a medium entrance angle inside the magnetic sector of 90°, ions entering with an angle <90° will follow a circular pathway larger than that of the center of the beam: Their transit time inside the magnetic sector will be longer, and consequently they will experiment with the magnetic field for a longer time. This approach will result in their stronger deflection. On the contrary, ions entering inside the magnetic sector with an angle higher than 90° (see Fig. 2.3) will follow a shorter pathway with a consequently lower transit time inside the magnetic field and a lower deflection. The results of this phenomena are that the diverging beam is focused on a well-defined point, symmetrical to that of the ion source exit slit. Then the magnetic sector leads to a direction focusing and it is for this reason that instruments based on the use of the magnetic sector are usually called "single-focusing" mass spectrometers.

But an ion beam is not only dishomogeneous in direction: It shows also a dishomogeneity in kinetic energy. In fact, if we look at Eq. 2.1, it assumes that all the ions experiment with the same acceleration voltage, V. The V value depends on the coordinates of the point from which the ion is extracted. Due to the ion source geometry, ions are generated at different points, and consequently experiment with different V values that reflects, on the basis of Eq. 2.1, on their kinetic energy. In other words, the ions emerging from an ion source will show a kinetic energy distribution. Usually, electrostatic sectors are employed to overcome this negative aspect, strongly reflecting on the mass resolution of the instrument. As shown in Fig. 2.4, they consist of two cylindrical plates on which potentials $+E/2$ and $-E/2$ are applied. The ions, interacting with the electrostatic field so generated, are forced to follow a circular pathway of radius R, due to the equality of centrifugal and centripetal forces.

$$\frac{mv^2}{R} = zE \qquad (2.4)$$

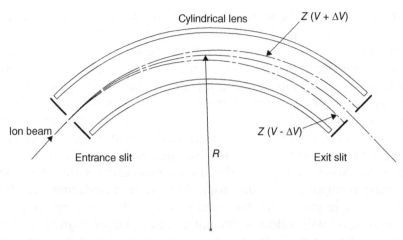

Figure 2.4. Scheme of the electrostatic sector.

If R is well defined by the use of two slits (before and after the electrostatic sector), only ions with a well-defined kinetic energy, obtained by Eq. 2.5

$$\varepsilon_k = \frac{RzE}{2} \tag{2.5}$$

pass through the electrostatic sector, while the others are filtered out by the device (those with higher ε_k will impact with the outside electrode, those with lower ε_k will discharge on the inside one). Just to give an order of magnitude, in high-resolution machines, the ions passing through the electrostatic sector are in the range of 1–10% of those entering inside it. Hence, the action of the electrostatic sector is a kinetic energy filtering of the ion beam.

Instruments using as analyzer a B,E arrangement are consequently called "double-focusing" mass spectrometers (see Fig. 2.5). Under these conditions, what are called "first-order" aberrations are overcome; however, further aberrations of the ion optics are still present, due to the fringing fields present either at the magnetic or electrostatic sector levels. To reduce these undesired effects, usually electrostatic lens are placed in the field-free region of the instrument.

Typical resolution values obtained by a commercially available double-focusing instrument are in the range 10,000–50,000. However, note that the highest resolution can be obtained only by narrow-width slit values and this necessarily reflects on the decrease of detected ions and the consequent drop of sensitivity. The ion transmission is not only affected by the slit width, but also by the use of the electrostatic sector

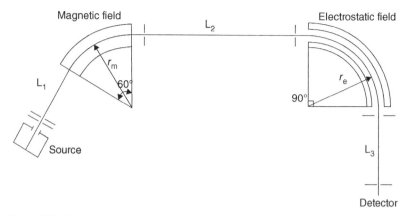

Figure 2.5. Scheme of a double-focusing mass spectrometer with BE geometry.

and the lens value setup. For the last parameter, a compromise between resolution and sensitivity is generally required.

2.2 QUADRUPOLE MASS FILTERS

The development of the quadrupole mass filter (Dawson, 1976) led to a revolution in the view held by the scientific community on MS. Until the end of 1960, mass spectrometers were considered highly complex devices, which needed the employment of expert personnel and a high budget for their acquirement. Magnetic sector instruments were large machines (covering some square meters of surface and with a weight usually in the tons range) that require (for their dimensions) more than one pumping line, and usually contained liquid nitrogen trap(s), diffusion pumps, and a rotating primary pump. All these points influenced the development of MS labs, where dedicated personnel (either at the scientific or technical level) gave support to the MS users and developed research areas in the MS fundamental, instrumental, and application fields.

The development of quadrupole mass filters and their commercial availability led to a drastic change in the view of MS. The analyzer, which in sector machines required ion pathways >1-m long (in some instrument 5–6 m!), was typically 20 cm long, consequently requiring a smaller pumping system: The result was a very compact machine with reduced surface dimensions (1 × 1.5 m), weight and price, and fully operated by data systems. This made it possible for mass spectrometers to enter into the analytical world, and no longer be present in the MS lab, and be directly employed by its users.

Quadrupole mass filters, as quadrupole ion traps, were created by the Bonn group, led by Prof. Wolfang Paul (Nobel Prize in Physics in 1989) (Paul, 1953, 1958, 1960) When the first quadrupole mass filters were commercialized, the scientific community, in particular the mass spectrometrists, were divided into two parties: The first party considered the new entry without any future, due to its performances, lower than those of sector instruments, while the second party looked in a positive way to the production of compact and low-cost machines, to be used in a more familiar way. Time is a good judge and the wide presence of quadrupole mass filter in the scientific and analytical world is proof that the second party had a correct view of what was required.

An exhaustive description of the physical aspects of the behavior of ions in a quadrupole field is beyond the aim of this book. For those interested in this aspect we suggest the March and Todd books, which describe in detail all the theoretical and practical aspects of quadrupole mass filters and ion traps very well. We will try to give a general idea of how the quadrupole mass filter operates, with the intent of making the reader conscious of the pros and cons of this instrumental approach.

A quadrupole mass filter consists of four hyperbolic rods on which a potential $U \pm V_0 \cos \omega t$ is applied (see Fig. 2.6). The hyperbolic shape lead to the production of a quadrupolar field, where the field intensity is linearly dependent on space. However, these conditions can be validly approximated by the use of cylindrical rods (see Fig. 2.7). The ions are injected in the z direction and experience a field due to the direct current (dc) and radio frequency (rf) potentials applied on

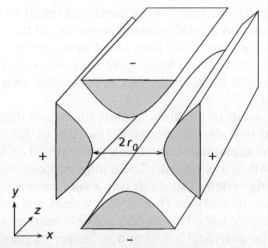

Figure 2.6. Quadrupole mass filter with hyperbolic rods.

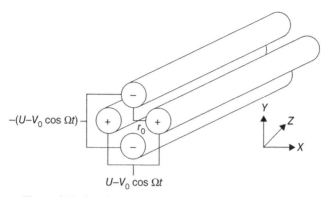

$-(U-V_0 \cos \Omega t)$

r_0

$U-V_0 \cos \Omega t$

Y

Z

X

Figure 2.7. Quadrupole mass filter with cylindrical rods.

the four rods. The motion equation can be determined by resolving a differential equation (Mathieu equation) (Mathieu, 1868), but for our purpose we can follow another simpler approach. What are the parameters affecting the ion motion inside the quadrupolar field? They are the mass of the ion m, its charge z, the potentials U and V, the frequency ω of the rf voltage, and the interrod distance r_0 (defined in Fig. 2.7). If we would like to analyze the interdependence of these six parameters, we would need, from the mathematical point of view, to use a hexa-dimensional space, which is difficult to be managed by our central nervous system. We are more familiar with two-dimensional (2D) structures. Consequently, we use the six parameters described above to define two new parameters, a and q

$$a = -\frac{8Uz}{mr_0^2\omega^2} \tag{2.6}$$

$$q = -\frac{4Vz}{mr_0^2\omega^2} \tag{2.7}$$

Then, with a and q we can define a bi-dimensional space by which we can study the interdependence of the six parameters affecting the behavior of an ion inside the quadrupolar field. First, we can define the "stability" region of the quadrupole mass filter (i.e., the values of a and q for which the motion of the ions in the x and y direction (see Fig. 2.7) do not exceed the r_0 value (in other words, the a and q values for which the ions do not crash on the rods!).

These regions, calculated on the basis of the motion equation are reported in Fig. 2.8. To obtain the ions transmission inside the quadrupole mass filter, the ions must be "stable" in both the x and y directions. Consequently, we must overlap the two stability diagrams of

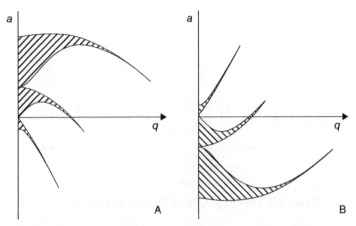

Figure 2.8. Stability diagrams in the q, a space corresponding to trajectories in the xy plane (A) and in the yz plane (B).

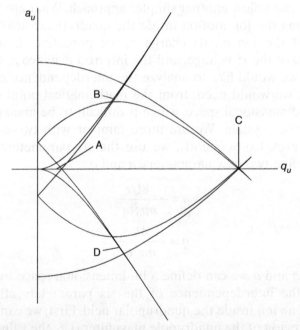

Figure 2.9. Overlapping of the two stability diagrams of Fig. 2.8, showing the stability regions in both xz and yz planes.

Fig. 2.8 and consider only the region of simultaneous stability (see Fig. 2.9). Four regions are put in evidence by this overlapping and the A region is the one usually employed for quadrupole operative conditions. This choice is due to the lower U,V potential values characteristic of this region, described in Fig. 2.10.

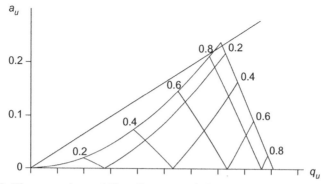

Figure 2.10. The operative stability diagram and the scan line defined for a specific U/V ratio.

Now, consider the ratio between a and q, as defined by Eqs. 2.6 and 2.7

$$\frac{a}{q} = \frac{8Uz}{mr_0^2\omega^2} \cdot \frac{mr_0^2\omega^2}{4Vz} = \frac{2U}{V} \quad \text{from which it follows that}$$

$$a = \frac{2U}{V}q \qquad (2.8)$$

This equation represents a straight line in the (a, q) space passing through the origin and whose slope is due to the U/V ratio (see Fig. 2.10). If the U/V ratio is carefully regulated, the straight line can intercept just the apex of the stability diagram.

Coming back to Eqs. 2.6 and 2.7, it is easy to recognize that, for constant values of V, U, r_0, and ω, the a and q values are inversely proportional to m/z. In other words we can consider that ions of increasing m/z ratio lie on the straight line defined by Eq. 2.8 for decreasing (a, q) values (see Fig. 2.11a). In these conditions all the ions, having a and q values outside the stability diagram, follow unstable trajectories and discharge on the quadrupole rods. But now if we increase both the U and V values (keeping constant the U/V ratio, r_0, and ω) the a and q values increase for all the ions. From a pictorial point of view, we can imagine that all ions start to travel along the straight line, as shown in Fig. 2.11b. When an ion has a and q values inside the stability diagram, its pathway will be stable. Consequently, it will pass through the quadrupole rods, while all the other ions will be "filtered"; that is, will follow unstable trajectories and discharge on the rods. Then, by scanning U and V, and keeping the U/V ratio constant, it is possible to make all the ions entering inside the quadrupole mass filter selectively stable. If a detector is mounted outside the rods, we can obtain the ion current due to each ionic species (i.e., the mass spectrum).

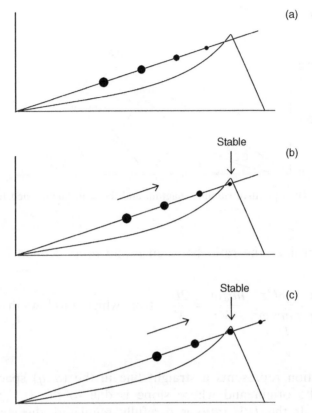

Figure 2.11. Change of the a, q values for ions of different m/z values by changing the U/V values at constant U/V ratio.

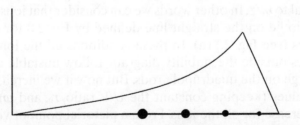

Figure 2.12. For $U = 0$, $a = 0$ and the stability diagram is the stability line on the q axis.

If $U = 0$, that is, if a potential $V \cos \omega t$ is only present on the rods, $a = 0$ and the bi-dimensional stability diagram became mono-dimensional, consisting of the q values between zero and the right limit of the bi-dimensional stability diagram (see Fig. 2.12). In this case, all the ions have q values inside the stability diagram, and consequently

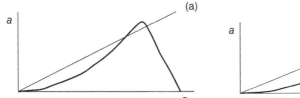

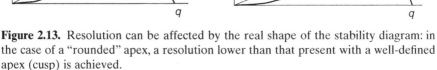

Figure 2.13. Resolution can be affected by the real shape of the stability diagram: in the case of a "rounded" apex, a resolution lower than that present with a well-defined apex (cusp) is achieved.

all follow stable trajectories. Under these conditions, the quadrupole is no more a mass filter, but is a highly effective ion lens, able to transmit ions from one region to another without (or with very small) ion loss. In these operative conditions, the quadrupole is used in many instrumental configurations.

Finally, it must be emphasized that by carefully playing with the U/V ratio (i.e., changing the slope of the straight line), it is possible to make a choice between resolution and sensitivity. In fact, by moving the straight line of Fig. 2.10 very close to the stability diagram apex, one can gain in resolution, but some ions of the same m/z values are lost, with a consequent sensitivity drop. Alternatively, by moving slightly down the straight line, a highly effective ion transmission is obtained, but under lower resolution conditions. Furthermore, note that resolution strongly depends on the quality of the quadrupolar field, in turn reflecting on the quality of the stability diagram. The use of rods of circular section leads to a good approximation of the quadrupolar field, but reflects on a stability diagram with a "rounded" apex (see Fig. 2.13a), while a hyperbolic-shaped rod reflects on a well-defined apex (Fig. 2.13b). Obviously, in the latter case the definition of the m/z value of the ions following stable trajectories inside the rods is much more accurate.

2.3   ION TRAPS

Ion traps (IT) (March and Todd, 2005) were (and still are!) very attractive devices for physical researches, being able to confine and store in a well-defined region of space, ions of interest on which to perform fundamental studies. However, they can be employed also as mass spectrometers and today they are among the more widely diffused mass analyzers, due on one hand to their low price and on the other hand to the ultrahigh performances of others. They can be classified in four main classes:

1. Three-dimensional (3D) quadrupole and high-field-order ion traps.
2. Linear ITs.
3. Ion cyclotron resonance FT–MS.
4. Orbitraps.

2.3.1 Three-Dimensional Quadrupole Ion Traps

The quadrupole ion trap (QIT) mass spectrometer consists of three hyperbolic-shaped electrodes arranged in a cylindrical geometry (see Fig. 2.14). By considering its axial geometry, we can move from the classical Cartesian coordinates to the polar ones. In fact, each point of the space inside the trap can be defined by the value of its axial (z) and radial (r) coordinates. If a potential $U + V \cos \omega t$ is applied at the intermediate electrode and the two end caps are grounded, a mathematical treatment analogous to that done for quadrupole mass filter can be employed. In this case, a and q values can be defined again as

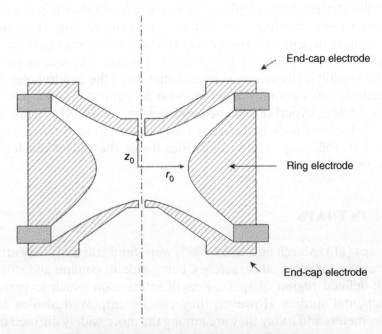

Figure 2.14. Scheme of a tridimensional ion trap.

$$a = -\frac{16zU}{mr_0^2\omega^2} \qquad (2.9)$$

$$q = \frac{8zV}{mr_0^2\omega^2} \qquad (2.10)$$

and stability regions can be calculated either in the z or r directions (see Fig. 2.15).

The condition for which ions can be stored inside the trap is that they follow stable pathways in both the z and r directions. In other words, their pathways cannot exceed the z_0 and r_0 values shown in Fig. 2.14. Consequently, we must consider the overlap of the two stability diagrams of Fig. 2.15 and consider the common regions in the a, q space, which lead to the operative diagram shown in Fig. 2.16.

What is the behavior of a trapped ion? An exhaustive description of the solution of the Mathieu equation and of the operative conditions of an ion trap can be found in the March and Todd books (March and Todd, 1995, 2005) on this argument. Just from a pictorial point of view we can say that an ion inside an ion trap follows some periodic fundamental trajectories with well-specified frequencies (called secular frequencies), on which some other periodic motions at higher frequencies (high-order frequencies) are superimposed (Nappi et al., 1997).

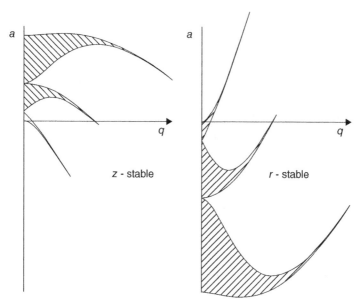

Figure 2.15. Stability diagrams in the (a, q) space showing the region of stability in the z and r directions.

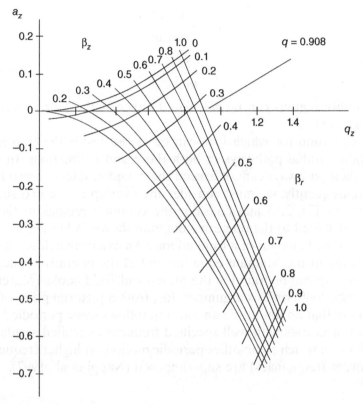

Figure 2.16. Stability diagram in the (a, q) space for the region of simultaneous stability in both the r and z direction.

Summarizing, it can be considered that ions inside the trap:

1. Have secular frequencies inversely related to their m/z values.
2. The "radius" of their periodic pathway is directly related to their m/z values.

Consider the case where $U = 0$; that is, any dc potential not given to the intermediate electrode. In this case, the stability diagram becomes a stability line, corresponding to q values between 0 and 0.908 (see Fig. 2.17). For suitable V, ω, r_0, and m/z values, all the ions follow trajectories confined inside the trap. By looking at Eq. 2.10, it follows that for V, ω and r_0 = const, ions of different m/z values exhibit q values inversely proportional to their m/z value, as schematized in Fig. 2.17.

Under these conditions, all the ions remain inside the trap, but if we increase the V value their q value increases, according to Eq. 2.10.

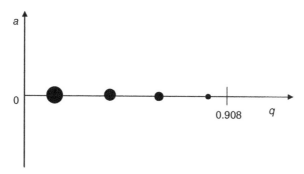

Figure 2.17. The stability diagram for $U = 0 \Rightarrow a = 0$ is a stability line on the q axis from $q = 0$ to $q = 0.908$. The q value of the trapped ion is inversely proportional to their m/z values.

When this value exceeds the $q = 0.908$ value, that is, out of the stability diagram, the ion motion becomes unstable in the z (axial) direction, and the ions are ejected from the trap. Then, by scanning the V value, it is possible to eject selectively all the ions from the trap and if an electron multiplier is mounted just outside, we can record the ion current of each ionic species originally trapped, thus obtaining the related mass spectrum.

The 3D quadrupole ion trap suffers from a severe limitation. If the number of trapped ions is too high, the electrical field due to the $V \cos \omega t$ potential is overlapped by that due to the ion cloud. The result is a drop in instrumental performances, particularly in mass resolution and linear response. To avoid this undesired phenomenon, a preliminary scan (not seen by the ion trap user) is performed and the ionization time (or the ion injection time) is optimized, thus confining the optimum number of ions inside the trap (see Fig. 2.18). This prescan leads to a well-controlled instrumental setup but, of course, it limits the sensitivity of the instrument. To overcome this problem, two different approaches can be employed: (1) increase the ion storage capacity of the trap by increasing the electric field strength; (2) increase the inner volume of the trap, so as to obtain a less dense ion cloud, which results in a decrease of space charge effects.

For the former approach, electrodynamic fields of intensity higher than the quadrupolar are employed. As summarized in Table 2.1, while a quadrupolar field increases linearly with the distance, for hexapoles it increases quadratically, and for octapolar it increases cubically. The last two conditions can be obtained by different electrode configurations (see Fig. 2.19), but components of higher fields can be simply obtained by varying the geometry of the ion trap, in particular the asymptotic cone

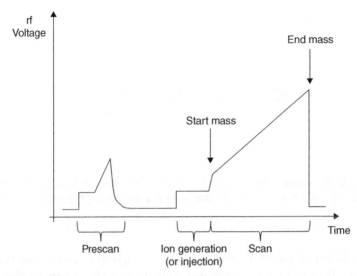

Figure 2.18. Typical scan function for acquiring the mass spectrum of trapped ions, showing the presence of the prescan for the evaluation of the best number of trapped ions.

TABLE 2.1. Pure Multipole Fields

Multipole Name	Potential Increase	Field Increases
Dipole (2 poles)	Linearly	Constant, no increase
Quadrupole (4 poles)	Quadratically	Linearly
Hexapole (6 poles)	Cubically	Quadratically
Octapole (8)	With 4th power	Cubically
Decapole (10)	With 5th power	With 4th power
Dodecapole (12)	With 6th power	With 5th power

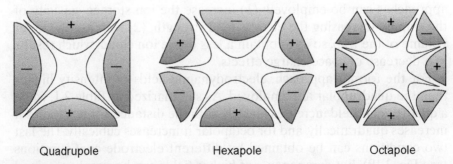

Quadrupole   Hexapole   Octapole

Figure 2.19. Schematic sections of electrode arrangements leading to quadrupolar, hexapolar, and octapolar fields.

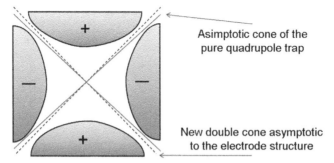

Figure 2.20. Higher order field components can be obtained by changing the geometry of a classical quadrupolar ion trap.

profile, as shown in Fig. 2.20 and/or the distances between the intermediate electrode and the end caps. The presence of field components higher than the quadrupole leads to the capacity of effectively storing a higher number of ions, with the subsequent increase of sensitivity. "High-capacity" ion traps are commercially available and exhibit a sensitivity one to two orders of magnitude higher than the quadrupolar ones.

2.3.2 Linear Ion Traps

Linear ITs have been developed to obtain a high ion storing efficiency. They are based on the use of a quadrupolar field, but their operative conditions are deeply different from those of quadrupole mass filters and quadrupole ion traps. As shown in Fig. 2.21, the first proposed configuration of this device is based on the use of a quadrupole mass

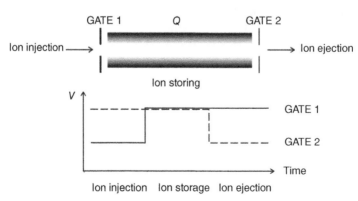

Figure 2.21. Scheme of a linear ion trap and voltage pulses applied on the two gates for ion injection, storage, and ejection.

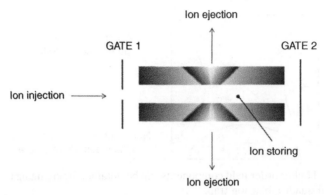

Figure 2.22. Scheme of a linear ion trap with orthogonal ion ejection.

filter (Q) mounted between two electrodes (Gates 1 and 2) (Campbell, 1998; Collins, 2001). The ions, generated inside the ion source are accelerated by interaction with the electrical field obtained by an electrode placed at a voltage V. They linearly move in the space with a speed

$$v = \sqrt{\frac{2zV}{m}}$$

For trapping the ions inside the quadrupolar field, in the first-stage, Gate 1 is placed at the voltage $V_1 = V$, while Gate 2 is placed at a voltage $V_2 > V_1$. Under these conditions, Gate 2 acts as an electrostatic mirror, so that the ions start to move in the opposite direction. When the number of stored ions is sufficiently high, the voltage of Gate 1 is placed at V_2: other ions do not enter in the quadrupole field and those present are trapped inside the quadrupole. Ions are ejected from the "linear IT" by decreasing the V_2 voltage.

The second configuration of a linear IT commercially available is schematized in Fig. 2.22. It operates, with respect to ion injection and storing, in the same way as described above, playing on the voltages applied to Gates 1 and 2, but in this case the ions are no longer ejected axially, but radially, thought two narrow slits present on two opposite quadrupole rods (Schwartz et al., 2002).

2.3.3   Digital Ion Trap

The ITs described in Section 2.3.2 operate with a sinusoidal electrical field in the rf region (V_{rf}) and the ions are ejected from the trap by increasing the V_{rf} value.

A deeply different approach is used in the digital IT systems (Ding et al., 2004). First, the trapping waveform is no longer sinusoidally shaped, but is generated by rapid switching between two well-defined voltage values. The timing of this switching (i.e., the frequency of the rectangular waveform) can be controlled with high precision by a suitable digital circuitry. With this approach, all the parameters [period (T), duty cycle (d), and voltage values(V_1, V_2)] are under digital control (see Fig. 2.23).

In the steady-trapping operation, a periodic rectangular wave voltage, generated between a high-voltage level V_1 and a low-voltage level V_2, is applied to the ring electrode of an IT. Under these conditions, the Mathieu equation cannot be used to determine the ion pathways inside the DIT and the matrix transform method must be utilized.

However, the Mathieu parameters a and q, defined by Eqs. 2.9 and 2.10, can still be employed for the description of the DIT theoretical stability diagram, considering the U and V values as the average values of the dc and ac components of the rectangular wave voltage applied to the intermediate electrode. They are defined as

$$U = dV_1 + (1-d)V_2 \tag{2.11}$$
$$V = 2(V_1 - V_2)(1-d)d \tag{2.12}$$

As shown in Fig. 2.23, the duty cycle d is described as the ratio between τ and the total period T of the rectangular wave. Unlike the sinusoidal wave, the rectangular wave can be generated with different pulsing times for V_1 and V_2 produced by the digital circuitry. In a DIT, the a_z value is a function of both the dc offset and the duty cycle d. In fact, the dc component U can be generated either by an imbalance between V_1 and V_2 or by variation of the duty cycle d.

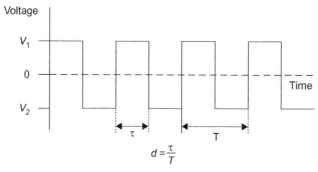

Figure 2.23. The digital waveform used for operating the digital ion trap.

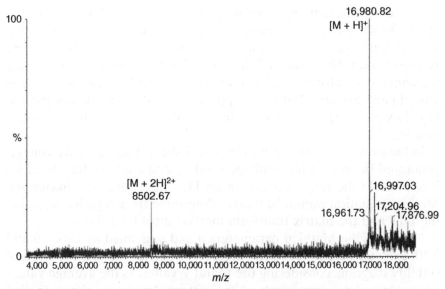

Figure 2.24. Mass spectrum of singly and doubly protonated horse heart myoglobin ions obtained by DIT.

For ITs driven by a sinusoidal waveform, the superimposition of a dc potential (U) on the main trapping field (V_{rf}) requires the use of an additional dc power supply. However, in the DIT, the dc component can be generated easily by varying the duty cycle of the rectangular waveform through appropriate variation of the parameter values entered into the control software of the mass spectrometer.

With this approach, first the V values (usually 10^3V) are lower than that employed in QIT (usually in the 10^4V range); second the ions are ejected from the trap by scanning the rectangular wave frequency.

The DIT scan function differs substantially from that used in the quadrupole IT experiments. Time is reported on the abscissa, while, on the ordinate, contrary to what was discussed for QIT (for which the V_{rf} voltage is reported), for DIT the period (T) of the square wave is reported. The DIT scan function used to obtain a complete mass spectrum is based on four separate steps: (1) a standby time at high T values, followed by (2) a stage devoted to ion introduction inside the trap, (3) a field adjusting phase, and (4) the analytical mass scan.

The results obtained by DIT are of high interest: mass resolution >17,000 and the capability for high mass range at low trapping voltage and fixed q_0 (see, e.g., the spectrum of horse heart myoglobin reported in Fig. 2.24) make this system deeply different from the ITs described in the previous paragraph. Unfortunately, DIT is not yet commercially available. Consequently, it becomes impossible to give an evaluation that arises by an extensive use in different application fields.

2.3.4 Fourier Transform–Ion Cyclotron Resonance

Fourier transform–ion cyclotron resonance (FT–ICR) (usually called, FT–MS) systems are the mass spectrometers exhibiting the highest resolution (up to 10^6) (Marshall and Schweikhard, 1992). They are based on the trapping of ions inside a magnetic field of high intensity. If an ion interacts with a strong magnetic field B, it follows a circular pathway of radius r, due to the equality of centrifugal and centripetal forces acting on it

$$\frac{mv^2}{r} = zvB \tag{2.13}$$

$$\frac{m}{z} = \frac{rB}{v} \tag{2.14}$$

For a circular pathway, $v = 2\pi r/T$, so that Eq. 2.14 becomes

$$\frac{m}{z} = \frac{BT}{2\pi} \quad \text{that is}$$

$$\frac{m}{z} = \frac{B}{2\pi\omega_c} \tag{2.15}$$

where ω_c is the induced cyclotron frequency.

Equation 2.15 showed the inverse relationship between the m/z value of an ion and the frequency ω_c of its circular motion inside the magnetic field of strength B.

If a simple plate arrangement, as shown in Fig. 2.25, is mounted inside the magnetic field, the phenomenon described above can be used to obtain a very effective mass analyzer. First, we can confine the ions in a very small region by imposing on the trapping plates a small voltage of the same sign as the trapped ion.

As shown by Eq. 2.15, ions of different m/z values exhibit different cyclotron frequencies. The radii of their circular motion are very small, but can be increased by the application of an "excitation" electrical field generated by an alternate voltage V_{ac} applied on the two side electrodes (see Fig. 2.25). For a well-defined V_{ac} frequency, only a well-defined ion resonates with it: The ion will acquire energy from the electrical field and the radius of its circular motion will increase (for

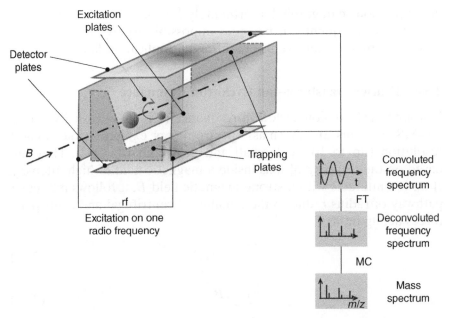

Figure 2.25. Scheme of a ICR cell and of the FT signal analysis.

the increase of its centrifugal force). Then it will pass very closely by the last two plates of the cell and will induce an alternating potential on them, which leads to an induced alternating current (ac). The frequency of this current will be exactly the same as the ω_c value of the excited ion and the accurate measurement of this electrical frequency will reflect on an equally accurate measurement of the m/z value (see Eq. 2.15.) The intensity of the induced current will be proportional to the number of ions passing close to the detector plates.

Hence, different from what has been described for the other mass analyzers, where the ion detection is obtained by a suitable detector mounted outside the analyzers themselves, the ICR system acts either as the mass analyzer or as the ion detector.

But why FT (Fourier transform)? The necessity of this data analysis becames essential when species of different m/z values are present inside the cell. In this case, the induced current will be a highly complex signal, due to overlapping of the alternating signals due to each single ionic species with a well-defined m/z value (see Fig. 2.26). The FT of the complex transient allows to achieve a highly accurate spectrum of the different frequencies of the ion motions inside the cell, which, by Eq. 2.15, gives rise to a highly accurate mass spectrum at the highest mass resolution available today.

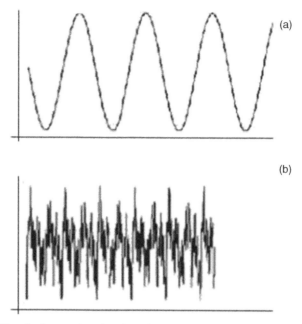

Figure 2.26. Signals detected at the detection plates: (a) for a single ionic species and (b) for many ionic species of different m/z value (and hence moving with different frequencies).

2.3.5   Orbitrap

The FT–ICR method described in Section 2.3.4 suffers only from one weak point: It requires magnetic fields with intensities > (or =) 3 Tesla. Consequently, cryomagnets are required, with high costs either for acquirement or for maintenance. The commercial availability of mass spectrometers exhibiting high performances, but low initial cost, modest maintenance cost, and reduced size, is surely of great interest, and the Orbitrap system (Hu et al., 2005) is the answer to this need.

The Orbitrap mass analyzer was invented by Makarov and can be considered the evolution of an early ion storage device, developed by Kingdom (Kingdom, 1923). It utilizes a purely electrostatic field for trapping the ions: either magnetic or electrodynamic fields (as those used for FT–ICR and IT, respectively) are no longer present. The Kingdom trap can be schematized, as shown in Fig. 2.27: it consists of a wire, a coaxial cylindrical electrode, and two end cap electrodes, each component electrically isolated from the others. A voltage is applied between the wire and cylinder. Under these conditions, if ions are injected into the device with a velocity perpendicular to the wire, those with appropriate perpendicular velocities will follow stable orbits

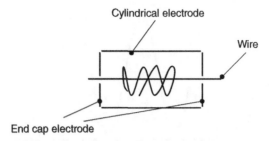

Figure 2.27. Scheme of the Kingdom ion trap.

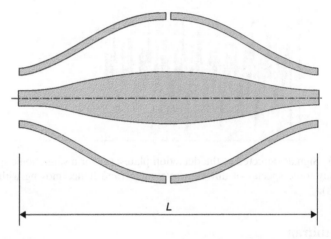

Figure 2.28. Shape of the electrodes of an Orbitrap. The length L is on the order of a few centimetres.

around the central wire. A weak voltage applied on the two end cap electrodes allows to confine ions in the center of the trap.

Makarov invented a new type of mass spectrometer by modifying the Kingdom trap with specially shaped outer and inner electrodes (see Fig. 2.28). Also, in this case a purely electrostatic field is obtained by a dc voltage applied to the inner electrode. Ions injected into the device undergo a periodic motion that can be considered the result of three different periodic motions: (1) rotation around the inner electrode; (2) radial oscillation; and (3) axial oscillations. These three components exhibit well-defined frequencies:

Frequency of rotation ω_φ $$\omega_\varphi = \frac{\omega_z}{\sqrt{2}} \sqrt{\left(\frac{R_m}{R}\right)^2 - 1} \qquad (2.16)$$

Frequency of radial oscillations ω_r $$\omega_r = \omega_z \sqrt{\left(\frac{R_m}{R}\right)^2 - 2} \qquad (2.17)$$

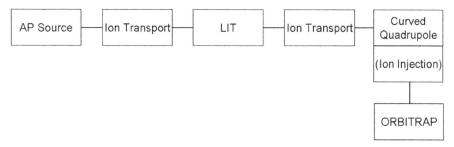

Figure 2.29. Scheme of a commercial Orbitrap instrument.

Frequency of axial oscillation ω_z $\qquad$ $\omega_z = \sqrt{\dfrac{k}{m/q}}$ $\qquad$ (2.18)

In particular, the ion motion in the z (axial) direction may be described as an harmonic oscillation and Eq. 2.18 showed the relationship between the axial frequency and the m/z (m/q) value of the trapped ion. By the same approach used for FT–ICR, in the case of Orbitrap ion detection is obtained by image current detection on the two outside electrodes, and by a FT algorithm the complex signal due to the copresence of ions of different m/z values (and hence exhibiting different ω_z values) is separated into its single m/z components. The typical mass resolution obtained by this analyzer is up to 10^5.

The commercially available instrument including Orbitrap technology is schematized in Fig. 2.29. Orbitrap represents the last step of a long journey carried our by the ions generated by an AP ion source (ESI, APCI, APPI, or APMALDI). The ions are first transported in a *linear ion trap*, by which MS/MS experiments can be effectively performed with high sensitivity. The ions (either precursor or collisionally generated fragments) are then transported to a "curved quadrupole" (c-trap) (Olsen et al., 2007), a highly effective device for ion storing, focusing, and ejection. By its action, a well-focalized ion beam is injected, through a further series of lenses, along the appropriate direction inside the Orbitrap and analyzed under high-resolution conditions, which are required by the analytical problem: As long as the trapping time is high the achieved resolution is obtained.

2.4   TIME OF FLIGHT

Time of flight is surely the simplest mass analyzer (Wollnik, 1993). In its basic form, it consists of an ion accelerator and a flight tube under vacuum. Magnetic, electrostatic, and electrodynamic fields are no longer present. In its "linear" configuration, TOF is based on the acceleration, by the action of suitable acceleration voltage V, of the ions generated inside the ion source. The potential energy is transformed into kinetic energy

$$zV = \frac{1}{2}mv^2 \qquad \text{from which}$$

$$v = \left(\frac{2zv}{m}\right)^{1/2} \tag{2.19}$$

Equation 2.19 shows that ions of different m/z values will follow, after acceleration, linear pathways with different speeds. In other words, the m/z values are inversely related to the squared speed.

If the ions follow the linear pathway inside a field-free region (drift tube) of length l, considering that $v = l/t \Rightarrow t = l/v$, it follows that

$$t = l\left(\frac{m}{2zV}\right)^{1/2} \tag{2.20}$$

This equation shows that ions of different m/z values reach the detector, placed at the end of the drift tube, at different times, proportional to the square root of their m/z value. By this experiment, we will obtain "arrival time" spectrum of the ions, which can be transformed into the mass spectrum by the relationship expressed by Eq. 2.20. For this reason, this device is called TOF.

The simplest TOF configuration (called "linear" for the linear pathways followed by ions) is reported in Fig. 2.30.

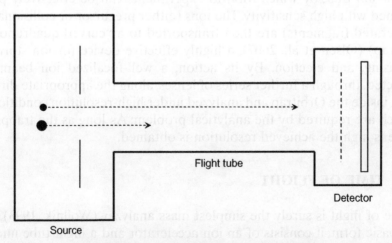

Flight tube

Detector

Source

Figure 2.30. Scheme of a linear TOF analyzer.

Different from what is present in the magnetic sector and quadrupole mass analyzer, TOF cannot operate in the continuous mode and an ion pulsing phase is required: The shorter the pulse, the better defined is the mass value and the peak shape. So, for example, when a TOF is linked to a MALDI source, if the analyzer is directly coupled to the source, ions will be produced during the time of laser shot irradiation (typically in the order of 100 ns) allowing to obtain a resolution in the range of 1000–2000. To obtain a shorter ion pulsing inside the TOF, a grid is usually mounted a few millimeters from the sample plate and placed, during the laser irradiation phase, at the same potential of the sample plate (see Fig. 2.31). Under these conditions, ions generated by MALDI cannot leave the source region and remain "trapped" in the region between the sample plate and the grid. When the laser irradiation has been stopped (and no more ions are generated) the voltage on the grid is switched off and the ions are accelerated inside TOF. The grid switching is usually carried out in the 10–30-ns range. Under these conditions, an increase in resolution of about one order of magnitude

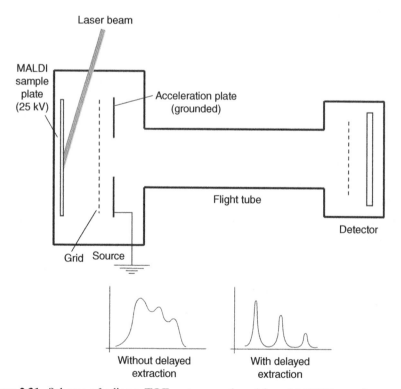

Figure 2.31. Scheme of a linear TOF system employed for a MALDI experiment with the grid to be used for delayed extraction of the ions from the source.

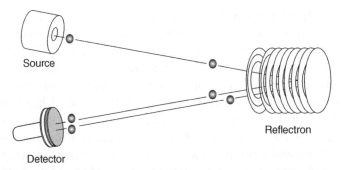

Source

Reflectron

Detector

Figure 2.32. Scheme of a TOF analyzer with the reflectron device, leading to a resolution increase.

is achieved (see lower part of Fig. 2.31). This approach is usually defined as the "delayed extraction method".

Furthermore, the ions emerging from the source are usually not homogeneous with respect to their speed (this effect mainly arises from the inhomogeneity of the acceleration field). Of course, a distribution of kinetic energy will reflect immediately on the peak shape and a wide kinetic energy distribution will lead to an enlarged peak shape, with the consequent decrease of resolution. To overcome this negative aspect, different approaches have been proposed. Those usually employed consists of a reflectron device. As shown in Fig. 2.32, the reflectron consists of a series of ring electrodes and a final plate. The plate is placed at a few hundred volts over the V values employed for ion acceleration. By using a series of resistors, the different ring electrodes are placed at decreasing potentials. When an ion beam with kinetic energy $E_k \pm \Delta E_k$ interacts with this field, the ion with excess kinetic energy $(E_k + \Delta E_k)$ will penetrate the field following a pathway longer than that followed by ions with mean kinetic energy E_k. In contrast, ions with a lower kinetic energy will follow a shorter pathway. This phenomenon leads to a thickening of the ion arrival time distribution with a consequent, significant increase in mass resolution.

Nowadays, TOF systems with resolutions >20,000 are commercially available.

REFERENCES

Beynon, J.H. (1960). Mass Spectrometry and its applications to organic chemistry, *Elsevier*, Amsterdam.

Campbell, J.M., Collins, B.A., and Douglas, D.J. (1998). A Linear Ion Trap Time-of-flight System with Tandem Mass Spectrometry Capabilities, *Rapid Commun. Mass Spectrom.*, **12**, 1463–1474.

Collins, B.A., Campbell, J.M., Mao, D., and Douglas, D.J. (2001). A Combined Linear Ion Trap Time-of-flight System with Improved performance and MS^n Capabilities, *Rapid Commun. Mass Spectrom.*, **15**, 1777–1795.

Dawson, P.H. (1976). Quadrupole Mass Spectrometry and its Applications, *Elsevier*, Amsterdam.

Ding, L., Sudakov, M., Brancia, F.L., Giles, R., and Kumashiro, S. (2004). A digital ion trap mass spectrometer coupled with atmospheric pressure ion source, *J Mass Spectrom.*, **39**, 471–448.

Hu, Q., Noll, R.J., Li, H., Makarov, A., Hardman, M., and Cooks, G.R. (2005). The Orbitrap: a new mass spectrometer, *J. Mass Spectrom.*, **40**, 430–443.

Kingdom, K.H. (1923). A Method for the neutralisation of electron space change by positive ionisation at very low gas pressure, *Phys. Rev.*, **21**, 408.

March, R.E. and Todd, J.F. (1995). Practical aspects of ion trap mass spectrometry Vols. I–III CRC Press, Boca Raton.

March, R.E. and Todd, J.F. (2005). Quadrupole ion trap Mass Spectrometry 2nd ed. Wiley–Interscience, New York.

Marshall, A.G. and Schweikhard, L. (1992). Fourier Transform Ion-Cyclotron resonance Mass-Spectrometry-Technique Developments, *Int. J. Mass Spectrom.*, **118**, 37–70.

Mathieu, E. (1868). Memoire sur le movement vibratoire d'une membrane de forme elliptique, *J. Math. Pures Appl.*, **13**, 137.

Nappi, M., Weil, C., Cleven, C.D., Horn, L.A., Wollnik, H., and Cooks, R.G. (1997). Visual representations of simulated three-dimensional ion trajectories in an ion trap mass spectrometer, *Int. J. Mass Spectrom. Ion Proc.*, **161**, 77–85.

Olsen, J.V., Macek, B., Lange, O., Makarov, A., Horning, S., and Mann, M. (2007). Higher-energy C-trap dissociation for peptide modification analysis, *Nature Methods*, **4**, 709–712.

Paul, W. and Steinwedel, H., *Ger. Pat.* 944, 900 (1956); *US Pat.* 2, 939, 952 (1960).

Paul, W. and Steinwedel, H.Z. (1953). *Naturforsch. A*, **8**, 44.

Paul, W., Reinhard, H.P., and von Zahn, U. (1958). *Z. Phys.*, **152**, 153.

Schwartz, J.C. and Senko, M.W. (2002). A Two-Dimensional Quadrupole Ion Trap Mass Spectrometer, *J. Am. Soc. Mass Spectrom.*, **13**, 659–669.

Wollnik, H. (1993). Time-of-flight Mass Analyzers, *Mass Spectrom. Rev.*, **12**, 89–114.

3

MS/MS METHODOLOGIES

The history of multiple mass spectrometry (MS/MS) (McLafferty, 1983) begins with studies from the 1960–1970s on metastable ions (Cooks et al., 1973). At that time the only instruments available were sector machines and the mass spectrum was in analogic form (i.e., just the recording of the electrical signal coming from the detector). The portion of a typical mass spectrum obtained by a B or EB sector instrument is reported in Fig. 3.1. The narrow peaks correspond to ions with a well-defined m/z value, but some wide Gaussian-shaped peaks of smaller intensity are well detectable at not entire m/z values. They originate from ions m_1^+ (defined metastable) that decompose

$$m_1^+ \rightarrow m_2^+ + m_3$$

in the region between the ion source and the magnetic field in the case of a single focusing (B) instrument or, in the case of an EB geometry, in the region between the electrostatic and magnetic sectors.

An ion that is produced along the pathway of its precursor maintains the precursor speed and, by taking into consideration the Eqs. 2.1 and 2.2, they exhibit a kinetic energy lower than that of all the ions ejected from the source.

Mass Spectrometry in Grape and Wine Chemistry, by Riccardo Flamini
and Pietro Traldi
Copyright © 2010 John Wiley & Sons, Inc.

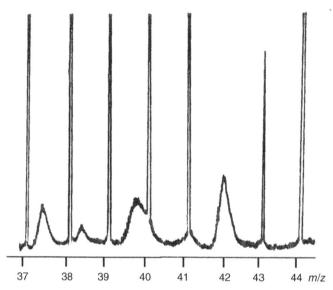

37   38   39   40   41   42   43   44 *m/z*

Figure 3.1. Typical "metastable" ions detected in a mass spectrum obtained by a magnetic sector instrument.

Consequently, they are detected at *m/z* values *m** lower than that of m_2 ions generated into the source. This value can be calculated by the relationship

$$m^* = \frac{m_2^2}{m_1} \qquad (3.1)$$

The wide peak shape can be explained by the fact that during the m_1^+ decomposition an amount of its internal energy is transformed into kinetic energy of m_2^+ This phenomenon is usually called "kinetic energy release" (Cooks et al., 1973) and can give important information on the structure of the precursor ions m_1^+. Then, m_2^+ does not exhibit the velocity v_1 of m_1^+, but a velocity distribution $v_1 \pm \Delta v$, where Δv is related to the kinetic energy released during the m_1 decomposition. Returning to Eq. 3.1, it is an equation with two unknowns (m_2 and m_1) and consequently in principle can have an infinite number of solutions. However, considering the ions detected in the mass spectrum, the couple of values m_1, m_2 leading to the detected *m**, can be determined with sufficient (but not absolute) certainty, thus allowing to gain important information on the decomposition pattern of the ionic species under study.

The *BE* double-focusing instrument (Fig. 2.5) turned out to be highly effective in performing studies on metastable ions. In fact, by this approach the precursor ion m_1^+ can be selected by a suitable B value and all its product ions m_{2i}^+ generated in the region between magnetic (B) and electrostatic (E) sectors can be easily separated by scanning the E potential. In fact, as reported in Eqs. 2.3 and 2.4, the electrostatic sector can be considered to be a kinetic energy analyzer and, by scanning E, it is possible to focus all the fragments ions m_{2i}^+ generated by m_1^+ in the field-free region on the detector. The spectrum so obtained is usually called the "mass analyzed ion kinetic energy spectrum" (MIKES) and the m_2^+ value can be calculated easily by the E_2 value, where it is focused on the detector:

$$\frac{m_2}{m_1} = \frac{E_2}{E_1}$$
$$m_2 = \frac{E_2}{E_1} m_1$$

(3.2)

The MIKE spectrum allows to obtain important structural information on m_1, which is both the decomposition pathways and the related kinetic energy released values strictly related to its structure.

However, note that the abundance of the ions detected by MIKE is two to three orders of magnitude lower than that observed in the usual mass spectrum. This analytically negative aspect is due to the origin of m_{2i}^+. In fact, the ions m_1^+ that decompose in the field-free region are only a small portion of all the ions generated inside the ion source.

If we consider the internal energy distribution of the ions, as shown by Fig. 3.2, for $E_{int} < E_1$ we will have stable ions, which will reach the detector undecomposed. For $E_{int} > E_2$, we will have unstable ions, which will decompose inside the ion source and whose fragments will be

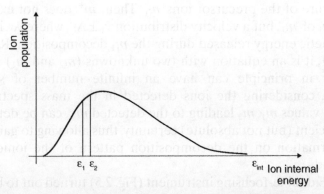

Figure 3.2. Internal energy distribution of ions generated by electron ionization.

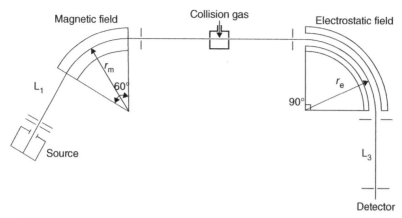

Figure 3.3. Double-focusing mass spectrometer with a collisional cell mounted in the second field-free region.

detected in the usual mass spectrum, while the ions m_1^+ with $E_1 < E_{int} < E_2$ will leave the source, but decompose along the pathway, leading to the fragment ions detected on the MIKE spectrum. Of course, the number of different kinds of ions will be the integer of the three portions of the distribution curve evidenced in Fig. 3.2, which accounts for the low intensity of MIKE spectra.

By considering the high structural diagnostic value of MIKE spectra, it became interesting to enhance the decomposition processes of selected ionic species in the field-free region and its collision with a target gas in a suitable cell was considered to be the most effective route (see Fig. 3.3). This represents the birth of the MS/MS methods (McLafferty, 1983).

The weak points of the MIKE approach were mainly two:

1. The low kinetic energy resolution, reflecting in low mass resolution.
2. The low sensibility of the method (generally 1–10% of precursor ions have effective collisions with the target gas and are able to produce fragment ions).

In order to obtain higher mass resolution and sensitivity, a new approach was investigated. The development of a Triple Quadrupole instrument, mainly due to the Cooks' group from Purdue University, led to an MS/MS instrument with high power and moved collisional experiments from the physicochemical environment to analytical chemistry.

All the MS/MS experiments consist of at least three different phases: (1) precursor ion selection; (2) collision of the precursor ion with target gas; (3) mass analysis of the collisionally generated fragments.

These three phenomena can take place in different space regions, and in this case the whole experiment is called "MS/MS in space". Otherwise, the three stages can occur in the same physical space, and consequently must be performed at different times. This approach is called MS/MS in time.

3.1   TRIPLE QUADRUPOLE

As reported in the previous paragraph, the triple quadrupole (QQQ) can be considered to be the first MS/MS instrument widely employed in the analytical field (Yost and Enke, 1983). In its simpler form, it consists in an arrangement, along the same axis, of three different quadrupole mass filters (see Fig. 3.4). The ionic specie of interest (M^+) is produced by the suitable ionization method inside the ion source S and is accelerated by the action of an electrostatic field (typical voltages applied to the acceleration plate are on the order of 10^2 V, one to two orders of magnitude lower than those used in the sector machines) and the ion beam so generated is focused just in the center of the quadrupole mass filter Q_1. If Q_2 and Q_3 operate in rf only, they behave as lenses, allowing the transfer of the ion from Q_1 to the detector: If, under these conditions, we perform a U and V scan (keeping the U/V ratio constant) in Q_1, we will obtain the usual mass spectrum, due to all the ionic species generated into the source. But if we choose the appropriate U, V values on Q_1 for a stable pathway of the M^+ ion, only this ion will pass through Q_1, while all the others will crash into the quadrupole rods.

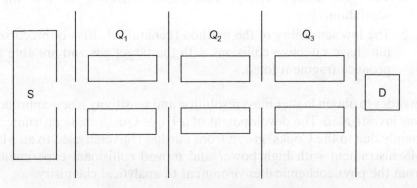

Figure 3.4. Scheme of a QQQ instrument.

Then, Q_1 allows to select the ion of interest. If a collision gas is injected into Q_2, (operating in rf only, i.e., with only V_{rf} and with $U = 0$), M^+ will collide with the target gas molecules. Its internal energy will increase, promoting the occurrence of the fragmentation processes.

The product ions so formed can be analyzed by scanning the U, V values imposed on Q_3. By using this description, it is easy to understand that the QQQ systems must be considered a "MS/MS in space" device.

The instrumental arrangement allows a wide series of collisional experiments to be performed, among which the most analytically relevant are (schematized in Fig. 3.5), the following:

1. Product ion scan: Identification of the decomposition products of a selected ionic species (Q_1 fixed, Q_2 in rf only mode, Q_3 scanned).
2. Parent ion scan: Identification of all the ionic species that produce the same fragment ion (Q_1 scanned, Q_2 in rf only mode, Q_3 fixed).
3. Neutral loss scan: Identification of all the ionic species that decompose through the loss of the same neutral fragment (Q_1 scanned, Q_2 in rf only mode, Q_3 scanned with a fixed difference with respect to Q_1).

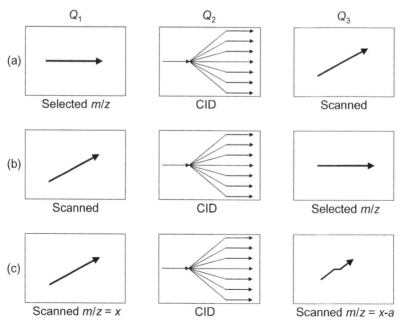

Figure 3.5. Collisional experiments that can be performed by a QQQ system.

The collisional phenomena occurring in a triple quadrupole (as well as in sector machines) lead to the production of an ion population with a wide internal energy distribution, due to the statistics of the preselected ion–target gas interactions. Hence, various decomposition channels, exhibiting different critical energies, can be activated. The resulting MS/MS spectrum is, in general, rich in peaks and, consequently, in analytical information.

What are the parameters in an MS/MS experiment that one can vary by use of a QQQ? Two parameters are (1) the nature of the target gas (the larger the target dimension, the higher the internal energy deposition on the preselected ion: In other words, Ar is more effective than He), and (2) its pressure (the higher the pressure, the higher the probability of multiple collisions leading to increased decompositions: Of course, the pressure must not exceed the limit compromising the ion transmission!). But, over all, the kinetic energy of colliding ions, which can be varied by suitable electrostatic lenses placed between Q_1 and Q_2, plays a fundamental role in MS/MS experiments.

The arrangement of the QQQs along the same axis can lead to same noise in the MS/MS spectra. In fact, by considering the collisionally induced decomposition

$$m_1^+ \rightarrow m_2^+ + m_3$$

we can manage, by playing with the fields of Q_1 and Q_3, the precursor (m_1^+) and product (m_2^+) ions, but not the neutral m_3 that will continue to follow the pathway of m_1^+ before its decomposition. Consequently, neutral species m_3 cannot be managed by an electrical field and can reach the detector, which leads to signals without any sense. To overcome this problem, QQQ of different geometrics have been proposed and are commercially available, as those reported in Fig. 3.6. The intermediate quadrupole Q_2 is no more linear, but has been substituted by

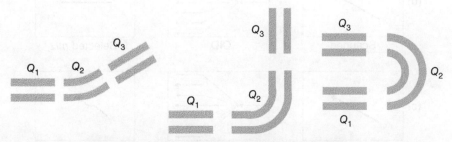

Figure 3.6. Different QQQ geometries available today.

curved quadrupoles. Since Q_1 and Q_3 are no longer on the same axis, m_3 cannot reach the detector, which leads to a sensible increase of the signal-to-noise (S/N) ratio.

3.1.1 Quadrupole Ion Traps

Ion traps (March and Todd, 2005) have *in themselves* the possibility of performing all three stages of MS/MS experiment (precursor ion selection, its collision with a target gas, analysis of the product ions) in the same physical space limited by the three-ion trap electrodes (intermediate ring and two end caps). This will operate as a "MS/MS in time" device. This result is achieved by varying, in a sequential way, the potential applied on the electrodes.

A typical sequence for these experiments is reported in Fig. 3.7. The ions are generated inside the ion trap (IT) (or injected into the trap after their outside generation) for a suitable time, chosen in order to optimize the number of trapped ions (an ion density that is too high leads to degraded data due to space-charge effects). The ions inside the trap exhibit motion frequency depending on their m/z values. The ion selection phase is achieved by the application, on the two end-caps, of a supplementary radiofrequency (rf) voltage with all the ion frequencies, except the ion of interest. Under these conditions, all of the undesidered ions are ejected from the trap and only those of interest remains trapped. The V_{rf} is changed, so that the selected ion reaches a

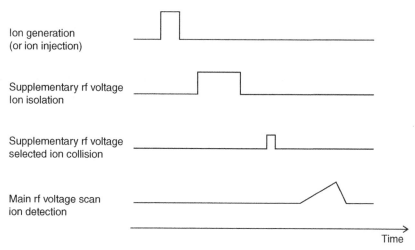

Figure 3.7. Sequence of pulses employed to perform MS/MS experiments by use of an ion trap.

well-defined q value. The collision of the preselected ion is again performed by its resonance with the supplementary radiofrequency (rf) field, whose frequency corresponds to that of the q value, but which has an intensity that can maintain the ion trajectory inside the trap walls. The ion collides with the He atoms present in the trap as a buffer gas and, once sufficient internal energy is acquired, it decomposes: The product ions so generated remain trapped and by using the main V_{rf} scan they can be ejected from the trap and detected.

Note that the collisional data obtained by an IT are quite different from those achieved by QQQ. In fact, in this case the energy deposition is a step-by-step phenomenon. Each time that the ion is accelerated by the supplementary radiofrequency (rf) field up and down inside the trap, it acquires, through collision with He atoms, a small amount of internal energy. When the internal energy necessary to activate the decomposition channel(s) at the lowest critical energy is reached, the ion fragments. In other words, while in the QQQ case the wide internal energy distribution from collisional experiments leads to the production of a large set of product ions, in the case of the IT only a few product ions are detected, originating from the decomposition processes at the lowest critical energy.

This aspect could be considered negative from an analytical point of view: In fact, a better structural characterization can be achieved by the production of a wider number of product ions. But it can be easily and effectively overcome by the ability of the IT to perform many MS/MS experiments. In fact, the sequence shown in Fig. 3.7 can be repeated by selection, among the collisionally generated product ions, of an ionic species of interest, its collision, and the detection of its product ions (MS³). This process can be repeated more times (MSⁿ), allowing on the one hand to draw a detailed decomposition pattern related to the lowest energy decomposition channels, and on the other hand to obtain fragment ions of high diagnostic value from a structural point of view (and hence analytically highly relevant). Hence, by use of an IT it is possible to perform MSⁿ experiments, that cannot be obtained by the QQQ approach.

3.1.2 Linear Ion Traps

As described in Section 3.1.1, a linear IT was introduced mainly for an increase in the number of trapped ion, which is reflected in a sensible increase of sensitivity. However, both of the instrumental configurations commercially available can be effectively employed for MS/MS experiments. Additionally, in this case, analogous to what was described

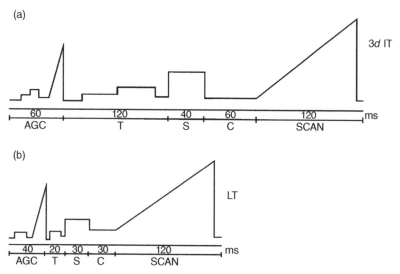

Figure 3.8. Scan function for MS/MS experiments performed by 3D ion trap (a) and linear ion trap (b).

for the 3D ITs, ion isolation and excitation are performed by the action of resonant rf fields. Product ions are ejected axially or radially from the trap (depending on the instrumental configuration) by using the mass selective instability mode of operation. Besides improving the trapping efficiency, fragmentation efficiency, and increased ion capacity (by linear traps), a scan speed higher than that employed in QIT can be employed. For example, with use of the QIT the sequence ion trapping–ion isolation–collision–fragment ion analysis requires 0.4 s. The same experiment performed by linear IT requires 0.24 s. (see Fig. 3.8). All these positive aspects reflect on the production of MS/MS spectra of quality higher than that obtained by 3D IT.

3.1.3 The MS/MS by a Digital Ion Trap

The three steps of an MS/MS experiment are performed by DIT using an approach substantially different from that employed in 3D IT or linear ITs (Ding, 2004). In those cases, the precursor ion isolation is performed by applying one or more dipole excitation waveforms, with a maximum isolation resolution of ~1300 (expressed as the isolation mass divided by the baseline width of the isolation window). In the case of DIT, ion isolation is performed by sequential forward and reverse scans, so as to eject all ions with m/z values lower and higher than that of interest, respectively. This method can provide precursor ion isolation with a resolution >3500.

In collisional activation conditions, DIT leads to differences in the relative efficiencies of the collisionally activated decomposition processes: In the low-mass range, DIT leads to an internal energy deposition higher than that observed by QIT; the opposite is true in the high-mass range.

3.1.4 The FT-MS (ICR and Orbitrap) for MS/MS Studies

Both ICR and Orbitrap cells operate under ultrahigh vacuum conditions. Then, if *even if* they could be used for ion selection (i.e., the first step of an MS/MS experiment), they cannot be used to perform collisional experiments. Consequently, the MS/MS systems commercially available based on ICR or Orbitrap devices requires the use of an external cell to perform MS/MS experiments.

The ion optics of three of these systems are reported in Fig. 3.9–3.11. In the first case, (Fig. 3.9) the ions, generated in the ion source are transported in a linear IT by a series of octapoles. In the trap, collisional experiments can be performed with high efficiency and the precursor and product ions can be transferred inside the ICR and analyzed under high-resolution conditions (Thermo Finningan).

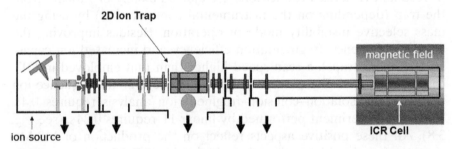

Figure 3.9. Scheme of a commercially available instrument (Thermo Finningan, technical literature) for MS/FTMS experiments.

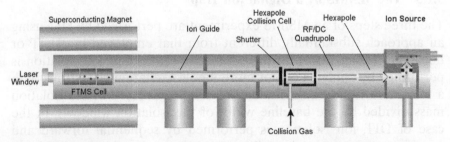

Figure 3.10. Scheme of a commercially available instrument (Varian, technical literature).

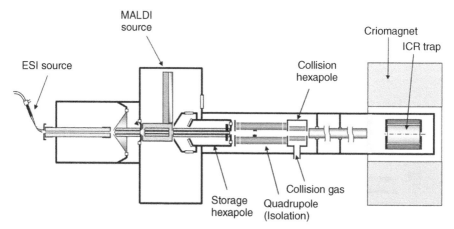

Figure 3.11. Scheme of a commercially available instrument (Bruker Daltonics, technical literature).

The second case (Fig 3.10) (Varian) is strongly analogous to that just described: The only difference lies in the collision cell, that in this case is hexapolar and uses a quadrupole mass filter for ion isolation.

Finally, the instrumental arrangement employed in the third case (Fig 3.11) (Bruker Daltonics) is more complex: The system allows to use both ESI and MALDI sources. Along the ion pathway, storage hexapole, quadrupole mass filter for ion selection, and collisional hexapole cells are present before the ion injection inside the ICR cell.

In all three instrumental configurations, it is possible to induce decomposition of the ion trapped inside the ICR cell not only by collisional experiments, but also by interaction with slow electrons (electron capture dissociation, ECD) or by irradiation with an infrared (IR) laser beam (infrared multi photon dissociations, IRMPD). Experiments of this type (mainly devoted to polypeptide identification) allow to maintain the high-vacuum conditions inside the ICR cell, which are necessary to achieve the high-resolution conditions.

In the case of Orbitrap, an ion optics strongly analogous to that reported in Fig. 3.12 is employed. The cryomagnet and the ICR cell are substituted by the curved quadrupole and the Orbitrap cell, as described in Section 2.3.5.

3.2   THE Q-TOF

Besides the FT–MS-based instruments, which are able to obtain specificity through either MS/MS or accurate mass measurements, another system is commercially available, based on the use of

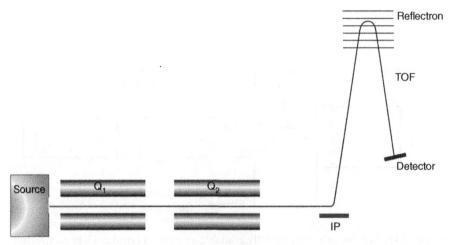

Figure 3.12. Scheme of a Q–TOF system.

quadrupole analyzers coupled with TOF. (Guilhaus et al., 2000; Constans, 2006).

Its basic structure is reported in Fig. 3.12 and can operate in TOF or in product ion scan mode. In the former, both Q_1 and Q_2 operate the rf only mode; in other words they transmit all the ions from the ion source to the ion pusher (IP). Once IP is reached, the ions are pulsed by the application of a suitable electrical field (typical voltage applied is on the order of 10^4 V) for 100 ns every 100 μs, in the TOF analyzer, in a direction orthogonal to the original pathway. By this experimental setup, the mass spectrum of all the ions generated into the ion source can be obtained, with resolution on the order of 15,000–20,000 and accuracy in the parts per million range.

In the product ion scan, mode Q_1 is used to select the ionic species of interest by applying the suitable U, V voltages on the rods. Where Q_2 is used as a collision cell and operating in the rf only mode, the selected ions collide with the target gas and the product ions are analyzed by TOF, thus obtaining accurate mass values.

The advantages of Q–TOF can be summarized as follows:

1. High efficiency in MS/MS experiments.
2. High mass accuracy for both normal mass spectra or product ion spectra.
3. A price surely lower than the FT/MS based instrument.

3.3   THE MALDI TOF-TOF

To achieve more specific and, consequently, more structurally diagnostic data, techniques based on MALDI followed by collisional activation have been developed. It was in 1992 that Spengler et al. showed that in MALDI a large fraction of the desorbed analyte ions undergo delayed fragmentation reactions (occurring during the flight) and that the m/z values of related decomposition ions can be determined by a reflectron time-of-flight (RETOF) analyzer. (Spengler et al., 1992). This technique was called *postsource decay* (PSD) and, until a few years ago, it was the only way to achieve structural information on MALDI generated ionic species of interest. It consists in ion selection by a suitable gating potential, which allows the selection and injection in the flight tube of only the ion of interest; for the mass analysis of its fragment ions the reflectron voltages have to be reduced stepwise. Typically, 10 spectral segments are recorded sequentially with enough overlap to lead to the complete product ion spectrum. The main problem related to this approach lies in mass calibration of the product ion spectrum. To avoid tedious and time-consuming procedures by model decomposition reactions (for which precursor and product ions are well known), computer software have been developed, based on the geometrical parameters and the electrical fields employed, as well as on the flight times of precursor and product ions (Kaufmann et al., 1993).

More recently, to improve the power of the MALDI–MS/MS approach, other two approaches have been proposed. The first is based on the reduction of the acceleration voltage of the MALDI source, so as to achieve an ion beam with a kinetic energy on the order of 8–10 keV. The selection of the ion of interest is performed by an electrostatic gating, by which ions with m/z values lower and higher than that of interest (i.e., with higher and lower speeds) are ejected from the usual ion pathway. After this selection, in a well-defined region placed immediately after the gating plate and consisting of a collision cell, the collision gas can be injected (typical pressure 5×10^{-5} mbar), so that the selected ion can interact with it over a pathway of ~20 cm. After the collision phenomena, the product ions (as well as the survived precursor ions) are further accelerated and analyzed by TOF. Hence, the collision cell can be considered as a supplementary ion source. In order to increase the mass resolution, a reflectron device is usually employed. This approach is called LIFT (Bruker Daltonics, Technical literature).

In the second case, called MALDI–TOF/TOF, an analogous (but significantly different in some aspects) configuration is employed: The collision cell is placed between two different regions; by a suitable lens systems the precursor ion selection is performed with high accuracy (Applied Biosystems, Technical literature). In this case, the ions leave the ion source after the interaction with the usual acceleration potential; only just before the collision cell region are they decelerated by a suitable electrostatic lens system (and by this approach the collision energy can be finely tuned). After collisions, they are further accelerated up to the usual kinetic energy.

Both systems have proved to be effective in the achievement of peptide sequence information without the problems of PSD related to mass calibration. Roughly, the two approaches seem, at first sight, strongly similar. However, deep differences are present in the ion path length before the collision cell and in the kinetic energy regimes of the colliding precursor ions.

Quite surprisingly in some experiments carried out in CID–LIFT conditions, strongly analogous spectra were obtained in the presence or absence of the target gas (Ar) inside the collision cell. The only way to rationalize the minor role of the collision gas is to assume that the internal energy content of the selected species is sufficiently high to promote extensive decomposition processes in the time window employed during the transit of the selected precursor ion inside the collision cell and the time that the contribution due to collisional phenomena becomes negligible (Moneti et al., 2007).

The occurrence of these "natural" decompositions has previously been related to the presence of "metastable" ions, analogous with what is usually observed in magnetic sector machines and analyzed, in the case of double focusing instruments, by MIKES or by linked scans described at the beginning of this chapter. In this case, the metastable ions represent only a minor percentage (~1%) of all the ions leaving the source, and represent ionic species with an internal energy content so as to promote decompositions in the time window corresponding to the field-free regions (FFR) of the instrument. However, in this case the situation is quite different from that observed in the LIFT experiments: In fact, the injection of a collision gas leads to an increase of the signals, but, overall, promotes the occurrence of new fragmentation channels due to an increase of internal energy deposition, typical of a collisional experiment.

The observed phenomenon was rationalized by considering the role of gas-phase collisions in the plume generated by laser irradiation. In this frame, two different kinetic energy regimes of the colliding ion must be considered: The first is due to the initial translational energy related to the ion sputtering prior to the acceleration phase, and the second is after ion acceleration and occurs in a larger volume due to the expansion of neutrals in the vacuum environment (Fig. 3.13).

In the former case, the kinetic energy of the sputtered species was evaluated to be on the order of several electronvolts, and considering the high density of the plume, multiple ion collision can take place with a consequent high internal energy deposition. When delayed extraction DE is employed, the neutral (and radical) species are rapidly pumped out (estimated speed 10^3 m/s) and ions can collide effectively with neutrals in a limited space and in a gas-phase environment, strongly reducing its density with respect to time. Practically speaking, after the DE time, the ejected ions are accelerated in vacuum. On the contrary, without (or with a small) DE time, the ion acquires a high translational energy by acceleration (in the order of 10^4 eV) and more effective collisions can take place inside the expanding neutral plume. This aspect has been described well by Kaufmann et al. (1993), proving that DE reduces to a large extent collisions in source activation.

Consider that the final effect of these collisionally induced internal energy deposition processes are the production of molecular species with a wide internal energy distribution, as shown in Fig. 3.14. Area A corresponds to the molecular species that have experienced a low internal energy uptake, while areas B and C correspond to molecular species experiencing internal energy deposition, so as to promote decomposition processes. In the case of C, the decomposition will take place inside the source prior to acceleration (and the decomposition products will be consequently detected in the usual MALDI spectrum),

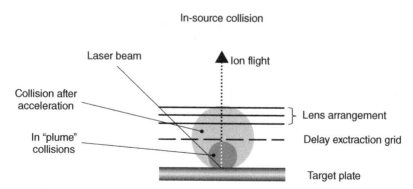

Figure 3.13. Interaction of the laser beam with the solid sample in a MALDI experiment, showing the production of a dense, expanding plume inside the ion source.

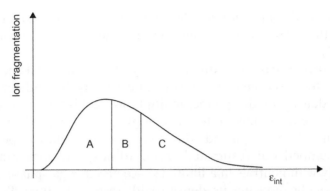

Figure 3.14. Internal energy distribution of ions generated in a MALDI source.

while in the case of B the internal energy deposition will be such as to activate decompositions inside the collision region of the LIFT cell and will be detected only in LIFT operative conditions. In this case the B region is much wider than that present in the metastable ion studies (compare Fig. 3.13 with Fig. 3.2) due to the effective energy deposition occurring in the plume.

REFERENCES

Bruker Daltonics, Technical literature.

Constans, A. (2006). Q-Tof mass spectrometer, *Scientist*, **20**, 73–78.

Cooks, R.G., Beynon, J.H., Caprioli, R.M., and Lester, G.R. (1973). Metastable Ions, *Elsevier*, Amsterdam.

Ding, L., Sudakov, M., Brancia, F.L., Giles, R., and Kumashiro, S. (2004). A digital ion trap mass spectrometer coupled with atmospheric pressure ion sources, *J. Mass Spectrom.*, **39**, 471–484.

Guilhaus, M., Selby, D., and Mlynski, V. (2000). Orthogonal acceleration time-of-flight mass spectrometry, *Mass Spectrom. Rev.*, **19**, 65–107.

Kaufmann, R., Spengler, B., and Lutzenkirchen, F. (1993). Mass spectrometric sequencing of linear peptides by product-ion analysis in a reflectron time-of-flight mass spectrometer using matrix-assisted laser desorption ionisation, *Rapid Commun. Mass Spectrom.*, **7**, 902.

March, R.E. and Todd J.F. (2005). Quadrupole ion trap Mass Spectrometry 2nd ed. Wiley–Interscience.

McLafferty, F.W. (1983). Tandem Mass Spectrometry, John Wiley & Sons, Inc., New York.

Moneti, G., Francese, S., Mastrobuoni, G., Pieraccini, G., Seraglia, R., Valitutti, G., and Traldi, P. (2007). Do collisions inside the collision cell play a relevant role in CID–LIFT experiments? *J. Mass Spectrom.*, **42**, 117–126.

Spengler, B., Kirsch, D., and Kaufmann, R. (1992). Fundamental aspects of post source decay in MALDI mass spectrometry, *J. Phys Chem.*, **96**, 9678.

Thermo Finnigan, Technical literature.

Varian, Technical literature.

Yost, R.A. and Enke, C.G. (1983). Tandem Mass Spectrometry, F.W. McLaffery (Ed.), John Wiley & Sons, Inc, New York, pp. 175–195.

PART II

APPLICATIONS OF MASS SPECTROMETRY IN GRAPE AND WINE CHEMISTRY

4

GRAPE AROMA COMPOUNDS: TERPENES, C₁₃-NORISOPRENOIDS, BENZENE COMPOUNDS, AND 3-ALKYL-2-METHOXYPYRAZINES

4.1 INTRODUCTION

Monoterpenes, C_{13}-norisoprenoids, and some benzenoid compounds are the most important grape aroma substances present in the pulp and skin of berries in both free and glycoside forms. Profiles of these compounds in the grape are mainly dependent on the variety, even if environmental variables and agricultural practices influence their contents (Marais et al., 1992). These compounds are transferred to the wine in winemaking and depend on the process used.

Monoterpenes in wines (principal structures are reported in Fig. 4.1) are mostly mono- and dihydroxylated alcohols and ethers (Strauss et al., 1986; Rapp et al., 1986; Versini et al., 1991; Winterhalter et al., 1998; Bitteur et al., 1990; Guth, 1995; Fariña et al., 2005). These compounds, associated with the floral aroma of aromatic grapes, such as *Muscat* and *Malvasie, Gewürztraminer*, and *Riesling*, are important variety markers (Rapp et al., 1978).

The content and profile of monoterpene compounds in wine change during winemaking and aging due to acid-catalyzed reactions (Rapp et al., 1986; Di Stefano, 1989). At the pH of must and wine, some monoterpenes are transformed into α-terpineol and 1,8-terpines,

Mass Spectrometry in Grape and Wine Chemistry, by Riccardo Flamini
and Pietro Traldi
Copyright © 2010 John Wiley & Sons, Inc.

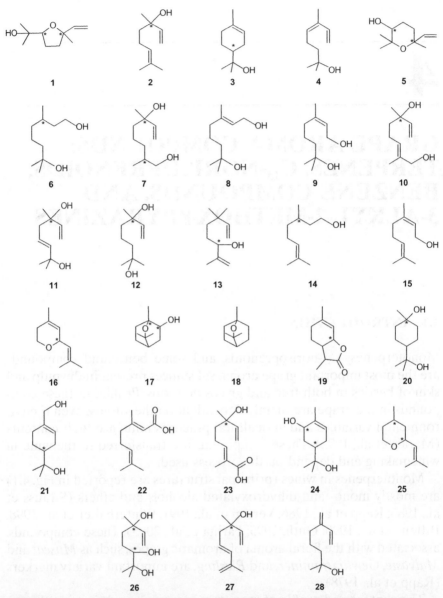

Figure 4.1. Principal monotepenes in grape and wine. (**1**) The *cis-* and *trans*-linalool oxide (5-ethenyltetrahydro-α,α,5-trimethyl-2-furanmethanol) (furanic form); (**2**) linalool (3,7-dimethyl-1,6-octadien-3-ol); (**3**) α-terpineol (α,α,4-trimethyl-3-cycloexene-1-methanol); (**4**) *cis-* and *trans*-ocimenol [(*E*- and *Z*-)2,6-dimethyl-5,7-octen-2-ol]; (**5**) *cis-* and *trans*-linalool oxide (6-ethenyltetrahydro-2,2,6-trimethyl-2H-pyran-3-ol) (pyranic form); (**6**) hydroxycitronellol (3,7-dimethyloctane-1,7-diol); (**7**) 8-hydroxydihydrolinalool (2,6-dimethyl-7-octene-1,6-diol); (**8**) 7-hydroxygeraniol [(*E*)-3,6-dimethyl-2-octene-1,7-diol]; (**9**) 7-hydroxynerol [(*Z*)-3,6-dimethyl-2-octene-1,7-diol]; (**10**) *cis-* and *trans*-8-hydroxylinalool [(*E*- and *Z*-) 2,6-dimethyl-2,7-octadiene-1,6-diol];

geraniol and nerol are transformed into linalool and α-terpineol (Di Stefano et al., 1992), some nonfloral diols and a triol are precursors of compounds characterized by aroma proprieties, such as neroloxide, roseoxide, anhydrofuranes, and anhydropyranes (Williams et al., 1980; Strauss et al., 1986), and (R)-(+)-citronellol is formed by enantio-specific reduction of geraniol and nerol by the yeasts (Gramatica et al., 1982; Di Stefano et al., 1992). Also, formation of highly odorant wine-lactone from (E)-2,6-dimethyl-6-hydroxyocta-2,7-dienoic acid, was observed (Winterhalter et al., 1997; 1998). β-Glycoconjugate monoterpenes in grape are mainly present as β-D-glucopyranoside, and as 6-O-α-L-arabinofuranosyl-, 6-O-β-D-apiofuranosyl-, 6-O-α-L-ramnopyranosyl-β-D-glucopyranosides (Ribéreau-Gayon et al., 1998). By performing an acid hydrolysis at pH 3.0 of β-D-glycosides extract from *Muscat of Alexandria* grape juice, formation of linalool, hotrienol, α-terpineol, geraniol, and nerol, was observed (Williams et al., 1982).

An important contribution to the aroma of aged wines is given by some C_{13}-norisoprenoids (Williams et al., 1989; 1992; Winterhalter et al., 1990; Winterhalter, 1992; Knapp et al., 2002; Versini et al., 2002). Compounds, such as 1,1,6-trimethyl-1,2-dihydronaphthalene (TDN), vitispiranes, actinidols, and β-damascenone confer kerosene, resinous–eucalyptus-like, woody, and rose-like scents, respectively, to the wine. Some noisoprenoids are important contributors to the typical aroma of *Chardonnay*, *Semillon*, *Sauvignon blanc* white wines, and red wines such as *Shiraz*, *Grenache*, *Merlot*, and *Cabernet Sauvignon* (Williams et al., 1989; Abbot et al., 1991; Gunata et al., 2002; Williams and Francis, 1996; Francis et al., 1998), and are the more characteristic aroma compounds of some grape varieties (Flamini et al., 2006). Structures of the principal norisoprenoids are shown in Fig. 4.2.

The most important flavoring benzenoid compounds of grapes are zingerone, zingerol, vanillin (vanilla note), ethyl- (flowery) and methyl- (dry herbs) vanillate, and methyl salicylate, whose structures are

◄

Figure 4.1. (*Continued*) (**11**) Ho-diendiol I (3,7-dimethyl-1,5-octadiene-3,7-diol); (**12**) endiol (3,7-dimethyl-1-octene-3,7-diol); (**13**) Ho-diendiol II (3,7-dimethyl-1,7-octadiene-3,6-diol); (**14**) citronellol (3,7-dimethyl-6-octen-1-ol); (**15**) nerol (Z), geraniol (E) (3,7-dimethyl-2,6-octadien-1-ol); (**16**) neroloxide; (**17**) 2-exo-hydroxy-1,8-cineol; (**18**) 1,8-cineol; (**19**) wine lactone; (**20**) cis- and trans-1,8-terpin; (**21**) p-menthenediol I (p-menth-1-ene-7,8-diol); (**22**) (E)-geranic acid (3,7-dimethyl-2,6-octadienoic acid); (**23**) (E)-2,6-dimethyl-6-hydroxyocta-2,7-dienoic acid; (**24**) (E- and Z)-sobrerol or p-menthenediol II (p-menth-1-ene-6,8-diol); (**25**) cis- and trans-rose oxide; (**26**) triol (2,6-methyl-7-octene-2,3,6-triol); (**27**) hotrienol [(5E)-3,7-dimethylocta-1,5,7-trien-3-ol]; (**28**) myrcenol (2-methyl-6-methylene-7-octen-2-ol).

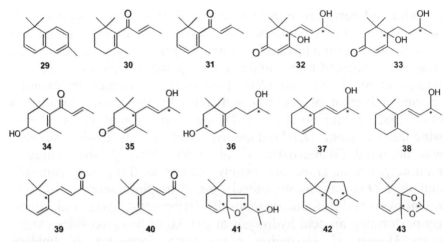

Figure 4.2. Principal norisoprenoid compounds in grape and wine. (**29**) TDN (1,1,6-trimethyl-1,2-dihydronaphthalene); (**30**) β-damascone; (**31**) β-damascenone; (**32**) vomifoliol; (**33**) dihydrovomifoliol; (**34**) 3-hydroxy-β-damascone; (**35**) 3-oxo-α-ionol; (**36**) 3-hydroxy-7,8-dihydro-β-ionol; (**37**) α-ionol; (**38**) β-ionol; (**39**) α-ionone; (**40**) β-ionone; (**41**) actinidols; (**42**) vitispiranes (spiro [4.5]-2,10,10-trimethyl-6-methylene-1-oxa-7-decene); (**43**) Riesling acetal (2,2,6-tetramethyl-7,11-dioxatricyclo[6.2.1.0$^{1.6}$] undec-4-ene).

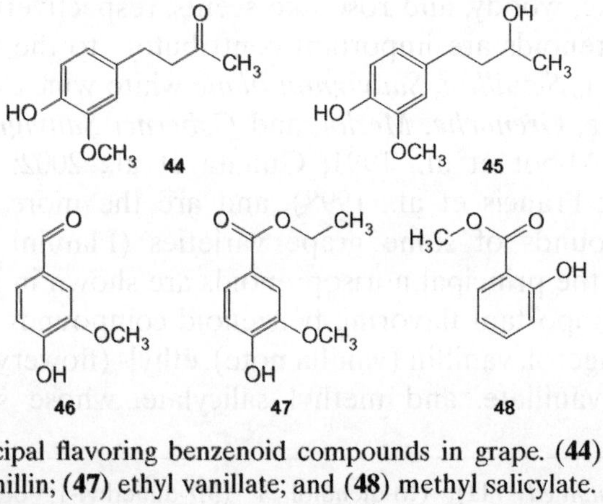

Figure 4.3. Principal flavoring benzenoid compounds in grape. (**44**) zingerone; (**45**) zingerol; (**46**) vanillin; (**47**) ethyl vanillate; and (**48**) methyl salicylate.

reported in Fig. 4.3 (Williams et al., 1983; 1989; Winterhalter et al., 1990; López et al., 2004).

3-Alkyl-2-methoxypyrazines are compounds present in skin, pulp, and bunch stems of grape, and contribute with very characteristic vegetative, herbaceous, bell pepper, or earthy notes to the aroma of *Cabernet Sauvignon, Sauvignon blanc, Semillon,* and other wines

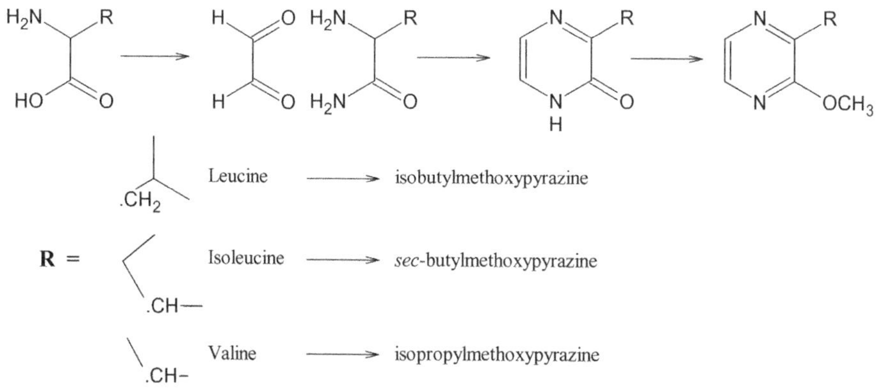

Figure 4.4. Biosynthetic pathways proposed for alkylmethoxypyrazines.

(Lacey et al., 1991; Allen and Lacey, 1993; Hashizume and Umeda, 1996; Hashizume and Samuta, 1997; 1999; Roujou de Boubee et al., 2002). The biosynthetic pathway of these compounds involves formation of the amide of the appropriate amino acid, formation of a pyrazinone, and methylation as shown in Fig. 4.4 (Murray and Whitfield, 1975).

In grape, the level of isobutylmethoxypyrazine decreases dramatically during ripening and generally the level of methoxypyrazines is higher in grapes from cooler regions. Also, exposure to light influences formation of these compounds and lower contents are observed in the more exposed berries (Hashizume and Samuta, 1999). 2-Methoxy-3-isobutylpyrazine, 2-methoxy-3-*sec*-butylpyrazine, and 2-methoxy-3-isopropylpyrazine are characterized by very low sensory thresholds, in water of 1–2 ng/L (Allen et al., 1991; Lacey et al., 1991). In wines, the level of isobutylmethoxypyrazine can be 10-fold of its sensory threshold, while *sec*-butylmethoxypyrazine and isopropylmethoxypyrazine are normally present in levels close to their sensory thresholds. A desirable content of these compounds in *Sauvignon blanc* and *Cabernet Sauvignon* wines could be between 8 and 15 ng/L.

Recently, rotundone was identified as a pepper aroma impact compound in *Shiraz* grapes (Siebert et al., 2008). Identification was achieved by performing GC–MS analysis of grape juice after purification by solid-phase extraction (SPE) using a styrene–divinylbenzene 500-mg cartridge and elution with *n*-pentane/ethyl acetate 9:1, followed by solid-phase microextraction (SPME) using a 65-µm polydimethylsiloxane–divinylbenzene (PDMS/DVB) fiber immersed in the sample for 60 min at 35 °C. d_5-Rotundone was used as an internal standard. The structure of the compound is reported in Fig. 4.5.

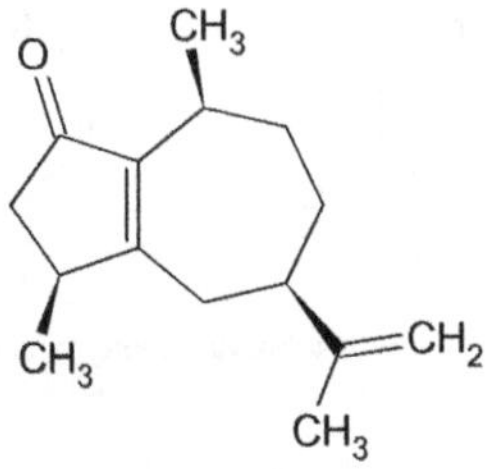

Figure 4.5. Structure of (−)-rotundone.

4.2 THE SPE–GC/MS OF TERPENES, NORISOPRENOIDS, AND BENZENOIDS

Many important fragrances are present in grapes in very low concentrations, but they are characterized by a very low sensory threshold. Isolation of glycosides by selective retention of the compounds on a solid-phase adsorbent is a technique commonly used. A method to extract glycoside precursors from grape juices and wines by using a 1-g C_{18} cartridge was proposed. Hydrophylic compounds are removed by water washing, free terpenes are recovered with dichloromethane, and the fraction containing glycosides is recovered with methanol (Williams et al., 1982, Di Stefano, 1991). Another method is based on the use of Amberlite XAD-2 resin, a sorbent characterized by an excellent capacity for adsorption of free terpenols from grape juice. In this case, free compounds are recovered from the stationary phase with pentane and glycosides with ethyl acetate (Gunata et al., 1985). Alternatively, free compounds can be eluted with a pentane–dichloromethane solution (Versini, 1987). Both Amberlite XAD-2 and Amberlite XAD-16 adsorbents also have the disadvantage of retaining free glucose in addition to glycosides. On the other hand, reversed-phase silica gel is particularly suitable for the isolation of glycosidic terpenes (Williams et al., 1995) and is commercially available in uniform, prepacked cartridges.

To perform sample preparation by solid-phase extraction using a sufficiently large cartridge allows us to concentrate the sample 1000-fold and to perform MS analysis operating in SCAN mode in order to use the mass spectra libraries for compound identification.

More recently, headspace and HS–SPME–GC/MS approaches for analysis of aroma in must and grape extracts were also proposed (López et al., 2004; Prosen et al., 2007; Sánchez-Palomo et al., 2005; Rosillo et al., 1999).

4.2.1   Preparation of Grape Sample

A satisfying method of sample preparation for GC/MS analysis of grape extract and used in several studies (Mateo et al., 1997; Chassagne et al., 2000; Flamini et al., 2001; 2006), was proposed by Williams et al. (1982) and Di Stefano (1991). Skins of 100 berries are separated from the pulp and are extracted with 35 mL of methanol for 4 h in the dark. Pulp and juice are reunited in a glass containing 100 mg of sodium metabisulfite.

> *Pulp.* After homogenization by Ultra-Turrax and centrifugation at 4000 g for 10 min, solid parts are washed with 50 mL of water, again centrifuged, and the clear liquid is reunited to the juice. The volume is adjusted to 250 mL by water and the solution is treated with 75 mg of pectolytic enzyme for 4 h at room temperature. The sample is centrifuged and kept frozen until analysis.

> *Skins.* After extraction, skins are homogenized with methanol, the solution is centrifuged, the solid residue is washed with 50 mL of water, and the supernatant is added to the organic phase. The volume is adjusted to 250 mL with water and the solution is treated with 2 g of insoluble poly(vinylpyrrolidone) (PVP) to reduce the polyphenolic content, which is finally filtered. The sample is kept frozen until analysis.

4.2.2   Analysis of Free Compounds

Analyses of skins and pulp can be carried out separately to study the aroma compounds in different parts of the berry, or a single sample can be analyzed in order to determine the mean contents in the grape. In the first case, 200 mL of extract are added to 200-μL 1-heptanol 180-mg/L water/ethanol 1:1 (v/v) solution as an internal standard, and the solution is passed through a 10-g C_{18} cartridge (e.g., Sep-Pak, Waters) previously activated by successive passage of 30 mL of dichloromethane, 30-mL methanol, and 30-mL water. In the latter case, a sample 100 mL pulp extract + 100 mL skin extract is prepared, added to the internal standard, and the resulting solution is passed through the 10-g C_{18} cartridge. After the sample loading, salts, sugars, and more polar compounds are removed by washing the cartridge with 50 mL of water, and the fraction containing free compounds is recovered by elution with 50 mL of dichloromethane. A second fraction containing glycoside compounds is recovered with 30 mL of methanol.

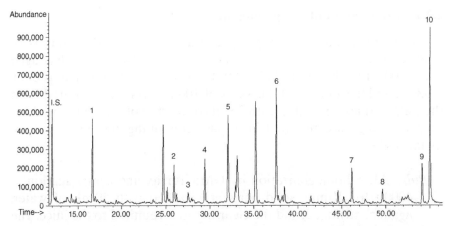

Figure 4.6. The GC/MS–EI (70 eV) chromatogram recorded in SCAN mode of free aroma compounds of a *Muscat* grape skins extract. I.S., internal standard (1-heptanol); peak 1. linalool; peak 2. *trans*-pyranlinalool oxide; peak 3. *cis*-pyranlinalool oxide; peak 4. nerol; peak 5. geraniol; peak 6. Ho-diendiol I; peak 7. Ho-diendiol II; peak 8. hydroxycitronellol; peak 9. 7-hydroxygeraniol; peak 10. (*E*)-geranic acid.

The dichloromethane solution is concentrated to 2–3 mL by distillation using a 40-cm length Vigreux column, and finally to 200 μL under a nitrogen flow prior to GC/MS analysis. The GC/MS profile of free aroma compounds of a *Muscat* grape skin extract is shown in Fig. 4.6.

4.2.3   Analysis of Glycoside Compounds

The methanolic solution is evaporated to dryness under vacuum at 40 °C, the residue is dissolved in 5 mL of a citrate–phosphate buffer (pH 5), then it is added to 200 mg of a glycosidic enzyme with strong glycosidase activity (e.g., AR 2000, Gist Brocades) and kept at 40 °C overnight (15 h). The next day the solution is centrifuged, added to 200 μL of a 1-octanol 180-mg/L solution as an internal standard, and the resulting solution is passed through a 1-g C_{18} cartridge previously activated by passage of 6-mL dichloromethane, 6-mL methanol, and 6-mL water. After cartridge washing with 5-mL water, the fraction containing the aglycones is eluted with 6 mL of dichloromethane, dehydrated with sodium sulfate, and concentrated to 200 μL with a nitrogen flow before analysis. A last fraction, containing the potentially aromatic precursor compounds, is recovered from the cartridge by elution with 5-mL methanol. The GC/MS profile of aglycones from hydrolysis of glycoside compounds of a *Prosecco* grape must is shown in Fig. 4.7.

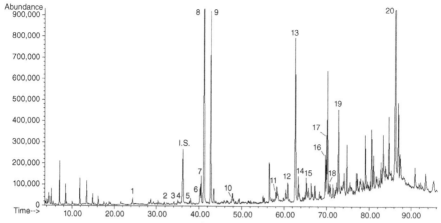

Figure 4.7. The GC/MS–EI (70 eV) SCAN mode chromatogram of aglycones formed by enzymatic hydrolysis of monoterpenol, norisoprenoid, and benzenoid glycosides of a *Prosecco* grape must. Peak 1. linalool; peak 2. α-terpineol; peak 3. *trans*-pyran linalool oxide; peak 4. methyl salicylate; I.S., internal standard (1-decanol); peak 5. nerol; peak 6. 2-*exo*-hydroxy-1,8-cineol; peak 7. geraniol; peak 8. benzyl alcohol; peak 9. β-phenylethanol; peak 10. endiol; peak 11. hydroxycitronellol; peak 12. *trans*-8-hydroxylinalool; peak 13. 7-hydroxygeraniol+*cis*-8-hydroxylinalool; peak 14. (*E*)-geranic acid; peak 15. 4-vinylphenol; peak 16. *p*-menthenediol I; peak 17. 3-hydroxy-β-damascone; peak 18. vanillin; peak 19. 3-oxo-α-ionol; peak 20. vomifoliol.

4.2.4 Analysis of Compounds Formed by Acid Hydrolysis

To reproduce changes in compounds occurring during ageing, hydrolysis of the extract is performed under similar acidic conditions of wines. The methanolic solution is evaporated to dryness under vacuum at 40 °C and the residue is dissolved in 10 mL of tartrate buffer at pH 3. After it is analyzed, the extract from enzymatic hydrolysis of the glycoside compounds fraction is added to this solution and, after addition of 1-g sodium chloride and 200 μL of a 180-mg/L solution of 1-decanol as an internal standard, the solution is heated to boiling and kept under reflux for 1 h. After cooling to room temperature, the solution is passed through a previously activated 360-mg C_{18} cartridge. The cartridge is washed by 3-mL water and the fraction containing volatile compounds is eluted with 4-mL dichloromethane. The solution is dehydrated over sodium sulfate and concentrated to 200 μL under a nitrogen flow before analysis. The GC/MS profile of compounds formed by acid hydrolysis of a *Raboso* grape skins extract is shown in Fig. 4.8.

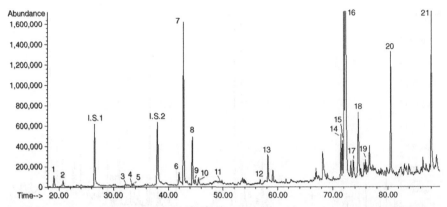

Figure 4.8. The GC/MS–EI (70 eV) SCAN mode chromatogram of compounds formed by acid hydrolysis of a *Raboso* grape skins extract. Peak 1. *trans*-furanlinalool oxide; peak 2. *cis*-furanlinalool oxide; I.S.1, internal standard (1-octanol); peak 3. (*Z*)-ocimenol; peak 4. (*E*)-ocimenol; peak 5. α-terpineol; I.S.2, internal standard (1-decanol); peak 6. 2-exo-hydroxy-1,8-cineol; peak 7. benzyl alcohol; peak 8. β-phenylethanol; peak 9. actinidols A; peak 10. actinidols B; peak 11. endiol; peak 12. eugenol; peak 13. vinylguaiacol; peak 14. *p*-menthenediol I; peak 15. 3-hydroxy-β-damascone; peak 16. vanillin; peak 17. methyl vanillate; peak 18. 3-oxo-α-ionol; peak 19. 3-hydroxy-7,8-dihydro-β-ionol; peak 20. homovanillic alcohol; peak 21. vomifoliol.

4.2.5 GC–MS

The analytical conditions commonly used for analysis of grape aroma compounds are reported in the Table 4.1. Figure 4.9 reports the mass spectra of the principal terpenols and norisoprenoids identified in grapes not reported in the main libraries commercially available.

4.3 THE SPME–GC/MS OF METHOXYPYRAZINES IN JUICE AND WINE

Accurate studies of grape juice and wine matrix effects in the headspace (HS)–SPME analysis of 3-alkyl-2-methoxypyrazines using a triphase fiber divinylbenzene–carboxen–polydimethylsiloxane (DVB/CAR/PDMS) was reported by Kotseridis et al. (2008). Also, PDMS/DVB and CAR/PDMS were selected as suitable fibers for analysis of wines (Sala et al., 2002; Galvan et al., 2008; Ryan et al., 2005). The optimized analytical conditions found are described in Table 4.2.

Increasing the ethanol content in the sample to 20% v/v induced an exponential decrease of recoveries. Best results were obtained by adjusting the sample at pH 6 and wine dilution 1:2.5 before extraction

TABLE 4.1. Common GC/MS Conditions Used for Analysis of Grape Aroma Compounds

GC column	Poly(ethylene)glycol (PEG) bound-phase fused-silica capillary (30 m × 0.25 mm i.d.; 0.25-μm film thickness)
Carrier gas	He Column head pressure 12 psi
Injector	Temperature 200°C, sample volume injected 0.5 μL, splitless injection
Oven program	60°C Isotherm for 3 min, 2°C/min to 160°C, 3°C/min to 230°C, 230°C Isotherm for 5 min
MS conditions	Ionization energy 70 eV, transfer line temperature 280°C, SCAN mode

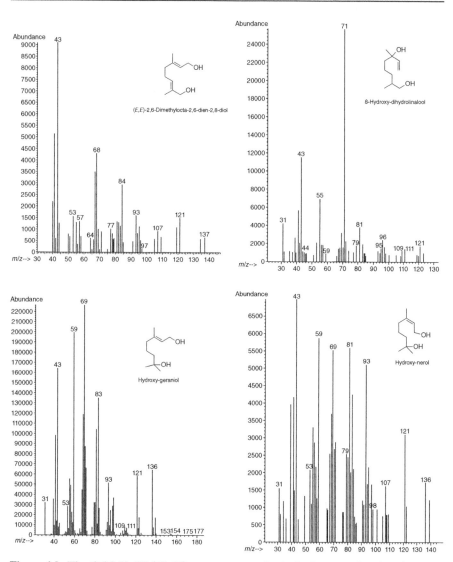

Figure 4.9. The GC/MS–EI (70 eV) mass spectra of principal terpenol and norisoprenoid compounds identified in grape and not reported in the main libraries commercially available.

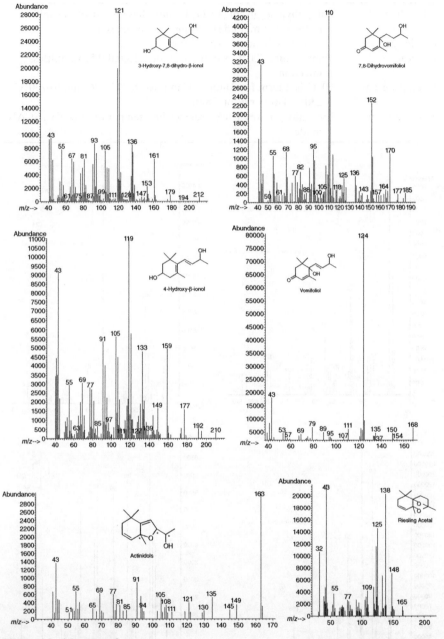

Figure 4.9. (*Continued*)

TABLE 4.2. Headspace-SPME Conditions for Analysis of 3-Alkyl-2-methoxypyrazines in Grape Juice and Wine[a]

Fiber	DVB/CAR/PDMS
	50/30-µm Coating thickness, 1-cm length
Sample volume	20 mL
Vial volume	40 mL
Addition to the sample	6 g NaCl (30% w/v)
Extraction temperature	Headspace: 50 °C for wine, 30 °C for juice
Extraction time	30 min under stirring
Desorption temperature	250 °C
Desorption time	5 min
Fiber cleaning	addition 5 min at 250 °C

[a]Kotseridis et al., 2008.

in order to reduce the ethanol content. By coupling SPME with the use of stable isotope-labeled internal standards, LODs < 0.5 ng/L in juice and 1–2 ng/L in wine were reported for 3-isobutyl-2-methoxypyrazine (IBMP), 3-sec-butyl-2-methoxypyrazine (SBMP), and 3-isopropyl-2-methoxypyrazine (IPMP), with recoveries from spiked wines ranging between 99 and 102% (Kotseridis et al., 2008).

As an alternative to dideuterated $[^2H_2]$-methoxypyrazine and $[^2H_2]$-IBMP (Kotseridis et al., 1999; Godelmann et al., 2008) and trideuterated $[^2H_3]$-IBMP, $[^2H_3]$-SBMP, and $[^2H_3]$-IPMP (Allen et al., 1994; 1995; Lacey et al., 1991; Roujou de Boubée et al., 2000; Kotseridis et al., 2008) isotope-labeled internal standards, ethoypyradines, which are commercially available, can be used (Sala et al., 2002; Hartmann et al., 2002). Various studies revealed that the optimum temperature for IBMP isolation lies between 30 and 40 °C, with fiber adsorption times between 30 and 40 min. Addition of NaCl at 30% maximizes the IBMP transfer into the HS. By comparing DVB/PDMS and DVB/CAR/PDMS fibers, the latter resulted in superior recoveries of IBMP, SBMP, and IPMP (Kotseridis et al., 2008).

The GC/MS conditions used for analysis of 3-alkyl-2-methoxypyrazines in juice and wine are reported in the Table 4.3; the SIM and MS/MS parameters in Table 4.4. Figure 4.10 shows the chromatogram from analysis of a 1:2.5 diluted wine spiked with methoxypyrazines at 0.25 ng/L each and two ethoxypyrazines at 45 ng/L each as internal standards, performed by an ion trap system using the chromatographic conditions reported in Table 4.3. Performing quantitative analysis on the main daughter ions produced by MS/MS of selected precursor ions allows to maximize the signal-to-noise ratio, which improves sensitivity and provides additional qualifier ion fragments.

TABLE 4.3. The GC/MS Conditions for Analysis of 3-Alkyl-2-methoxypyrazines in Juice and Wine

GC column	5% Diphenyl–95% dimethlypolysiloxane bound-phase fused-silica capillary (30 m × 0.25 mm i.d.; 0.25-µm film thickness) or similar
Carrier gas	He Column 1.2 mL/min
Injector	250 °C
Oven program	40 °C for 5 min, 10 °C/min to 230 °C, held 3 min
MS conditions	Ionization energy 70 eV, transfer line temperature 250 °C, SIM (single quadrupole) or MS/MS

TABLE 4.4. The GC/MS–EI (70 eV) Parameters for Single Quadrupole and Ion Trap (IT) Analysis of 3-Alkyl-2-methoxypyrazines[a]

			SIM Quantification Qualifier		IT–MS/MS	
Analyte	MW	GC Retention Time (min)	Ion	Ion	Precursor Ion	Daughter Ions
			m/z			
3-Ethyl-2-methoxypyrazine	138.17	11.7	138	123	138	123;119,109
3-Isopropyl-2-methoxypyrazine	152.20	12.4	137	152	137	109;105;81
3-sec-Butyl-2-methoxypyrazine	166.22	13.6	138	124	138	123;119;81
3-Isobutyl-2-methoxypyrazine	166.22	13.7	124	109	124	109;94;81
Internal Standards						
[²H₃]-IPMP	155.21		140	155		
[²H₃]-SBMP	169.24		141	127		
[²H₃]-IBMP	169.24		127	112		
3-Ethyl-2-ethoxypyrazine	152.20	12.8	123	95	152	152;124;95
3-Isopropyl-2-ethoxypyrazine	166.22	13.4	123	151	166	166;151;123

[a]Analytical conditions reported in Table 4.3.

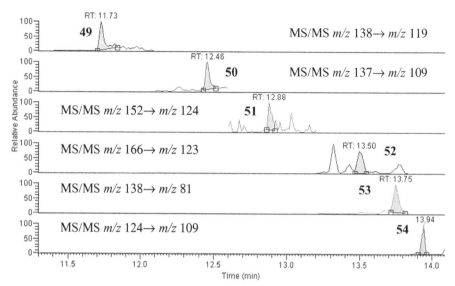

Figure 4.10. Extracted ion chromatograms of GC/IT–MS/MS analysis of a 1:2.5 diluted wine spiked with four methoxypyrazines at 0.25 ng/L and two ethoxypyrazines at 45 ng/L as internal standards. The chromatographic conditions are described in Table 4.3. (**49**) 3-Ethyl-2-methoxypyrazine (*m/z* 119); (**50**) 3-isopropyl-2-methoxypyrazine (IPMP) (*m/z* 109); (**51**) 3-ethyl-2-ethoxypyrazine (internal standard 1) (*m/z* 124); (**52**) 3-isopropyl-2-ethoxypyrazine (I.S. 2) (*m/z* 123); (**53**) 3-*sec*-butyl-2-methoxypyrazine (SBMP) (*m/z* 81); (**54**) 3-isobutyl-2-methoxypyrazine (IBMP) (*m/z* 109).

REFERENCES

Abbott, N.A., Coombe, B.G., and Williams, P.J. (1991). The contribution of hydrolyzed flavor precursors to quality differences in Shiraz juice and wines. An investigation by sensory descriptive analysis, *Am. J. Enol. Vitic.* **42**, 167–174.

Allen, M.S., Lacey, M.J., and Boyd, S. (1994). Determination of methoxypyrazines in red wines by stable isotope dilution gas chromatography-mass spectrometry, *J. Agric. Food Chem.* **42**, 1734–1738.

Allen, M.S., Lacey, M.J., and Boyd, S. (1995). Methoxypyrazines in red wines: occurrence of 2-methoxy-3-(1-methylethyl)pyrazine, *J. Agric. Food Chem.*, **43**, 769–772.

Allen, M.S., Lacey, M.J., Harris, R.L.N., and Brown, W.V. (1991). Contribution of methoxypyrazines to Sauvignon blanc wine aroma, *Am. J. Enol. Vitic.*, **42**(2), 109–112.

Allen, M.S. and Lacey, M.J. (1993). Methoxypyrazine grape flavour: influence of climate, cultivar and viticulture, *Wein-Wissenschaft.*, **48**, 211–213.

Bitteur, S.M., Baumes, R.L., Bayonove, C.L., Versini, G., Martin, C., and Dalla Serra, A. (1990). 2-*exo*-Hydroxy-1,8-cineole: a new component from grape var. Sauvignon, *J. Agric. Food Chem.*, **38**(5), 1210–1213.

Chassagne, D., Alexandre, H., Massoutier, O., Salles, O., and Feuilljat, M. (2000). The aroma glycosides composition of Burgundy Pinot noir must, *Vitis*, **39**(4), 177–178.

Di Stefano, R., Cravero, M.C., and Gentilini, N. (1989). Metodi per lo studio dei polifenoli dei vini, *L'Enotecnico*, **5**, 83–89.

Di Stefano, R., Maggiorotto, G., and Gianotti, S. (1992). Transformation of nerol and geraniol induced by yeasts (Italian), *Riv. Vitic. Enol.*, **45**(1), 43–49.

Di Stefano, R. (1991). Proposition d'une methode de preparation de l'echantillon pour la determination des terpenes libres et glycosides des raisins et des vins, *Bull. O.I.V.*, **64**(721–722), 219–223.

Fariña, L., Boido, E., Carrau, F., Versini, G., and Dellacassa, E. (2005). Terpene compoundsas possible precursors of 1,8-cineole in red grapes and wines, *J. Agric. Food Chem.*, **53**(5), 1633–1636.

Flamini, R., Dalla Vedova, A., and Calò, A. (2001). Studio sui contenuti monoterpenici di 23 accessioni di uve *Moscato*: correlazione tra profilo aromatico e varietà, *Riv. Vitic. Enol.* (**2/3**), 35–49.

Flamini, R., Dalla Vedova, A., Panighel, A., Biscaro, S., Borgo M., and Calò, A. (2006). Caratterizzazione aromatica del Torbato (V. vinifera) e studio degli effetti dell'accartocciamento fogliare sui composti aromatici delle sue uve, *Riv. Vitic. Enol.*, **59**(1), 13–26.

Francis, I.L., Kassara, S., Noble, A.C., and Williams, P.J. (1998). The contribution of glycoside precursors to Cabernet savignon and Merlot aroma, In *Chemistry of wine flavor*, A.L. Waterhouse and S.E. Ebeler (Eds.), ACS Symp. Series 714, *Am. Chem. Soc.*, Washington DC.

Galvan, T.L., Kells, S., and Hutchison, W.D. (2008). Determination of 3-alkyl-2-methoxypyrazines in Lady Beetle-infested wine by solid-phase microextraction headspace sampling, *J. Agric. Food Chem.*, **56**, 1065–1071.

Godelmann, R., Limmert, S., and Kuballa, T. (2008). Implementation of headspace solid-phase-microextraction-GC-MS/MS methodology for determination of 3-alkyl-2-methoxypyrazines in wine, *Eur. Food Res. Technol.*, **227**(2), 449–461.

Gramatica, P., Manitto, P., Ranzi, B.M., Dalbianco, A., and Francavilla, M. (1982). Stereospecific reduction of geraniol to (R)-(+)-citronellol by Saccharomyces cerevisiae, *Experientia*, **38**, 775–776.

Gunata, Y.Z., Bayonnove, C.L., Baumes, R.L., and Cordonnier, R.E. (1985). The aroma of grapes I. Extraction and determination of free and glicosidically bound fractions of some grape aroma components, *J. Chromatogr.*, **331**, 83–90.

Gunata, Z., Wirth, J.L., Guo, W., and Baumes, R.L. (2002). C13-norisoprenoid aglycon composition of leaves and grape berries from Muscat of Alexandria and Shiraz cultivars, In: *Carotenoid-derived aroma compounds*, ACS Symp. Series 802, *Am. Chem. Soc.*, Washington DC.

Guth, H. (1995). Potente Aromastoffe von Weißweinen unterschiedlicher Rebsorten—Identifizierung und Vergleich, *Lebensmittelchemie*, **49**, 107.

Hartmann, P.H., McNair, H.M., and Zoecklein, W. (2002). Measurement of 3-alkyl-2-methoxypyrazine by headspace solid-phase microextraction in spiked model wines, *Am. J. Enol. Vitic.*, **53**(4), 285–288.

Hashizume, K. and Samuta, T. (1999). Grape maturity and light exposure affect berry methoxypyrazine concentration, *Am. J. Enol. Vitic.*, **50**(2), 194–198.

Hashizume, K. and Umeda, N. (1996). Methoxypyrazine content of Japanese red wines, *Biosci. Biotechnol. Biochem.*, **60**, 802–805.

Hashizume, K. and Samuta, T. (1997). Green odorants of grape cluster stem and their ability to cause a wine stemmy flavor, *J. Agric. Food Chem.*, **45**, 1333–1337.

Knapp, H., Straubinger, M., Stingl, C., and Winterhalter, P. (2002). Analysis of norisoprenoid aroma precursors. In *Carotenoid-derived aroma compounds*, ACS Symp. Series 802, *Am. Chem. Soc.*, Washington DC.

Kotseridis, Y.S., Baumes, R.L., Bertrand, A., and Skouroumounis G.K. (1999). Quantitative determination of 2-methoxy-3-isobutylpyrazine in red wines and grapes of Bordeaux using a stable isotope dilution assay, *J. Chromatogr. A*, **841**, 229–237.

Kotseridis, Y.S., Spink, M., Brindle, I.D., Blake, A.J., Sears, M., Chen, X., Soleas, G., Inglis, D., and Pickering, G.J. (2008). Quantitative analysis of 3-alkyl-2-methoxypyrazines in juice and wine using stable isotope labelled internal standard assay, *J. Chromatogr. A*, **1190**, 294–301.

Lacey, M.J., Allen, M.S., Harris, R.L.N., and Brown, W.V. (1991). Methoxypyrazines in Sauvignon blanc grapes and wines, *Am. J. Enol. Vitic.*, **42**(2), 103–118.

López, R., Ezpeleta, E., Sánchez, I., Cacho, J., and Ferreira, V. (2004). Analysis of the aroma intensities of volatile compounds released from mild acid hydrolysates of odourless precursors extracted from Tempranillo and Grenache grapes using gas chromatography-olfactometry, *Food Chem.*, **88**, 95–103.

Marais, J., Versini, G., van Wyk, C.J., and Rapp, A. (1992). Effect of region on free and bound monoterpene and C13-norisoprenoids in Weisser Riesling wines, *S. Afric. J. Enol. Vitic.*, **13**, 71–77.

Mateo, J.J., Gentilini, N., Huerta, T., Jiménez, M., and Di Stefano, R. (1997). Fractionation of glycoside precursors of aroma in grapes and wine, *J. Chromatogr. A*, **778**, 219–224.

Murray, K.E. and Whitfield, F.B. (1975). The occurrence of 3-alkyl-2-methoxypyrazines in raw vegetables, *J. Sci. Food Agric.*, **26**, 973–986.

Prosen, H., Janeš, L., Strlič, M., Rusjan, D., and Kočar, D. (2007). Analysis of free and bound compounds in grape berries using headspace soli-phase microextraction with GC-MS and preliminary study of solid-phase extraction with LC-MS, *Acta Chim. Slov.*, **54**, 25–32.

Rapp, A., Hastrich, H., Engel, L., and Knipser, W. (1978). Possibilities of characterising wine quality and wine varieties by means of capillar chromatography. In G. Charalambous (Ed.) *Flavor of food and beverages, Academic Press*, New York.

Rapp, A., Mandery, H., and Niebergall, H. (1986). Neue Monoterpendiole in Traubenmost und Wein sowie in Kulturen von Botrytis cinerea, *Vitis*, **25**, 79–84.

Ribéreau-Gayon, P., Glories, Y., Maujean, A., and Dubourdieu, D. (1998). Traité d'Oenologie 2. Chimie du vin—stabilisation et traitements, *Dunod*, Paris.

Rosillo, L., Rosario Salinas, M., Garijo, J., and Alonso, G.L. (1999). Study of volatiles in grapes by dynamic headspace analysis. Application to the differentiation of some Vitis vinifera varieties, *J. Chromatogr. A*, **847**, 155–159.

Roujou de Boubee, D., Van Leeuwen, C., and Dubourdieu, D. (2000). Organoleptic impact of 2-methoxy-3-isobutylpyrazine on red Bordeaux and Loire wines. Effect of environmental conditions on concentrations in grapes during ripening, *J. Agric. Food Chem.*, **48**(10), 4830–4834.

Roujou de Boubee, D., Cumsille, A.M., Pons, D., and Dubordieu, D. (2002). Location of 2-methoxy-3-isobutylpirazine in Cabernet sauvignon bunches and its extractability during vinification, *Am. J. Enol. Vitic.*, **53**, 1–5.

Ryan, D., Watkins, P., Smith, J., Allen, M., and Marriott, P. (2005). Analysis of methoxypyrazines in wine using headspace solid phase microextraction with isotope dilution and comprehensive two-dimensional gas chromatography, *J. Sep. Sci.*, **28**, 1075–1082.

Sala, C., Mestres, M., Martí, M.P., Busto, O., and Guasch, J. (2002). Headspace solid-phase microextraction analysis of 3-alkyl-2-methoxypyrazines in wines, *J. Chromatogr. A*, **953**, 1–6.

Sanchéz-Palomo, E., Díaz-Maroto, M.C., and Pérez-Coello, S. (2005). Rapid determination of volatile compounds in grapes by HS-SPME coupled with GC-MS, *Talanta*, **66**, 1152–1157.

Strauss, C.R., Wilson, B., Gooley, P.R., and Williams, P.J. (1986). Role of monoterpenes in grape and wine flavour. In T.H. Parliment and R. Croteau (Eds.), Biogenertion of aromas, ACS Symp. Series 317, *Am. Chem. Soc.*, Washington DC.

Siebert, T.E., Wood, C., Elsey, G.M., and Pollnitz, A.P. (2008). Determination of rotundone, the pepper aroma impact compound, in grapes and wine, *J. Agric. Food Chem.*, **56**, 3745–3748.

Versini, G., Carlin, S., Dalla Serra, A., Nicolini, G., and Rapp, A. (2002). Formation of 1,1,6-trimethyl-1,2-dihydronaphthalene and other norisoprenoids in wine. Considerations on the kinetics. In Carotenoid-derived Aroma Compounds, ACS Symp. Series 802, *Am. Chem. Soc.*, Washington DC.

Versini, G., Rapp, A., Reniero, F., and Mandery, H. (1991). Structural identification and presence of some *p*-menth-1-enediols in grape products, *Vitis*, **30**, 143–149.

Versini, G., Rapp, A., Scienza, A., Dalla Serra, A., and Dell'Eva, M. (1987). Nuovi componenti monoterpenici e nor-isoprenici complessati identificati nelle uve. In: *Le Sostanze Aromatiche dell'Uva e del Vino—Primo Simposio Internazionale*, 25–27 giugno S. Michele All'Adige (TN), pp. 71–92.

Williams, P.J. and Francis, I.L. (1996). Sensory analysis and quantitative determination of grape glycosides—the contribution of these data to winemaking and viticulture. In G.R. Takeoka, R. Teranishi, P.J. Williams, and A. Kobayashi (Eds.), Biotechnology for improved foods and flavors, ACS Symposium Series 637, *Am. Chem. Soc.*, Washington DC.

Williams, P.J., Cynkar, W., Francis, I.L., Gray, J.D., and Coombe, B.G. (1995). Quantification of glycosides in grapes, juices and wines through a determination of glycosyl glucose, *J. Agric. Food Chem.*, **43**, 121–128.

Williams, P.J., Sefton, M., and Francis, I.L. (1992). Glycosidic precursors of varietal grape and wine flavour. In Flavor precursors—Thermal and enzymatic conversions, R. Teranishi, G.R. Takeoka, and M. Güntert (Eds.), ACS Symp. Series 490, *Am. Chem. Soc.*, Washington DC.

Williams, P.J., Sefton, M.A., and Wilson, B. (1989). Nonvolatile conjugates of secondary metabolites as precursors of varietal grape flavor components. In: Flavor chemistry trends and developments, R. Teranishi, R.G. Buttery, and F. Shahidi (Eds.), ACS Symp. Series 388, *Am. Chem. Soc.*, Washington DC.

Williams, P.J., Strauss, C.R., and Wilson, B. (1980). Hydroxylated linalool derivatives as precursors of volatile monoterpenes of Muscat grapes, *J. Agric. Food Chem.*, **28**, 766–771.

Williams, P.J., Strauss, C.R., Wilson, B., and Massy-Westropp, R.A. (1982). Use of C18 reversed-phase liquid chromatography for the isolation of monoterpene glycosides and nor-isoprenoid precursors from grape juice and wines, *J. Chromatogr.*, **235**(2), 471–480.

Williams, P.J., Strauss, C.R., Wilson, B., and Massy-Westropp, R.A. (1983). Glycosides of 2-phenylethanol and benzyl alcohol in Vitis vinifera grapes, *Phytochemistry*, **22**(9), 2039–2041.

Winterhalter, P., Messerer, M., and Bonnländer, B. (1997). Isolation of the glucose ester of the (*E*)-2,6-dimethyl-6-hydroxyocta-2,7-dienoic acid from Riesling wine, *Vitis*, **36**, 55–56.

Winterhalter, P., Baderschneider, B., and Bonnländer, B. (1998). Analysis, structure, and reactivity of labile terpenoid aroma precursors in Riesling wines. In Chemistry of wine flavor, Waterhouse, A.L. and Ebeler, S.E. (Eds.), ACS Symp. Series 714, *Am. Chem. Soc.*, Washington DC.

Winterhalter, P. (1992). Oxygenated C13-norisoprenoids—Important flavor precursors. In: Flavor precursors—Thermal and enzymatic conversions, R. Teranishi, G.R. Takeoka, and M. Güntert (Eds.), ACS Symp. Series 490, Washington DC.

Winterhalter, P., Sefton, M.A., and Williams, P.J. (1990). Two-dimensional GC-DCCC analysis of the glycoconjugates of monoterpenes, norisoprenoids, and shikimate derived metabolites from Riesling wine, *J. Agric. Food Chem.*, **38**(4), 1041–1048.

5

VOLATILE AND AROMA COMPOUNDS IN WINES

5.1 HIGHER ALCOHOLS AND ESTERS FORMED FROM YEASTS

5.1.1 Introduction

The main volatiles in wines are the higher aliphatic alcohols, ethyl esters, and acetates formed from yeasts during fermentation. Acetates are very important flavors characterized by fruity notes, C_4–C_{10} fatty acid ethyl esters manly confer fruity scents to the wine. Other wine aroma compounds are C_6 alcohols, such as 1-hexanol and *cis*- and *trans*-3-hexen-1-ol, 2-phenylethanol, and 2-phenylethyl acetate. Contents of these compounds in wine are linked to the winemaking processes used: fermentation temperature, yeast strain type, nitrogen level in must available for yeasts during fermentation, clarification of wine (Rapp and Versini, 1991). Much literature on the wine aroma compounds was reported in reviews by Schreier (1979) and Rapp (1988).

5.1.2 SPME–GC/MS Analysis of Higher Alcohols and Esters

Analysis of volatile compounds in wine is usually performed by gas chromatography (GC) coupled with flame ionization (FID) or GC/

Mass Spectrometry in Grape and Wine Chemistry, by Riccardo Flamini and Pietro Traldi
Copyright © 2010 John Wiley & Sons, Inc.

MS. The first methods of sample preparation were performed by liquid–liquid extraction using solvents, such as dichloromethane/pentane 2:1 (v/v), Freon 11, or a Freon 11–dichloromethane 9:1 (v/v) mixture (Drawert and Rapp, 1968; Hardy, 1969; Rapp et al., 1978; Marais, 1986). Alternatively, liquid–liquid discontinuous extraction with 1,1,2-trichloro-1,2,2-trifluoroethane (Freon 113 or Kaltron), was proposed (Ferreira et al., 1993; Rapp et al., 1994; Genovese et al., 2005).

Different methods of solid-phase extraction (SPE) of wine volatiles were developed by using Amberlite XAD-2 polystyrenic resins (Gunata et al., 1985; Versini et al., 1988), reverse-phases C_{18} (Williams, 1982; Gianotti and Di Stefano, 1991; Di Stefano, 1991; Zulema et al., 2004; Ferreira et al., 2004), and more recently, highly cross-linked hydroxylated polystyrenic phases (e.g., ENV+, Ferreira et al., 2004; Boido et al., 2003) and highly cross-linked ethylvinylbenzene–divinylbenzene copolymers (e.g., LiChrolut EN, López et al., 2002; Ferreira et al., 2004; Genovese et al., 2005) stationary phases.

In general, SPE provides high recoveries of most fermentative volatiles in wine (80–100%), but requires longer times and is quite solvent consuming. On the other hand, the main advantage of this approach is to allow separation of the fraction of glycoside compounds that can be analyzed as aglycones after an enzymatic hydrolysis.

Solid-phase microextraction (SPME) of wine was developed by both headspace (HS) (Vas et al., 1998) and liquid-phase sampling (De la Calle et al., 1996). Exhaustive overviews on materials used for the extraction–concentration of aroma compounds in wines were published from Ferreira et al. (1996), Cabredo-Pinillos et al. (2004), and Nongonierma et al. (2006).

Headspace is useful for the trace analysis of compounds having a high affinity for the fiber phase and that can be enriched in the HS of the sample. The use of a multiphase fiber is a very interesting and low time-consuming approach. It also considers the possibility of sampling automation using a GC/MS system coupled with a statistical method for treatment of fragment abundance (Kinton et al., 2003; Cozzolino et al., 2006).

Many qualitative and comparative studies of volatiles in wines by HS–SPME were preformed (Favretto et al., 1998; Marengo et al., 2001; Vas et al., 1998; Begala et al., 2002; Demyttenaere et al., 2003; Rocha et al., 2006). The calibration curves of the common volatiles in wines can be calculated because of standards that are commercially available and the method is suitable for quantitative analyses. Alternatively, quantitative GC/MS can be performed by using deuterated standards, for example, by synthesis of ethyl esters by reaction of the corresponding organic acid with d_5-ethanol (Siebert et al., 2005). Polydimethylsiloxane (PDMS) fibers have high affinity for high molecular weight nonpolar compounds, such as ethyl esters and higher alcohols (Bonino et al., 2003, Vianna and Ebeler, 2001; Martí et al., 2003), a mixed-fiber carbowax–divinylbenzene (CW/DVB) for C_6 and higher aliphatic alcohols and monoterpenols (Bonino et al., 2003). The triphase CAR/PDMS/DVB fiber is suitable for sampling of both lower molecular weight and more polar compounds (Howard et al., 2005), overcoming the lack of selectivity toward some compounds of the one- or two-phase fibers (Ferreira and de Pinho, 2003). The HS–SPME–GC/MS chromatogram relative to analysis of wine volatiles by triphase fiber is shown in Fig. 5.1, the experimental conditions are reported in Table 5.1. A list of wine volatiles detectable by this method is reported in Table 5.2.

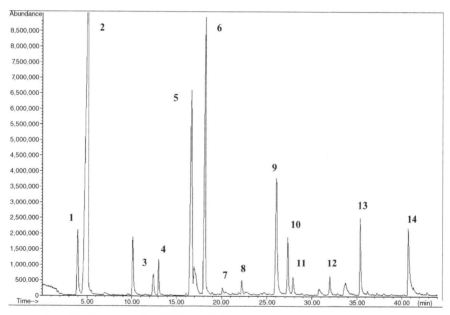

Figure 5.1. HS (headspace)–SPME–GC/MS chromatogram recorded in the analysis of a *Gewürztraminer* wine volatiles performed using a CAR–PDMS–DVB fiber and the experimental conditions reported in Table 5.1. (**1**) ethyl hexanoate; (**2**) 2- and 3-methyl-1-butanol (isoamyl alcohols); (**3**) ethyl lactate; (**4**) 1-hexanol; (**5**) ethyl octanoate; (**6**) 1-heptanol (internal standard); (**7**) benzaldehyde; (**8**) linalool; (**9**) ethyl decanoate; (**10**) diethyl succinate; (**11**) α-terpineol; (**12**) 2-phenylethyl acetate; (**13**) 2-phenylethanol; (**14**) octanoic acid.

TABLE 5.1. Experimental Conditions for HS–SPME–GC/MS Analysis of Volatiles in Wine (Howard et al., 2005)

Sample volume	6 mL of wine diluted 1:1 (v/v)
Vial volume	10 mL
SPME fiber	CAR/PDMS/DVB 50/30-μm coating thickness, 1 cm length
Addition to the sample	2.1 g Na_2SO_4
Extraction temperature	Headspace at 40 °C
Extraction time	30 min under stirring
Desorption temperature	250 °C
Desorption time	5 min
GC column	Poly(ethylene)glycol (PEG) bound-phase fused-silica capillary (30 m × 0.25 mm i.d.; 0.25-μm film thickness)
Internal standards	50-μL methyl heptanoate (or 4-methyl-2-pentanol) 1 ppm solution for higher alcohols
	Ethyl nonanoate 0.1 ppm for esters, 1,6-heptadien-4-ol 50 ppb for monoterpenols
Carrier gas	He Column head pressure 12 psi
Injector	Splitless
Oven program	35 °C Isotherm for 5 min, 3 °C/min to 210 °C, 210 °C isotherm for 10 min
MS conditions	EI (70 eV), SCAN range m/z 40–450

Alternatively, the HS–SPME of higher alcohols and, in particular, of aliphatic esters in wine can be efficiently performed by a 100-μm PDMS fiber (Francioli et al., 1999; Vas et al., 1998; Pozo-Bayón et al., 2001; Marengo et al., 2001; Vianna and Ebeler, 2001; Demyttenaere et al., 2003; Alves et al., 2005). In this case, a volume of 4 mL of wine is added to sodium chloride (1–2 g) and of an internal standard (e.g., 1-heptanol or 2-octanol). The solution is transferred in a 10-mL vial, kept at 25–30 °C for 10 min, and the sampling is performed by exposing the fiber, under stirring for 30 min, to the HS of the sample. By this approach, tertiary alcohols (e.g., linalool, geraniol, and citronellol) are well extracted, but low absorption on the fiber is observed for 2-phenylethanol and 1-hexanol. The absorption of 2-phenylethyl acetate and diethyl succinate (Versini et al., 2008) are also found to be lower. A recent study of varietal compounds in grape homogenate (Sanchéz-Palomo et al., 2005) described a more reproducible adsorption efficiency by operating HS–SPME/GC–MS with the other apolar phase PDMS/DVB by satisfying adsorption on the fiber of monoterpenols and benzenoids including 2-phenylethanol. A list of wine volatiles detectable by HS–SPME using 100-μm PDMS fiber is reported in Table 5.3.

TABLE 5.2. Wine Volatiles Detectable by HS–SPME Using a CAR–PDMS–DVB Fiber and Their Principal *m/z* Signals[a]

Compound	*m/z*	Compound	*m/z*
(E)-2-Nonenal	70;41;83	Ethyl octanoate	88;101;127
(E,E)-2,4-Decadienal	81;41;39	Ethyl propanoate	29;57;27
1,1,6-Trimethyl-1,2-dihydronaphthalene	157;142;172	Ethyl lactate	45;29;75
1-Hexanol	56;43;69	Geranyl ethylether	69;93;121
1-Octen-3-ol	72;57;85	Hexanoic acid	60;73,87
2-Methyl-1-butanol	57;41;70	Hexyl acetate	43;56;61
2-Octanone	43;58;71	Isoamyl acetate	43;70;55
2-Phenylethanol	91;92;122	Isoamyl alcohol	55;42;70
2-Phenylethyl acetate	104;43;91	Isoamyl octanoate	70;127;43
3-Methyl-1-butanol	55;42;70	Isobutyl alcohol	43;41;42
Acetic acid	43;45;60	Linalool	71;93;55
cis-3-Hexenol	67;41;55	Linalyl ethylether[b]	71;43;99
cis-Furanlinalool oxide	59;94;111	Methyl decanoate	74;87;155
Decanoic acid	60;73;129	Methyl heptanoate (I.S.)	74;43;87
Diethylsuccinate	101;129;55	Methyl hexanoate	74;87;99
Ethyl 2-methylbutanoate	102;85;74	Methyl octanoate	74;87;127
Ethyl 3-hexenoate	69;41;68	Octanoic acid	60;73;101
Ethyl 9-decenoate	41;55;88	Propanol	31;29;42
Ethyl acetate	61;70;73	trans-Furanlinalool oxide	59;43;68
Ethyl butanoate	71;43;88	Vitispiranes	192;177;121
Ethyl decanoate	88;61;155	α-Ionone	121;93;192
Ethyl dodecanoate	88;101;183	α-Terpineol	59;93;136
Ethyl hexanoate	88;99;60	β-Damascenone	69;121;190
Ethyl isobutanoate	43;29;71	β-Ionone	177;178;135
Ethyl isovalerate	29;57;88	Furfural	96;95;39

[a]EI 70 eV. Versini et al., 2008; Ferreira and de Pinho, 2003; Bosch-Fusté, 2007.
[b]Data kindly provided by Prof. R. Di Stefano.

TABLE 5.3. Wine Volatiles Detectable by HS–SPME Using a 100-μm PDMS Fiber and Their Principal *m/z* Signals[a]

Compound	*m/z*	Compound	*m/z*
(E)-2-Nonenal	70;41;83	Ethyl hexanoate	88;99;60
(E)-Cinnamaldehyde	131;103;51	Ethyl isobutanoate	43;29;71
(E)-β-Ocimene	93;91;73	Ethyl isovalerate	29;57;88
(E,E)-2,4-Decadienal	81;41;39	Ethyl lactate	45;29;75
(Z)-3-Hexen-1-ol	67;41;55	Ethyl nonanoate	88;101;73
(Z)-β-Ocimene	93;91;79	Ethyl octanoate	88;101;127
1,2-Dihydro-1,1,6-trimethylnaphthalene	157;142;172	Ethyl propanoate	29;57;27
1,3-Butanediol	43;45;57	Ethyl sorbate	67;95;41

TABLE 5.3. (*Continued*)

Compound	m/z	Compound	m/z
1-Hexanol	56;43;69	Geraniol	69;41;93
1-Octen-3-ol	72;57;85	Geranyl ethylether	69;93;121
2,2,6-Trimethyl-6-vinyltetrahydropyran	43;71;68	Hexanoic acid	60;73,87
2,3-Butanediol	45;57;75	Hexyl acetate	43;56;61
2,6-Di-terbutyl-4-methylphenol	205;57;220	Ho-trienol	71;82;67
2-Ethyl-1-hexanol	57;41;55	Isoamyl acetate	43;70;55
2-Methyl-1-butanol	57;41;70	Isoamyl decanoate	70;43:155
2-Methyl-1-butyl acetate	43;87;70	Isoamyl octanoate	70;127;43
2-Methyl-1-butyl hexanoate	70;43;99	Isobutyl acetate	43;56;73
2-Methyl-1-propanol	33;74;55	Isobutyl octanoate	57;127;41
2-Methylbutyl octanoate	127;43;60	Isopentyl hexanoate	70;43;99
2-Phenylethanol	91;92;122	Lavandulyl acetate	69;81;95
2-Phenylethyl acetate	104;43;91	Limonene	68;93;67
3,7-Dimethyl-1,5,7-octatrien-3-ol	71;82;43	Linalool	71;93;55
3-Methyl-1-butanol	55;42;70	Linalyl ethylether[b]	71;43;99
3-Methyl-1-butyl hexanoate	70;43;99	Methyl decanoate	74;87;155
4-Ethylphenol	107;122;77	Methyl octanoate	74;87;127
4-Ethylphenyl acetate	91;29;164	Methyl *trans*-geranate	69;41;114
cis-Furanlinalool oxide	59;94;111	Nerol oxide	68;41;83
Citronellol	69;81;123	Nerolidol	69;93;107
Citronellyl acetate	69;43;95	Neryl ethylether[b]	41;69;93
Decanal	57;95;112	Octanoic acid	60;73;101
Decanoic acid	60;73;129	*p*-Cymene	119;134;91
Diethylsuccinate	101;129;55	Sorbic acid	97;112;67
Ethyl 2-hexenoate	97;55;73	*trans*-Furanlinalool oxide	59;43;68
Ethyl 2-methylbutanoate	102;85;74	Vitispiranes	192;177;121
Ethyl 2-methylpropanoate	43;71;116	α-Ionone	121;93;192
Ethyl 3-methylbutanoate	88;57;70	α-Terpenyl ethylether[b]	59;136
Ethyl 7-octenoate	55;88;124	α-Terpineol	59;93;136
Ethyl 9-decenoate	41;55;88	α-Terpinolene	93;136;121
Ethyl acetate	43;61;45	β-Damascenone	69;121;190
Ethyl butanoate	71;43;88	β-Ionone	177;178;135
Ethyl decanoate	88;101;155		
Ethyl dodecanoate	88;101;183		
Ethyl heptanoate	88;43;113		

[a]EI-70 eV. Demyttenaere et al., 2003; Ferreira and de Pinho, 2003; Versini et al., 2008; Marengo et al., 2001.
[b]Data kindly provided by Prof. R. Di Stefano.

5.2   VOLATILE SULFUR COMPOUNDS IN WINES

5.2.1   Introduction

Heavier sulfur compounds have a detrimental effect on the wine aroma. Off-flavor sulfur compounds in wine are divided into light (boiling point <90 °C) and heavy (b.p. > 90 °C) thiols, sulfides, thioesters, and heterocyclic compounds (Ribereau-Gayon et al., 1998; Mestres et al., 2000). Garlic odor is associated with *trans*-2-methylthiophan-3-ol and 4-methylthiobutan-l-ol; onion to 2-methyltetrahydrothiophenone, cabbage to methionol, and cauliflower to 2(methylthio)-ethanol. Other sulfur compounds with b.p. < 90 °C are often characterized by strong off-flavors and very low sensory thresholds (Rapp et al., 1985). These compounds are formed through both enzymatic processes during fermentation of yeasts and nonenzymatic chemical, photochemical, and thermal reactions occurring in wine making and during aging (Mestres et al., 2000).

3-Mercaptohexan-1-ol (3-MH) and 4-methyl-4-mercaptopentan-2-one (4-MP, at ppt levels characterized by a box-tree-like aroma typical of *Sauvignon blanc* wines) are present in grape mainly as (*S*)-cysteine conjugates (Ribereau-Gayon et al., 1998; Flanzy, 1998), in wines 3-mercaptohexyl acetate (3-MHA) is present also (Swiegers and Pretorius, 2007; Swiegers et al., 2007). Both 3-MH and 3-MHA are typical of *Sauvignon blanc* wines and are characterized by a tropical fruit-like scent. Other sulfur compounds that may be present in wines are ethylmercaptan (EtSH), dimethyl sulfide (DMS, grassy/truffle-like note), diethyl sulfide (DES), dimethyl disulfide (DMDS), diethyl disulfide (DEDS), methyl thioacetate (MTA), ethyl thioacetate (ETA), 2-mercaptoethanol (ME), 2-(methylthio)-1-ethanol (MTE), 3-(methylthio)-1-propanol (MTP), 4-(methylthio)-1-butanol (MTB, earthy-like scent), benzothiazole (BT) and 5-(2-hydroxyethyl)-4-methylthiazole (HMT), since DMS, MTB, MTE, and BT are characteristic of *Merlot* wines (Versini et al., 2008). Also, 2-methyl-3-furanthiol (MF, a very odoriferous compound with an odor threshold of 0.4–1.0 ppt) was found in wines (Bouchilloux et al., 1998). Structures of these compounds are reported in Fig. 5.2.

In general, DMS increases during aging (Simpson, 1979; Segurel et al., 2005), instead methyl and ethyl thioacetates hydrolyze, and there is an increase of thiols and disulfides observed (Rauhut, 1996). Another sulfur compound found in wine is bis(2-hydroxyethyl)disulfide, which was found in extracts of wines, in particular from *Vitis labrusca* grapes or its hybrids (Anocibar Beloqui et al., 1995). This compound is a precursor of H_2S and 2-mercaptoethanol, two substances with a very strong rotten egg and unpleasant poultry-like odor, respectively. After

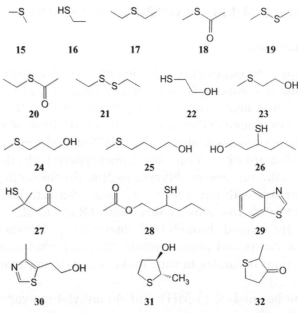

Figure 5.2. Volatile sulfur compounds of wines: (**15**) dimethyl sulfide, (**16**) ethylmercaptan, (**17**) diethyl sulfide, (**18**) methyl thioacetate, (**19**) dimethyl disulfide, (**20**) ethyl thioacetate, (**21**) diethyl disulfide, (**22**) 2-mercaptoethanol, (**23**) 2-(methylthio)-1-ethanol, (**24**) 3-(methylthio)-1-propanol, (**25**) 4-(methylthio)-1-butanol, (**26**) 3-mercaptohexan-1-ol, (**27**) 4-methyl-4-mercaptopentan-2-one, (**28**) 3-mercaptohexanol acetate, (**29**) benzothiazole, (**30**) 5-(2-hydroxyethyl)-4-methylthiazole, (**31**) *trans*-2-methylthiophan-3-ol, (**32**) 2-methyltetrahydrothiophen-3-one.

addition of 5-g Na₂SO₄ to 50mL of the sample, extraction was performed with 2 × 5mL of ethyl acetate. The GC/MS–EI (70eV) mass spectrum of bis(2-hydroxyethyl)disulfide is reported in Fig. 5.3.

5.2.2 HS–SPME–GC/MS Analysis of Volatile Sulfur Compounds

The carboxen–polydimethylsiloxane–divinylbenzene (CAR/PDMS/DVB) 50:30µm and 2-cm length resulted in the more efficient fiber for the extraction of sulfur compounds with the simple sampling conditions (e.g., ionic strength, sample temperature, and adsorption time) (Fedrizzi et al., 2007a). A suitable HS–solution volumes ratio is 1:2 (Mestres et al., 2000). Figure 5.4 reports the HS–SPME–GC/MS chromatograms relative to analyses of compounds used as internal standards (a) and of analytes (b) using the SPME conditions reported

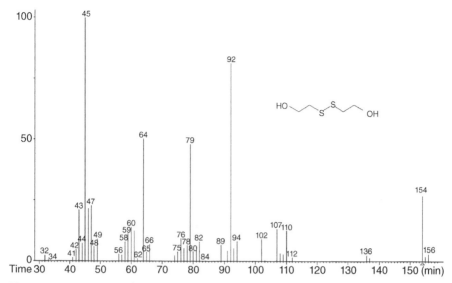

Figure 5.3. GC/MS–EI (70 eV) mass spectrum of bis(2-hydroxyethyl)disulfide, a precursor of H_2S (strong odor of rotten eggs) and of 2-mercaptoethanol (poultry-like aroma) in wine.

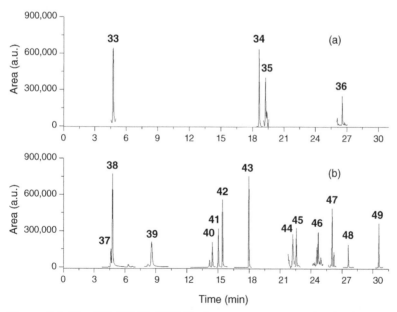

Figure 5.4. The HS–SPME–GC/MS chromatograms recorded in SIM mode in the analysis of compounds reported in Table 5.5: (a) internal standards (**33**. d_6-DMS, m/z 68; **34**. DPDS, m/z 108; **35**. MT, m/z 71; **36**. MTH, m/z 148), (b) analytes (**37**. EtSH, m/z 62; **38**. DMS, m/z 62; **39**. DES, m/z 75; **40**. MTA, m/z 90; **41**. DMDS, m/z 94; **42**. ETA, m/z 104; **43**. DEDS, m/z 122; **44**. ME, m/z 78; **45**. MTE, m/z 92; **46**. MTP, m/z 106; **47**. MTB, m/z 120; **48**. BT, m/z 135; **49**. HMT, m/z 122). The SPME conditions are reported in Table 5.4. (Reprinted from Rapid Communications in Mass spectrometry 21, Fedrizzi et al., Concurrent quantification of light and heavy sulphur volatiles in wine by headspace solid-phase microextraction coupled with gas chromatography/mass spectrometry, p. 710, Copyright © 2007, with permission from John Wiley & Sons, Ltd.)

TABLE 5.4. The SPME Conditions Used for Analysis of Sulfur Compounds[a]

Sample volume	20 mL
Vial volume	30 mL
SPME fiber	CAR/PDMS/DVB 30:50 μm—2 cm
Salt addition	5 g $MgSO_4 \cdot 7H_2O$
Extraction	Headspace at 35 °C
Extraction time	30 min under stirring

[a]Fedrizzi et al., 2007a.

TABLE 5.5. The *m/z* Signals Used for GC/MS Analysis of Sulfur Compounds in Wines[a]

Sulfur Compounds	Quantifier Ion	Qualifier Ions
Dimethyl sulfide (DMS)	76	62 45,47
Ethylmercaptan (EtSH)	61	62 47,61
Diethyl sulfide (DES)	52	75 61,90
Methyl thioacetate (MTA)	42	90 43,47
Dimethyl disulfide (DMDS)	94	64,79
Ethyl thioacetate (ETA)	104	43,60
Diethyl disulfide (DEDS)	122	66,94
2-Mercaptoethanol (ME)	78	47,60
2-(Methylthio)-1-ethanol (MTE)	92	47,61
3-(Methylthio)-1-propanol (MTP)	106	58,61
4-(Methylthio)-1-butanol (MTB)	120	61,102
Benzothiazole (BT)	135	69,108
5-(2-Hydroxyethyl)-4-methylthiazole (HMT)	112	85,143
Internal Standards		
d_6-DMS	68	50,66
Dipropyl disulfide (DPDS)	108	66,150
4-Methylthiazole (MT)	71	39,99
3-(Methylthio)-1-hexanol (MTH)	148	61,75

[a]Chromatograms are showed in Fig. 5.4. Below, the internal standards are reported. (From Versini et al., 2008).

in Table 5.4. The MS fragments recorded in singular ion monitoring (SIM) mode used for detection of analytes and internal standards are listed in the Table 5.5.

The use of a 2-cm length fiber provides a higher sensitivity for all analytes. For most compounds, the best salting-out effect is achieved by addition of $MgSO_4 \cdot 7H_2O$ 1.0 M. A temperature of 35 °C is the best compromise between efficient sampling of less volatile compounds and reduced desorption of the higher volatile compounds.

Recoveries also depend on the wine matrix: white and red wines differ for volatiles and polyphenols, and contents of alcohol and sugars. For quantitative analysis, internal standards have to be spiked to the sample in suitable concentration, such as dimethyl sulfide-d_6 (d_6-DMS) 25 µg/L, dipropyl disulfide (DPDS) 25 µg/L, 4-methylthiazole (MT) 10 µg/L, and 3-(methylthio)-1-hexanol (MTH) 50 µg/L. A white wine matrix can be used [e.g., 10% v/v ethanol, sugar content <4 g/L, and a polyphenolic content of 115 mg/L expressed as (+)-catechin] for preparation of standard solutions to calculate the calibration curve for analytes, containing total SO_2 corrected to 100 g/L and previously treated twice with charcoal 3 g/L to remove sulfur and less polar volatile compounds (higher alcohols are not removed).

The GC/MS (EI 70 eV) analysis is performed by using a PEG bound-phase fused silica (30 m × 0.32 mm i.d.; 0.25-µm film thickness) capillary column (carrier gas He at a flow rate of 1.2 mL/min) recording the MS signals in SIM mode (transfer line 220 °C, MS source 150 °C). GC conditions: injector temperature 250 °C in splitless mode for 1 min, oven temperature program from a 35 °C isotherm for 5 min, 1 °C/min to 40 °C, 10 °C/min to 250 °C (Fedrizzi et al., 2007a).

5.2.3 HS–SPME–GC/MS Analysis of 3-MH and 3-MHA

Several separation and preconcentration methods for analysis of 3-MH, 3-MHA, and 4-MP, were reported (Tominaga et al., 1998; Schneider et al., 2003; Mateo-Vivaracho et al., 2007). Recently, HS–SPME–GC/MS method with limits of quantification (LOQs) of a dozen ng/L (close to the sensory thresholds of 3-MH and 3-MHA) was proposed (Fedrizzi et al., 2007b). This method does not include determination of 4-MP because the sensory threshold of this compound in wines (~0.4 ng/L) is too low. The method has to be calibrated using a white wine pretreated with charcoal, corrected to a total SO_2 100 g/L, and spiked with different quantities of standard solutions. The optimized HS–SPME conditions (temperature, sampling time, solution pH) are reported in Table 5.6.

Alternatively, 3-MH and 3-MHA can be determined by solid-phase extraction (SPE) using a cross-linked styrene–divinylbenzene polyhydroxylated polymer cartridge (ENV⁺ 1 g) and reaching LODs comparable to those provided by the HS–SPME method (Boido et al., 2003; Fedrizzi et al., 2007b), or by Purge and Trap (PT). Before performing SPE, the sample pH is adjusted to 7.0 to reduce the free fatty acids absorption on the resin. A volume of 100 mL of wine is diluted 1:1 v/v with water and passed through the cartridge. Analytes are recovered

TABLE 5.6. Optimized HS–SPME Conditions for Analysis of 3-MH and 3-MHA in Wine[a]

Sample volume	20 mL
Vial volume	30 mL
Solution pH	7.0
SPME fiber	CAR/PDMS/DVB 30:50 µm—2 cm
Salt Addition	5 g $MgSO_4 \cdot 7H_2O$
Extraction	Headspace at 40 °C
Extraction time	40 min under stirring

[a]Proposed by Fedrizzi et al. (2007b).

TABLE 5.7. Optimized experimental conditions used for Purge and Trap analysis of 3-MH and 3-MHA in wine[a]

Sample volume	5 mL
Sample temperature	45 °C
Solution pH	7.0
Purge flow time	15 min
Purge flow rate	50 mL/min
Trap temperature	38 °C
Desorption time	2 min

[a]Versini et al., 2008.

by 30 mL of dichloromethane and the solution is added to 60 mL of *n*-pentane to obtain a pentane–dichloromethane 2:1 (v/v) solution with a lower boiling point. After drying over Na_2SO_4 the solution is concentrated to 100–150 µL by distillation using a Vigreux column.

For PT extraction of 3-MH and 3-MHA, the sample pH is adjusted to 7.0 to reduce the free fatty acids stripping and to avoid 3-MHA hydrolysis. The optimized PT conditions are reported in Table 5.7 (Versini et al., 2008).

Recent studies showed that the wine matrix strongly influences the apparent partition between the liquid phase and the SPME fiber coatings (Fedrizzi et al., 2007a; 2007b). Wine components may potentially affect the sampling: remarkable sugar content may induce the signals to rise, and polyphenols and esters can participate in competitive adsorption processes (Murray, 2001; Lestremau et al., 2004) and redox reactions (Murat et al., 2003; Blanchard et al., 2004). Consequently, a reliable HS–SPME method needs the use of a suitable internal standard, such as 6-mercaptohexan-1-ol (6-MH).

Comparison among the HS–SPME, SPE, and PT methods, coupled with GC/MS–SIM analysis, showed that PT is more sensitive than HS–SPME in both 3-MHA (LOD ~0.036 and 0.057 µg/L, respectively) and 3-MH (LOD 0.048 and 0.069 µg/L, respectively) analysis, and SPE LODs are comparable with the HS–SPME ones. Purge and Trap overcomes the matrix effects observed for HS–SPME and SPE. The latter is more suitable for analysis of volatile compounds by providing more exhaustive extraction (Versini et al., 2008).

For GC/MS analysis of 3-MH and 3-MHA, a fused-silica apolar column (10 m × 0.32 mm i.d.; 0.25-µm film thickness) connected to the PEG-like polar column can be used with the following operative conditions: carrier gas He (flow rate 1.2 mL/min), transfer line temperature 220 °C, MS source temperature 150 °C, GC injector temperature 250 °C, injection in splitless mode, oven temperature program from a 35 °C isotherm for 5 min, 1 °C/min to 40 °C, 10 °C/min to 250 °C.

5.2.4 Analysis of Wine Mercaptans by Synthesis of Pentafluorobenzyl Derivatives

Polyfunctional thiols in wine can be determined also by synthesis of their pentafluorobenzyl derivatives (Mateo-Vivaracho et al., 2007). A volume of 6 mL of wine is extracted for 15 min at low temperature (<10 °C) using 1.5 mL of benzene containing 4-methoxy-α-toluenethiol as an internal standard for 2-methyl-3-furanthiol (MF), 2-furfurylthiol as an internal standard for 2-furanmethanethiol (FFT), 3-MHA and 3-MH, 3-mercapto-3-methylbutyl formate as an internal standard for 4-MP. The extract (0.9 mL) is then added to 50 µL of two benzene solutions containing pentafluorobenzyl bromide (PFBBr) 2 g/L and a strong alkali secondary reagent (1,8-diazabicyclo[5.4.0]undec-7-ene or triethylamine) to improve the reaction speed. Reaction is carried on at 4 °C for 40 min, then the mixture is washed with a water/methanol (5:1) solution containing HCl 0.5 M (3 × 1 mL). The derivatives are analyzed by GC/MS negative-ion chemical ionization using methane as a reagent gas. A 5% phenyl–95% dimethyl polysiloxane capillary column (20 m × 0.15 mm i.d., 0.15-µm film thickness) can be used with the following temperature program: a 70 °C isotherm for 3 min, heated to 140 °C at 20 °C/min, to 180 °C at 15 °C/min, to 210 °C at 30 C/min, to 300 °C at 250 °C/min, kept isotherm for 10 min. To perform the SIM mode analysis allows us to determine FFT (at m/z 113 and 274), 4-MP (m/z 131 and 194), 3-MHA (m/z 175, 194, and 213), and 3-MH (m/z 133 and 194) with LODs below their odor detection threshold:

0.5 ng/L for FFT, 0.1 ng/L for MP, 0.6 ng/L for 3-MHA, and 7 ng/L for 3-MH (inconsistent results were obtained for MF). Before analysis, excess of PFBBr reagent can be removed by a SPE clean up of the reaction mixture, or using a suitable temperature program of GC injector.

Synthesis of PFB derivatives was also performed by direct reaction on an SPME fiber PDMS/DVB (65-μm thickness) previously adsorbed with the reagent by two successive exposures for 5 min at 55 °C to vapors of (a) 10 mL of tributylamine and (b) 10 mL of PFBBr water/acetone (9:1) 200-mg/L solutions. The fiber is then exposed to 10-mL wine HS in a 20-mL vial at 55 °C for 10 min. By GC–NICI–MS analysis in SIM mode, low LODs were achieved for FFT and 3-MHA (0.05 and 0.03 ng/L, respectively); LODs of 0.11, 0.5, and 0.8 ng/L, were reported for MF, 4-MP, and 3-MH, respectively (Mateo-Vivaracho et al., 2006).

5.3   CARBONYL COMPOUNDS IN WINES AND DISTILLATES

5.3.1   Introduction

Among the number of compounds formed by bacteria and yeasts during fermentation, carbonyl compounds are important in determining the chemical composition of wine. After fermentation of grape must by the yeasts, malolactic fermentation (MLF) is another important industrial scale process operated by lactic bacteria aimed at improving organoleptic characteristics and to confer microbiological stability to quality wines (Davis et al., 1985). The main change occurring in this process is transformation of L(−)-malic acid in L(+)-lactic acid. In addition, profound changes of the carbonyl compounds profile of wine occur (Sauvageot and Vivier, 1997). These compounds contribute to the aroma complexity of the wine: the pungent note is associated with a high acetaldehyde level (Di Stefano and Ciolfi, 1982); the herbaceous odor to aliphatic aldehydes, such as hexanal, (E)-2-hexenal, (E)-2-heptenal, octanal, and (E)-2-octenal (de Revel and Bertrand, 1994; Allen, 1995); decanal and (E)-2-nonenal are associated with "sawdust" or "plank" odor (Chatonnet and Dubourdieu, 1996; 1998). Also, glyoxal, methylglyoxal, and hydroxypropandial are produced by *Leuconostoc oenos* bacteria during MLF (de Revel and Bertrand, 1993; Guillon et al., 1997; Guillon, 1997). These dicarbonyl compounds are characterized by very low odor thresholds, and are associated with browning processes due to reactions with amino acids (de Revel and Bertrand, 1993; Guillon et al., 1997). Aldehydes, such as formaldehyde, acetaldehyde, acrolein, and benzaldehyde, are reputed to be carcinogens (Nascimento et al., 1997).

Diacetyl is thought to add complexity to wine, but if present in high concentration (>5 mg/L) it can be overpowering and confer a distinct butterlike undesirable note. The sensory threshold of diacetyl in wines ranges between 0.2 and 2.8 mg/L (Martineau et al., 1995). Acetoin and diacetyl are produced by yeasts with alcoholic fermentation and their levels generally further increase two to threefold after MLF (Davis et al., 1985). A HS–SPME–GC/MS method for analysis of diacetyl in wine was reported, and an LOD of 0.01 mg/L was reached by using d_6-diacetyl as an internal standard (Hayasaka and Bartowsky, 1999). After sodium chloride addition to the wine, a polydimethylsiloxane and carbowax–divinylbenzene fiber coating was exposed for 10 min to a HS of 3-mL sample in a 15-mL vial at 40 °C. Molecular ions of diacetyl (m/z 86) and diacetyl-d_6 (m/z 92) were recorded in SIM mode and the compound was quantified on the analyte ion response relative to the internal standard ratio. The use of deuterated internal standards ensure the robustness of the method in that quantitative results are not affected by changes in the HS–SPME parameters. On the other hand, diacetyl is usually present in wines in considerable concentration, and the use of deuterated (quite expensive) standards is not indispensable.

A dramatic increase of glycolaldehyde was observed as a consequence of MLF. The amount of glycolaldehyde that resulted was associated with the amounts of glyoxal present in the wine, and glyoxal decreased as glycolaldehyde increases, probably due to reduction mechanisms promoted by bacteria. Glycolaldehyde is very effective in inducing (+)-catechin oxidation and wine browning, inferring it could play a significant role in the color stability of white wines (Flamini and Dalla Vedova, 2003).

A list of carbonyl compounds identified in *Chardonnay* and *Cabernet Sauvignon* wines (before and after MLF) and grappa samples (spirit produced by distillation of fermented grape marc), are reported in Table 5.8 (Flamini et al., 2002a; 2005a).

For analysis of α-dicarbonyl compounds, the synthesis of 1,2-diaminobenzene derivatives can be used (Guillou et al., 1997). Figure 5.5 shows the GC/MS–EI (70 eV) mass spectrum of 3-hydroxy-2-oxopropanal quinoxaline derivative.

TABLE 5.8. Mean Contents of Carbonyl Compounds Identified in Chardonnay and Cabernet Sauvignon Wines[a] and Grappa Samples[b]

	Wine				Grappa	
	Cabernet Sauvignon		Chardonnay		Cabernet Sauvignon	Chardonnay
Compound (μg/L)[c]	Before MLF	After MLF	Before MLF	After MLF	(μg/100 mL a.e.)	(μg/100 mL a.e.)
(E)-2-Hexenal	3.43	0.58	0.82	1.09	71.2	73.3
(E)-2-Nonenal	trace	0.18	0.18	0.29	8.7	12.8
(E)-2-Octenal					n.f.	4.4
(E)-2-Pentenal					15.7	28.5
(E,E)-2,6-Nonadienal	13.78	0.27	n.f.	n.f.		
2,3-Butanedione	250	2820	50	940	179.2	189.4
3-Hydroxy-2-butanone	1620	4560	5030	10910	892.1	384.9
Acetaldehyde (mg/L)	52.01	5.02	46.54	67.04	33.2	20.8
Benzaldehyde					1117.2	699.5
Butyraldehyde	12.36	19.99	8.48	5.08		
Decanal	0.34	1.56	1.84	4.13		
Glycolaldehyde[d]	48.43	391.6	38.51	110.87	6.1	6.0
Glyoxal	41.33	106.01	49.11	95.92	0.2	0.1
Heptanal	3.36	0.56	0.35	1.99	52.3	116.2
Hexanal	94.63	91.67	10.79	89.94	1093.5	1563.2
Isovaleraldehyde+					50.6	148.6
2-Methylbutyraldehyde[d]	25.69	45.51	45.85	24.46	11.5	18.2
Methylglyoxal	10.96	14.37	34.73	71.55	0.3	0.2
Nonanal	1.30	1.85	1.05	2.92	11.2	32.5
Octanal	0.82	0.77	0.39	0.80	4.7	13.6
Propanal					53.8	90.4
Vanillin					1.30	1.8

[a]Before and after Malolactic fermentation.
[b]Not found, n.f., a.e., anhydrous ethanol.
[c]Amounts expressed as internal standard o-chlorobenzaldehyde (I.S.).
[d]Quantified on basis of one of the two PFBOA syn–anti oxime peaks.

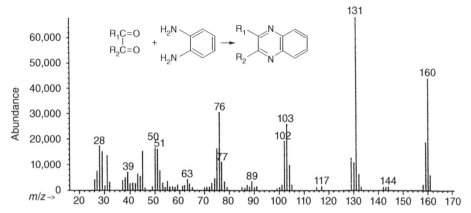

Figure 5.5. Mass spectrum of quinoxaline derivative of 3-hydroxy-2-oxopropanal in wine formed by reaction with 1,2-diaminobenzene. (Reprinted from *Journal of Agricultural and Food Chemistry 45*, Guillou et al., Occurrence of hydroxypropanedial in certain musts and wines, pp. 3383, 3385, Copyright © 1997, with permission from American Chemical Society).

5.3.2 The GC/MS Analysis of Wine Carbonyl Compounds by Synthesis of PFBOA Derivatives

The GC/MS analysis of O-pentafluorobenzyl (PFB) oximes formed by reaction of carbonyl compounds with O-(2,3,4,5,6-pentafluorobenzyl)-hydroxylamine (PFBOA), is a satisfactory method for many carbonyl compounds in term of selectivity, and for high sensitivity achieved while operating in SIM mode and recording the base peak signal at m/z 181 characteristic of all PFB-oximes (Cancilla and Que Hee, 1992; Lapolla et al., 2003; 2006). Several studies on carbonyl compounds in hydro-alcoholic matrices, such as wine, model wine solutions, spirits, and beer, were performed by this method (Vanderlinde et al., 1992; Vidal et al., 1992; de Revel and Bertrand, 1993; 1994; Guillou, 1997; Guillou et al., 1997; Flamini et al., 2002a; 2002b; Flamini and Dalla Vedova, 2003, Ochiai et al., 2003; Vesely et al., 2003; Flamini et al., 2005a; 2005b). By performing direct derivatization at room temperature of aqueous sample, ketones (e.g., 2-heptanone, 2-nonanone, 2-undecanone, and 2-tridecanone in Cognac) require longer reaction times with respect to aldehydes (Vidal et al., 1993). To overcome this problem, a pentane–ethyl ether liquid–liquid extraction of the sample followed by derivatization of the extract at 65 °C for 30 min can be performed. Unfortunately, extraction with organic solvent may induce losses of the more polar compounds. The simple procedure and the good linearity and sensitivity observed for most carbonyl compounds, suggests direct derivatization of aqueous samples is the more convenient procedure.

o-Chlorobenzaldehyde can be added to the sample as an internal standard and quantitative results can be expressed as micrograms per liter (µg/L) of I.S., or by calculation of the calibration curve of each compound using 1-octanol, lindane, pivaldehyde, acetophenone, or o-chlorobenzaldehyde as an internal standard (de Revel et al., 1993; Guillou et al., 1997; Flamini et al., 2002a; 2002b; Flamini and Dalla Vedova, 2003; Ojala et al., 1994; Vidal et al., 1993). In general, for each compound containing one carbonyl group the two geometrical oximes, syn and anti, form (except for formaldehyde); in the case of diacetyl, three chromatographic peaks, corresponding to the (Z,Z), (E,E), and $(Z,E) + (E,Z)$ isomers, form. Figure 5.6 shows the reaction scheme of aldehyde PFB-oxime synthesis.

The large amount of pyruvic acid present into the wine subtracts considerable PFBOA reagent at the reaction with the other carbonyl compounds. The pyruvate-PFBOA-derivatives formed and the not reacted PFBOA reagent (at the end of reaction) leave the GC column as broad peaks covering large parts of the chromatogram. Therefore, it may be useful to remove this compound from wine prior to derivatization by passing the sample through a 10-g strong anionic exchange column preconditioned by successive passages of 20-mL HCl 1% (v/v) and 100-mL sodium fluoride 0.5 M solutions. Finally, the stationary phase is rinsed with water. Before passage through the column, the sample is adjusted to pH 7.5 with 1 mL 4 M NaOH. The column is washed with 10-mL ethanol and the two fractions eluted are collected and combined. After addition of o-chlorobenzaldehyde as an internal standard (200 µL of 0.1-mM ethanol solution), the resulting solution is adjusted to pH 3.0 with HCl and 20 mg of PFBOA are added. The reaction is carried out by stirring for 1 h at

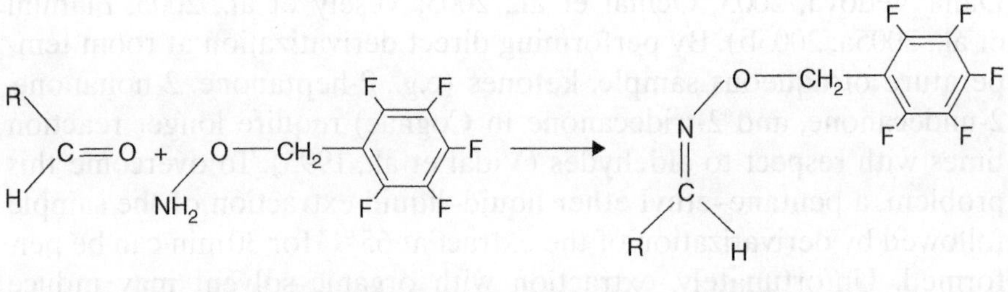

PFBOA   (syn + anti)

Figure 5.6. Synthesis of aldehyde pentafluorobenzyl (PFB) oximes.

room temperature. The PFB–oximes are recovered by a 3×3-mL liquid–liquid extraction with ethyl ether–hexane 1:1 (v/v), the resulting organic solution is dried over Na_2SO_4, and reduced to 0.4 mL before analysis (Flamini et al., 2002a). The GC/MS conditions used for analysis of PFBOA derivatives are reported in Table 5.9.

Main carbonyl compounds of interest usually are present into the wine in low levels and very sensitive methods are necessary for quantitative analysis. Consequently MS is performed in SIM mode recording m/z 181, which corresponds to the pentafluorobenzyl ion (base peak of PFBOA-derivatives mass spectra). Saturated and unsaturated aldehyde derivatives can be distinguished from the other derivatives by recording the signals at m/z 239 and 250, respectively. The former is typical of non-α-substituted compounds and probably corresponds to the N-vinyl pentafluorobenzyl oxime cation (formed by 1-vinyl neutral aliphatic chain loss); the latter is typical of α,β-unsaturated compounds and corresponds to the probable loss of the aliphatic radical chain with formation of the isoxazoline ring. Chromatograms recorded at m/z 181, 239, and 250 are shown in Fig. 5.7. Retention time and characteristic m/z signals of PFBOA derivatives of carbonyl compounds in wines, are reported in Table 5.10.

A GC/MS–EI specific method for analysis of acetaldehyde in wine and hydro-alcohol matrixes was proposed by synthesising PFBOA oximes (Flamini et al., 2002a). Quantitative analysis is done on the sum of the syn and anti acetaldehyde O-PFB-oximes peak areas recorded in SCAN mode. By using analytical conditions similar to those reported in Table 5.9, but keeping the initial oven temperature of 60 °C for 1 min and increasing the temperature at a rate of 5 °C/min to 130 °C, then 10 °C/min to 210 °C, retention time of the internal standard 1-octanol is ~10.3 min, and the two acetaldehyde O-PFB-oxime peaks fall in the 7.8–8.0-min range of the chromatogram. The main advantages of this

TABLE 5.9. The GC/MS Conditions Used for Analysis of PFBOA Derivatives

GC column	PEG bound-phase fused-silica capillary (30 m × 0.25 mm i.d.; 0.25-μm film thickness)
Carrier gas	He Column head pressure 12 psi
Injector	250 °C, Sample volume injected 0.5 μL, splitless injection
Oven program	60 °C Isotherm for 5 min, 3 °C/min to 210 °C, 210 °C isotherm for 5 min
MS conditions	Ionization energy 70 eV, transfer line temperature 280 °C, SIM mode acquisition

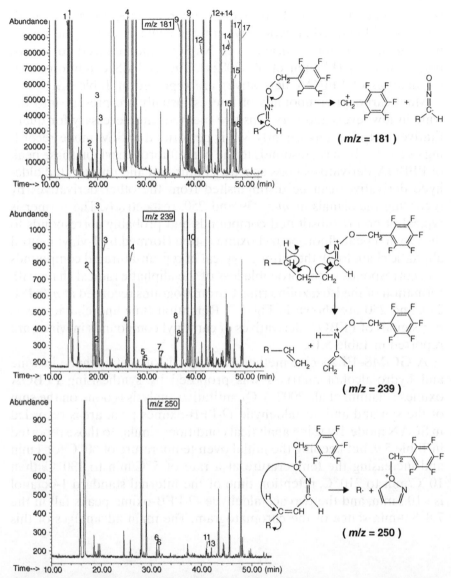

Figure 5.7. Carbonyl compound analysis of a *Cabernet Sauvignon* wine: chromatograms are recorded in SIM mode on signals at *m/z* 181, 239, and 250. The chromatogram at *m/z* 181 shows signals of all PFBOA derivatives. The chromatogram at *m/z* 239 shows signals of saturated aldehyde derivatives and the chromatogram at *m/z* 250 shows the signals of α,β-unsaturated aldehyde derivatives. On the right, the schemes for formation of these ions are showed. (1) acetaldehyde, (2) butyraldehyde, (3) isovaleraldehyde+2-methylbutyraldehyde, (4) hexanal, (5) heptanal, (6) (*E*)-2-hexenal, (7) octanal, (8) nonanal, (9) acetoin, (10) decanal, (11) (*E*)-2-nonenal, (12) glycolaldehyde, (13) (*E,Z*)-2,6-nonadenal, (14) diacetyl, (15)IS, (16) methylglyoxal, (17) glyoxal. (Reprinted from Vitis 41, 2002, Flamini et al., Changes in carbonyl compounds in Chardonnay and Cabernet Sauvignon wines as a consequence of malolactic fermentation, p. 110, with permission).

TABLE 5.10. Characteristic *m/z* Signals and Retention Times (RT) of syn, anti, and syn+anti PFBOA Derivatives of Carbonyl Compounds[a]

Compound	RT1 (min)	RT2 (min)	RT3 (min)	Characteristic Signals (*m/z*)	RF
Propanal	13.93	14.28		236;253	0.5
2-Methyl butyraldehyde	16.99	17.10		100;239;281	1.2
3-Methyl butyraldehyde	17.88	18.70		100;239;281	1.2[b]
Pentanal	20.28	20.71		100;239;281	1.2[b]
(*E*)-Crotonaldehyde	22.73	24.00		84;250;265	1.7[b]
Hexanal	23.75	24.16		114;239;295	0.2
(*E*)-2-Pentenal	26.06	26.71		98;250;279	1.7[b]
Heptanal	27.10	27.45		128;239;309	0.3
(*E*)-2-Hexenal	29.00	29.48		112;250;293	1.7
Octanal	30.47	30.75		142;239;323	0.8
(*E*)-2-Heptenal	32.86	33.13		126;250;307	1.7[b]
Nonanal	33.74	33.95		156;207;239;337	0.8
(*E*)-2-Octenal	36.27	36.36		140;250;321	1.7[b]
3-Hydroxy-2-Butanone	35.13	36.94		240	0.9
Decanal	36.87	37.05		170;239;351	0.6
(*E*)-2-Nonenal	38.91			154;250;335	0.7
(*E,E*)-2,6-Nonadienal	39.85	39.96		152;250;333	0.3
Glycoladehyde	39.71	41.16		225;255	2.1
(*E,Z*)-2,6-Nonadienal	40.40			152;250;333	0.4
Benzaldehyde	40.74	41.00		301	1.2
2,3-Butanedione	41.29	43.41	45.72	476	3.5
o-Chlorobenzaldehyde (IS)	45.05	45.81		300	1
Methylglyoxal	45.71	46.35		462	4.2[c]
Glyoxal	47.32	47.60		448	4.9
p-Anisaldehyde	52.69	53.09		331	0.2
Cinnamaldehyde	53.24	53.73		326;327	0.2
4-Hydroxybenzaldehyde	57.80			317	0.4
Vanillin	63.67			347	0.2

[a]Response factor (RF) relative to I.S. *o*-chlorobenzaldehyde calculated as: $[RF_X = (A_X/C_X)/(A_{IS}/C_{IS})]$.
[b]The RF calculated for 2-methylbutanal and (*E*)-2-hexenal.
[c]The RF calculated as average of glyoxal and 2,3-butanedione response factors. (From Rivista di Viticultura e di Enologia 2005 (1), Flamini et al., Study of carbonyl compounds in some Italian marc distillate (grappa) samples by synthesis of *O*-(2,3,4,5,6-pentafluorobenzyl)-hydroxylamine derivatives, p. 56).

method are the possibility of using one of the PEG capillary columns commonly used for analysis of wine volatiles and satisfying linearity and a fairly good reproducibility with limited cost and time of sample preparation.

5.3.3 HS–SPME–GC/MS of PFBOA Derivatives

A specific SPME method for analysis of acetaldehyde, diacetyl, and acetoin PFBOA derivatives for monitoring the wine MLF, was proposed (Flamini et al., 2005b). The volume of a 100-μL sample is transferred to a 4-mL vial and, after addition of 200 μL of an internal standard *o*-chlorobenzaldehyde 4-mg/L ethanol–water 1:1 (v/v) solution, and of 1 mL of PFBOA 2 g/L of aqueous solution, the volume is adjusted to 2 mL with water. In general, the best conditions to achieve the highest reaction yields for the synthesis of PFBOA derivatives are to carry on the reaction for 1 h at room temperature. In this case, the highest sensitivity was achieved by addition of 50-mg sodium chloride to the sample before performing HS–SPME for 15 min (see Table 5.11). Since these compounds are usually present in wines at the parts per million levels, a good compromise between method sensitivity and analysis time is to perform the reaction for 20 min at 50 °C, and extraction by exposing a 65-μm poly(ethylene glycol)/divinylbenzene (PEG/DVB) fiber to the sample headspace for 5 min at 50 °C, after addition of 50-mg NaCl to the sample. In general, the PFBOA residue confers higher volatility and a good affinity for the PEG/DVB fiber to the molecule. This method has good linearity and is sufficiently accurate and repeatable for all three analytes. Presently, this type of fiber is commercially not readily available. As an acternative, a 65-μm poly(dimethylsiloxane)–divinylbenzene (PDMS/DVB) fiber can be used (Vesely et al., 2003; Carlton et al., 2007).

The GC–MS of PFBOA derivatives of these three compounds also was performed by positive chemical ionization (PICI) using methane as the reagent gas. Abundant formation of the $[M+H]^+$ ion of acetaldehyde derivatives at m/z 240, diacetyl-mono derivatives at m/z 282 and of the internal standard *o*-chlorobenzaldehyde derivatives at m/z 336, and abundant formation of the $[M+H-18]^+$ ion of acetoin derivatives at m/z 282, were observed. Figure 5.8 shows the mass spectra of acetaldehyde, diacetyl monooxime, acetoin, and *o*-chlorobenzaldehyde PFBOA derivatives. The analytical conditions are reported in Table 5.12.

Collision induced dissociation (CID) experiments, by selecting the $[M+H]^+$ ion of diacetyl-mono derivatives at m/z 282 and $[M+H-18]^+$ ion of acetoin derivatives at m/z 266, allows us to confirm the compound identification. The principal daughter ions produced are reported in the Table 5.13.

Quantitative analysis is performed on the signal of $[M+H]^+$ ions at m/z 240 for acetaldehyde, m/z 282 for diacetyl, and m/z 336 for the internal standard, and on the signal at m/z 266 of $[M+H-18]^+$ ion for

TABLE 5.11. Normalized Peak Areas of *m/z* Signals Used for Quantification of Acetaldehyde, Diacetyl, and Acetoin PFBOA Derivatives Calculated with Respect to I.S.[a]

TABLE 5.11A[b]

PFBOA Derivative	*m/z*	Ion	1 h on Fiber Reaction at 50 °C	
			SPME 15 min	SPME 15 min + NaCl 50 mg
Acetaldehyde	240	$[M+H]^+$	75	31
Diacetyl	282	$[M+H]^+$	3	5
Acetoin	266	$[M+H-18]^+$	5	5

TABLE 5.11B[c]

PFBOA Derivative	*m/z*	Ion	50 °C		25 °C
			20-min Reaction	40-min Reaction	1-h Reaction
			SPME 15 min	SPME 15 min	SPME 15 min
Acetaldehyde	240	$[M+H]^+$	61	28	96
Diacetyl	282	$[M+H]^+$	100	24	65
Acetoin	266	$[M+H-18]^+$	40	31	8

TABLE 5.11C[d]

PFBOA Derivative	*m/z*	Ion	20-min Reaction at 50 °C		
			SPME 5 min	SPME 15 min	SPME 30 min
Acetaldehyde	240	$[M+H]^+$	68	61	13
Diacetyl	282	$[M+H]^+$	49	100	18
Acetoin	266	$[M+H-18]^+$	29	40	45

TABLE 5.11D[e]

PFBOA Derivative	*m/z*	Ion	1-h Reaction at 25 °C	
			SPME 15 min	SPME 15 min + NaCl 50 mg
Acetaldehyde	240	$[M+H]^+$	96	100
Diacetyl	282	$[M+H]^+$	65	100
Acetoin	266	$[M+H-18]^+$	8	100

[a]The more intense area is equal to 100, the others were calculated referring to it.
[b]Study of the effect of salt addition before SPME on the intensity of signals after 1 h of on-fiber derivatization.
[c]Study of the in-solution reaction time and temperature effects.
[d]Study of the SPME-time effect.
[e]Study of the effect of salt addition before SPME after 1 h of in-solution derivatization. (Reprinted from *Journal of Mass Spectrometry* 40, Flamini et al., Monitoring of the principal carbonyl compounds involved in malolactic fermentation of wine by synthesis of *O*-(2,3,4,5,6-pentafluorobenzyl) hydroxylamine derivatives and solid-phase-micro-extraction positive-ion-chemical-ionization mass spectrometry analysis, p. 1560, Copyright © 2005, with permission from John Wiley & Sons, Ltd.

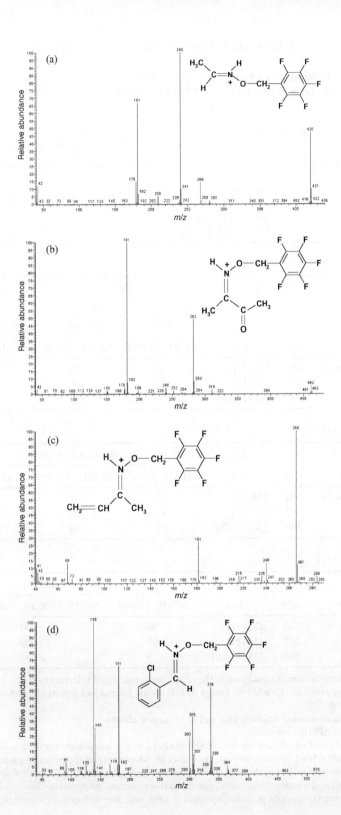

TABLE 5.12. Derivatization and HS–SPME–GC/MS–PICI Conditions for Analysis of Acetaldehyde, Diacetyl, and Acetoin PFBOA Derivatives in Wine[a]

Sample volume	100 µL
Vial volume	4 mL
Derivatization conditions	200 µL IS o-chlorobenzaldehyde 4 mg/L ethanol/water solution, 1 mL of PFBOA 2-g/L water solution, volume adjusted to 2 mL by water, reaction at 50 °C for 20 min
SPME fiber	PEG/DVB 65-µm coating thickness, 1 cm length
Addition to the sample	50 mg NaCl
Extraction temperature	50 °C
Extraction time	5 min under stirring
Desorption temperature	240 °C
Desorption time	1 min
Fiber cleaning	Additional 5 min at 250 °C
GC column	PEG bound-phase fused-silica capillary (30 m × 0.25 mm i.d.; 0.25-µm film thickness)
Carrier gas	He, Column head pressure 16 psi
Injector	Temperature 240 °C, sample volume injected 0.5 µL, splitless injection
Oven program	60 °C isotherm for 5 min, 3 °C/min to 210 °C, 210 °C isotherm for 5 min
MS–IT conditions	PICI mode using methane as reagent gas (flow 1 mL/min), ion source at 200 °C, dumping gas 0.3 mL/min, simultaneous SCAN (range m/z 40–660, 1.67 scan/s) and MS/MS
CID experiments	Collision gas He, excitation voltage 225 mV (precursor ion m/z 282 for diacetyl-mono derivatives, m/z 266 for acetoin derivative)
Quantitative	Recording signal at m/z 240 for acetaldehyde, m/z 282 for diacetyl, m/z 266 for acetoin, m/z 336 for o-chlorobenzaldehyde (I.S.)

[a]Flamini et al., 2005b.

Figure 5.8. The GC/MS–positive ion chemical ionization mass spectra of (a) acetaldehyde, (b) diacetyl monooxime, (c) acetoin, and (d) o-chlorobenzaldehyde PFBOA derivatives. The analytical conditions are reported in Table 5.12. (Reprinted from *Journal of Mass Spectrometry* 40, Flamini et al., Monitoring of the principal carbonyl compounds involved in malolactic fermentation of wine by synthesis of O-(2,3,4,5,6-pentafluorobenzyl) hydroxylamine derivatives and solid-phase-micro-extraction positive-ion-chemical-ionization mass spectrometry analysis, p. 1560, Copyright © 2005, with permission from John Wiley & Sons, Ltd.)

TABLE 5.13. Principal Daughter Ions Produced by CID of Precursors: [M+H]⁺ of Diacetyl mono-PFBOA-oximes (m/z 282), [M+H-18]⁺ of Acetoin PFBOA-Oximes (m/z 266) and [M+H]⁺ of o-Chlorobenzaldehyde PFBOA-Oximes (m/z 336)[a]

	Precursor Ion (P)		Fragment Ion					
	$[M+H]^+$	$([M+H]-H_2O)^+$	$(P-H_2O)^+$	$(P-CH_2O)^+$	$(P-CH_3O)^+$	$(P-C_2H_2O)^+$	$(P-CH_3OF)^+$	$(P-C_7H_3OF_5)^+$
Mono-PFB-diacetyl	**282**		264	252		240		
PFB-Acetoin		**266**	248		235		216	
PFB-o-Chlorobenzaldehyde	**336**		318		305			138

[a]The analytical conditions are reported in Table 5.12. (Reprinted from *Journal of Mass Spectrometry* 40, Flamini et al., Monitoring of the principal carbonyl compounds involved in malolactic fermentation of wine by synthesis of O-(2,3,4,5,6-pentafluorobenzyl) hydroxylamine derivatives and solid-phase-micro-extraction positive-ion-chemical-ionization mass spectrometry analysis, p. 1561, Copyright © 2005, with permission from John Wiley & Sons, Ltd.)

acetoin. This method has a lower sensitivity in the detection of acetoin, probably due to higher polarity and lower volatility of its oximes. Also, the higher steric hindrance of acetoin might affect the derivatization reaction yield.

Figure 5.9 shows the GC/MS chromatograms recorded in the analysis of a *Merlot* wine (**a**) at the beginning of MLF and (**b**) after 5 days of fermentation.

In addition, SPME methods of direct on-fiber derivatization with PFBOA (Vesely et al., 2003; Carlton et al., 2007) and direct derivatization on an SPE cartridge using a styrene–vinylbenzene polymer stationary phase (Ferreira et al., 2004; Culleré et al., 2004), were proposed. Derivatization at 50 °C on a PEG–DVB fiber previously adsorbed on PFBOA showed very low reaction yield for acetoin and diacetyl, probably due to high polarity, and consequently the low volatility, of these compounds (Table 5.11A) (Flamini et al., 2005b).

5.4   ETHYL AND VINYL PHENOLS IN WINES

5.4.1   Introduction

Due to the difficulty of cleaning and sterilizing wooden barrels during storage, the wine can undergo attack by slow-growing species, such as *Brettanomyces bruxellensis*, *Brettanomyces anomalus*, *Saccharomyces bailli*, and certain genera of lactic bacteria (Fugelsang, 1998). This growing may be promoted by the high pH and low sulfur dioxide (SO_2) level of wine. Both *Brettanomyces* and *Dekkera* can grow during wine aging (Froudiere and Larue, 1988) and even after bottling. These yeasts are rarely found during alcoholic fermentation of must (Wright and Parle, 1973), and few studies reported an abnormal presence of *Brettanomyces* on grape clusters (Pretorius, 2000).

Volatile phenols originate from hydroxycinnamic acids (ferulic, *p*-coumaric, or caffeic acid) by the action of hydroxycinnamate decarboxylase enzyme, which turn the hydroxycinnamics acid into vinylphenols (Albagnac, 1975; Grando et al., 1993). Then, these compounds are reduced to ethyl derivatives by vinylphenol reductase enzymes characteristic of species, such as *Dekkera bruxellensis*, *Dekkera anomala*, *Pichia guillermondii*, *Candida versatilis*, *Candida halophila*, and *Candida mannitofaciens* (Edlin et al., 1995; 1998; Dias et al., 2003; Chatonnet et al., 1992; 1995; 1997; Dias et al., 2003), apart from very small quantities produced by some yeasts and lactic acid bacteria under peculiar growth conditions (Chatonnet et al., 1995; Barata et al., 2006;

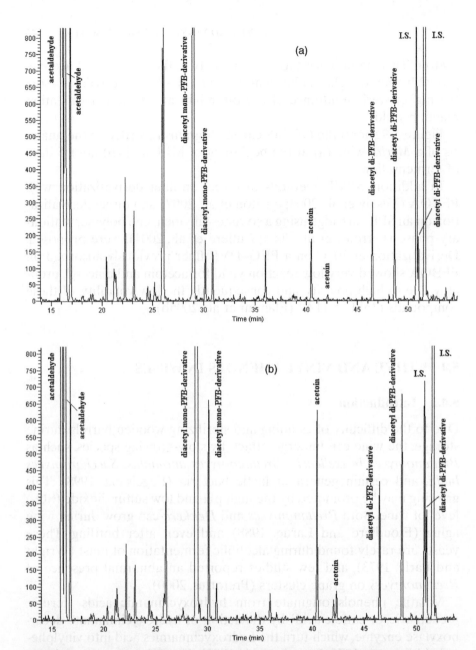

Figure 5.9. Analysis of a *Merlot* wine at the beginning of MLF (**a**) and after 5 days of fermentation (**b**): reconstructed ion chromatograms (RIC) of [M+H]⁺ species of acetaldehyde–PFB derivatives at *m/z* 240, diacetyl mono–PFB derivatives at *m/z* 282, diacetyl di-PFB-derivatives at *m/z* 477 and *o*-chlorobenzaldehyde–PFB derivatives (I.S.) at *m/z* 336, and of [M+H-18]⁺ ion of acetoin–PFB derivatives at *m/z* 266. The analytical conditions are reported in Table 5.12. (Reprinted from *Journal of Mass Spectrometry* 40, Flamini et al., Monitoring of the principal carbonyl compounds involved in malolactic fermentation of wine by synthesis of *O*-(2,3,4,5,6-pentafluorobenzyl) hydroxylamine derivatives and solid-phase-micro-extraction positive-ion-chemical-ionization mass spectrometry analysis, p. 1563, Copyright © 2005, with permission from John Wiley & Sons, Ltd.)

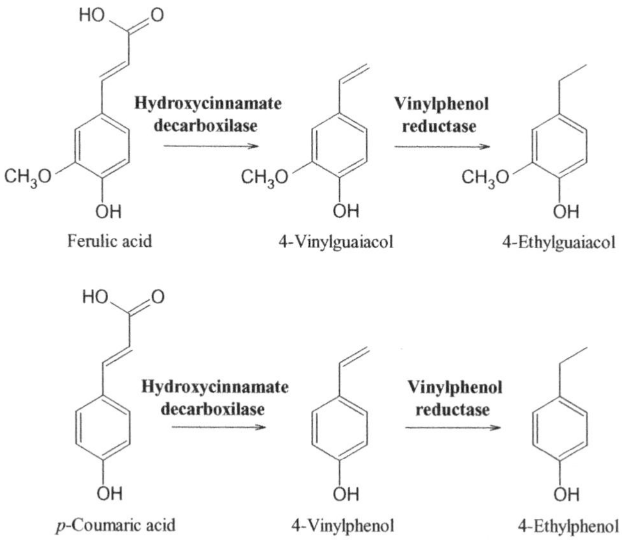

Figure 5.10. Formation of ethylphenols from hydroxycinnamic acids.

Couto et al., 2006). Figure 5.10 shows a scheme for formation of ethylphenols from hydroxycinnamic acids.

Volatile phenols greatly influence the aroma of wine, the most important are 4-vinylphenol (4-VP), 4-vinylguaiacol (4-VG), 4-ethylphenol (4-EP), and 4-ethylguaiacol (4-EG) (Chatonnet et al., 1992). The 4-EP compound was reported in wine for the first time in 1967 by Webb and co-workers and its presence, together with the other phenols cited, was confirmed in 1970 by Dubois and Brulé (Webb, 1967; Dubois and Brulè, 1970).

Oligomer proanthocyanidins inhibit *Saccharomyces cerevisiae* cinnamate decarboxylase (Chatonnet et al., 1990), justifying the very low amounts of vinylphenols in red wines; on the contrary, *Brettanomyces* decarboxylase is not inhibited by proanthocyanidines (Chatonnet et al., 1993). Formation of stable vinylphenol–anthocyanin adducts can arrest the successive formation of ethylphenols in red wines (Fulcrand et al., 1996). Also, derivates of vinylcatechol (Schwarz et al., 2003; Hayasaka and Asenstorfer, 2002), were found in wine.

White wines can contain vinylphenols in concentrations up to several hundreds of a microgram per liter, but they usually lack ethylphenols. On the contrary, in red wine ethylphenols can reach some milligrams per liter (Chatonnet et al., 1992; 1993; Chatonnet, 1993).

In red wines, high levels of 4-EP are associated with disagreeable odors described as "phenolic", "leather", "horse sweat", "stable", or "varnish" (Etievant, 1991; Chatonnet et al., 1992; 1993; Rodrigues et al., 2001). The odor threshold limits in wine reported in the literature for 4-ethylphenol and 4-ethylguaiacol are 440 and 33 µg/L, respectively.

Vinylphenols are liable to give sensory characteristics generally classified among the "off flavors" and generally described as phenolic, medicinal, pharmaceutical, smoky, spicy, and clovelike (Montedoro and Bertuccioli, 1986; Rapp and Versini, 1996). The 4-VP is said to negatively affect and mask the fruity scent of white wines, even if present in concentrations lower than the sensory threshold (Dubois, 1983; Chatonnet, 1993), conferring odors resembling "band-aid" and gouache (van Wyk and Rogers, 2000). The latter can be considered less negative, and with 4-VG contributes to the floral aroma of *Chardonnay* wines (Versini et al., 1992). The 4-VG gives the spicy note of *Gewürztraminer* wines (Versini, 1985), as well as the typical characteristics of some beers from Belgium and Bavaria, for example, Lambic and Weizen (Narziss et al., 1990; Coghe et al., 2004).

5.4.2   Analysis of Ethylphenols

The very low odor threshold limits of 4-EG and 4-EP in wine need sensitive analytical methods, including prior concentration steps. By using GC/MS analysis, detection limits of a few microgram per liter and wide linearity range can be achieved (Pollnitz et al., 2000; Martorell et al., 2002). Several methods of sample preparation were proposed: liquid–liquid extraction (Chatonnet et al., 1992; Versini, 1985; Chatonnet et al., 1995; Pollnitz et al., 2000; Chatonnet and Boidron, 1988; Rocha et al., 2004); SPE (Aznar et al., 2001; Dominguez et al., 2002; López et al., 2002); SPME and HS–SPME (Martorell et al., 2002; Ferreira et al., 1996; 1998; Monje et al., 2002; Castro Mejías et al., 2003); and stir bar sorptive extraction (SBSE) (Díez et al., 2004).

Also, use of the deuterated internal standard 2,3,5,6-[$^2H_4$]-4-ethylphenol (d_4-4-ethylphenol) to perform an accurate quantitative GC/MS analysis of 4-EP and 4-EG, was proposed. In this case, liquid–liquid extraction was performed by diethyl ether/pentane 1:2 (v/v), SIM mode analysis was performed by recording the signals at m/z 111 and 126 for [$^2H_4$]-4-EP, m/z 107 and 122 for 4-EP, m/z 122, 137, and 152 for 4-EG. Quantitative analysis is performed on the ions at m/z 126, 122, and 152 (Pollnitz et al., 2000).

A method of dispersive liquid–liquid microextraction (this technique is based on the use of a ternary component solvent system composed of an extraction and a dispenser solvent) coupled to GC/MS was recently proposed. This method provided limits of detection (LOD) and of quantification (LOQ) of 28 and 95 µg/L for 4-EG, and 44 and 147 µg/L for 4-EP, respectively (Farina et al., 2007). The HS–SPME approach is a valid sample preparation method because it is simple, rapid, solventless, easy to be automatized, and has a minimal sample manipulation.

Sampling using a 100-µm PDMS fiber by extraction of 25 mL of wine added of a suitable amount of NaCl to get a 6 M solution at 25 °C for 60 min, allows LODs and LOQs of 1 and 5 µg/L for 4-EG, and 2 and 5 µg/L for 4-EP, respectively (Martorell et al., 2002). Also, the HS–SPME of NaCl saturated samples using a 85-µm polyacrylate (PA) fiber for 40 min at 55 °C provided LODs in the low microgram per liter range (Monje et al., 2002).

A method based on the use of multiple-headspace (MHS) SPME using a carbowax–divinylbenzene (CW/DVB) fiber (three consecutive extractions of the same sample to minimize the possible matrix effects), showed LODs (S/N = 3) of 0.06 µg/L for 4-EG and 4-EP, 0.20 µg/L for 4-VG, and 0.12 µg/L for 4-VP, below the sensory thresholds of these compounds in wines (Pizarro et al., 2007).

The HS–SPME are usually carried out under nonequilibrium conditions. The distribution constants of analytes between the fiber and the sample, between the HS and the sample, and the volume of the three phases (sample, headspace, and coating) must be constant, like the other SPME extraction parameters (sample agitation, fiber exposure time, etc.). The total analyte area (A_T) corresponding to a cumulative extraction yield after multiple extractions can be determined as the sum of the areas obtained for each individual extraction when each is exhaustive, or expressed as:

$$A_T = \sum A_i = A_1/(1-\beta)$$

Here A_i is the peak area obtained in the ith extraction, A_1 is the peak area obtained after the first extraction, and β is a constant calculated from the linear regression of the following equation:

$$\ln A_i = (i-1)\times \ln\beta + \ln A_1$$

The experimental conditions for MHS–SPME–GC/MS/MS of volatile phenols in wines are reported in Table 5.14. The MS/MS parameters and method performances are reported in Table 5.15.

Also, a HS–SPME–GC/MS method for analysis of 4-ethylcatechol in wine was performed using a triphase divinylbenzene–carboxen–polydimethylsiloxane (DVB/CAR/PDMS) fiber after derivatization of the sample with acetic anhydride (Carrillo and Tena, 2007).

TABLE 5.14. Multiple Headspace (MHS) SPME (n = 3) and GC/MS/MS Conditions for Analysis of Volatile Phenols in Wine[a]

Sample volume	5 mL
Vial volume	20 mL
SPME fiber	CW/DVB 70 μm
Addition to the sample	1.17 g NaCl
Extraction	60 °C for 50 min under stirring, 3 extractions
Desorption	220 °C for 2 min at 200 °C
Internal standard	3,4-Dimethylphenol (100 μL of a 100-mg/L solution)
GC column	PEG bound-phase fused-silica capillary (30 m × 0.25 mm i.d.; 0.25-μm film thickness)
Carrier gas	He (1 mL/min)
Oven program	35 °C for 2 min, 20 °C/min to 170 °C, 170 °C for 1 min, 3 °C/min to 210 °C held for 15 min
MS–IT conditions	EI (70 eV), ion source temperature 200 °C
CID experiments	Precursor ions isolation window 3 amu, scan time at 0.46 s/scan

[a]Reported by Pizarro et al. (2007).

TABLE 5.15. Multiple Headspace SPME–GC/MS/MS Method (n = 3) for Analysis of Volatile Phenols in Wines: MS/MS Parameters and Performances[a]

Compound	Column RT (min)	Quantification Ions (m/z)	CID Parameters	
			Storage Level (m/z)	Amplitude (V)
4-EG	12.9	137 + 152	75	76
4-EP	15.2	107 + 122	60	65
4-VG	15.5	150 + 135	75	62
4-VP	19.3	120 + 91	65	62

	Linear Range (μg/L)	Correlation Coefficient (r)	LOQ S/N = 10 (μg/L)	LOD S/N = 3 (μg/L)
4-EG	2.74–706	0.997	0.18	0.06
4-EP	2.76–1714	0.995	0.20	0.06
4-VG	3.60–762	0.994	0.66	0.20
4-VP	2.80–760	0.999	0.40	0.12

[a]Pizarro et al., 2007.

Derivatization of 4 mL of wine is carried out at 70 °C for 1 min in a 20-mL sealed vial after addition of 140 μL of acetic anhydride, 1 mL of K_2CO_3 5.5% solution, and 0.9 g NaCl. Sampling is performed by exposing the fiber at 70 °C for 70 min (desorption at 270 °C for 7 min). A PEG fused-silica capillary column was used for GC/MS–IT EI (70 eV) analysis of derivatives, which performed quantification in selected ion storage (SIS) mode on the signals m/z 123 + 138 for 4-ethylcatechol, and m/z 107 + 122 for internal standard 3,4-dimethylphenol. The LODs reported are 4 µg/L for 4-ethylcatechol (with a LOQ 6 µg/L), and 2 and 17 µg/L for 4-EG and 4-EP, respectively.

5.5   2′-AMINOACETOPHENONE IN WINES

There are countries (e.g., Italy) where winemaking does not use *V. vinifera* (hybrid) grapes and commercialization of their wines is not permitted. 2′-Aminoacetophenone (AAP) was identified as the cause of the aging note, so-called "hybrid note" or "foxy smelling", typical of *Labruscana* grapes and also found in some *V. vinifera* wines (e.g., *Müller-Thurgau, Riesling*, and *Silvaner*) (Rapp et al., 1993).

Wines showing this note, described as "acacia blossom", "naphthalene note", "furniture polish", "fusel alcohol", "damp cloth", have caused a considerable amount of rejections. Developing of AAP in grape is promoted by several factors, such as reduced nitrogen fertilization in combination with hot and dry summers, and the risk increases in wines from grapes harvested early. Hormone plant indole-3-acetic acid (IAA) is an important precursor of AAP in nonenzymatic processes. Pyrrole ring cleavage of IAA by superoxide radicals generated by aerobic oxidation of sulfite during storage of sulfurized wines, leads to formation of *N*-formyl-2′-aminoacetophenone (FAP), which is further decomposed to AAP. The scheme of AAP formation is shown in Fig. 5.11 (Hoenicke et al., 2002a; 2002b).

Analysis of fermented synthetic media revealed that AAP is also a secondary metabolite of *S. cerevisiae* yeasts together with *o*-aminopropiophenone and 3-(*o*-aminophenyl)-prop-1-en-3-one (Ciolfi et al., 1995). Methylanthranilate also contributes to the typical hybrid–foxy taint of American hybrid and wild vine wines. It was found in some *V. vinifera* white wines in concentration >0.3 µg/L (Rapp and Versini, 1996).

For analysis of AAP in wine, an SPE sample preparation method can be performed by using highly cross-linked ethylvinylbenzene–divinylbenzene copolymers (e.g., LiChrolut EN) cartridges. A volume of 50 mL of wine spiked with an internal standard (e.g., acetophenone-d_8 or d_3-AAP) is passed through the cartridge, then analytes are recovered with 1 mL of dichloromethane. The organic phase is washed with 1 mL of a sodium hydrogen carbonate solution, then dried over anhydrous sodium sulfate. The residual solution is added to 100-µL

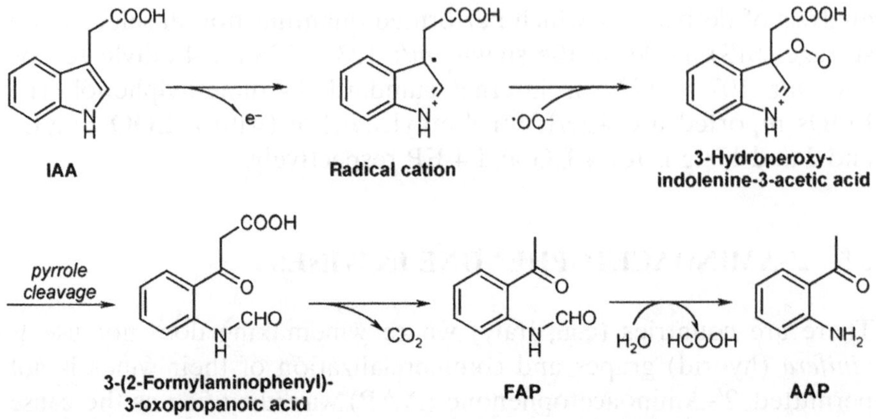

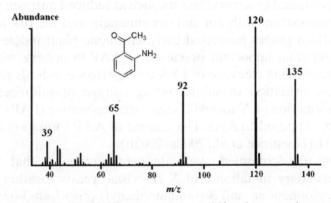

Figure 5.11. Mechanism of AAP formation proposed by Hoenicke (2002b). (Reprinted from *Journal of Chromatography* A 1150, Schmarr (2007) Analysis of 2-aminoaceto-phenone in wine using a stable isotope dilution assay and multidimensional gas chromatography–mass spectrometry, p. 79, Copyright © 2006, with permission from Elsevier.)

Figure 5.12. The MS–EI (70 eV) mass spectrum of 2′-aminoacetophenone (AAP).

n-heptane as a cosolvent, then dichloromethane is evaporated at room temperature in order to have a residual extract in *n*-heptane (Schmarr et al., 2007).

Alternatively, a direct-immersion SPME method with DVB/CAR/PDMS fiber (50/30 μm, 2 cm length) has been proposed. An aliquot of 15 mL of wine is transferred into a 20-mL vial and equilibrated at 30 °C for 5 min, then the fiber is immerged into the solution for 30 min under stirring. The fiber is then desorbed into the GC injection port at 250 °C (Fan et al., 2007).

The GC/MS analysis can be performed using a 5% diphenyl-dimethylpolysiloxane column (30 m × 0.32 mm i.d., 0.25 μm). The oven temperature program starts at 40 °C and is increased to 250 °C at 10 °C/min. These methods allow detection of AAP in wine up to 1 ng/L (Fan et al., 2007). The MS–EI (70 eV) mass spectrum of AAP is shown in Fig. 5.12.

REFERENCES

Albagnac, G. (1975). La décarboxylation des acides cinnamiques substitués par les levures, *Ann. Technol. Agric.*, **24**, 133–141.

Allen, M. (1995). What level of methoxypyrazines is desired in red wines? The flavour perspective of the classic red wines of Bordeaux, *Aust. Grapegrower Winemaker*, **381**, 7–9.

Alves, R.F., Nascimento, A.D.M., and Nogueira, J.M.F. (2005). Characterization of the aroma profile of Madeira wine by sorptive extraction techniques, *Anal. Chim. Acta*, **546**, 11–21.

Anocibar Beloqui, A., de Pinho, P.G., and Bertrand, A. (1995). Bis-(hydroxyethyl)disulfide a new sulfur compound found in wine, *Am. J. Enol. Vitic.*, **46**(1), 84–87.

Aznar, M., Lopez, R., Cacho, J.F., and Ferreira, V. (2001). Identification and quantification of impact odorants of aged red wines from Rioja. GC-olfactometry, quantitative GC-MS, and odor evaluation of HPLC fractions, *J. Agric. Food Chem.*, **49**, 2924–2929.

Barata, A., Nobre, A., Correia, P., Malfeito-Ferreira, M., and Loureiro, V. (2006). Growth and 4-ethylphenol production by the yeast Pichia guillier-mondii in grape juices, *Am. J. Enol. Vitic.*, **57**, 133–138.

Begala, M., Corda, L., Podda, G., Fedrigo, M.A., and Traldi, P. (2002). Headspace solid-phase microextraction gas chromatography/mass spectrometry in the analysis of the aroma constituents of "Cannonau of Jerzu" wine, *Rapid Commun. Mass Spectrom.*, **16**, 1086–1091.

Blanchard, L., Darriet, P., and Dubourdieu, D. (2004). Reactivity of 3-mercaptohexanol in red wine: impact of oxygen, phenolic fractions, and sulfur dioxide, *Am. J. Enol. Vitic.*, **55**(2), 115–120.

Boido, E., Loret, A.L., Medina, K., Fariña, L., Carrau, F., Versini, G., and Dellacassa, E. (2003). Aroma composition of *Vitis vinifera* Cv. Tannat: the typical red wine from Uruguay, *J. Agric. Food Chem.*, **51**(18), 5408–5413.

Bonino, M., Schellino, R., Rizzi, C., Aigotti, R., Delfini, C., and Baiocchi, C. (2003). Aroma compounds of an Italian wine (*Ruché*) by HS–SPME analysis coupled with GC–ITMS, *Food Chem.*, **80**, 125–133.

Bosch-Fusté, J., Riu-Aumatell, M., Guadayol, J.M., Caixach, J., López-Tamames, E., and Buxaderas, S. (2007). Volatile profiles of sparkling wines obtained by three extraction methods and gas chromatography–mass spectrometry (GC–MS) analysis, *Food Chem.*, **105**, 428–435.

Bouchilloux, P., Darriet, P., and Dubourdieu, D. (1998). Identification d'un thiol fortement odorant, le 2-methyl-3-furanthiol, dans les vins, *Vitis*, **37**(4), 177–180.

Cabredo-Pinillos, S., Cedron-Fernandez, T., and Saenz-Barrio, C. (2004). Comparison of different extraction methods applied to volatile compounds in wine samples previous to the determination by gas chromatography, *Anal. Lett.*, **37**(14), 3063–3084.

Cancilla, D.A. and Que Hee, S.S. (1992). *O*-(2,3,4,5,6-pentafluorophenyl) methylhydroxylamine hydrochloride: a versatile reagent for the determination of carbonyl-containing compounds, *J. Chromatogr.*, **627**, 1–16.

Carlton, W.K., Gump, B., Fugelsang, K., and Hasson, A.S. (2007). Monitoring acetaldehyde concentrations during micro-oxygenation of red wine by headspace solid-phase microextraction with on-fiber derivatization, *J. Agric. Food Chem.*, **55**, 5620–5625.

Carrillo, J.D. and Tena, M.T. (2007). Determination of ethylphenols in wine by in situ derivatisation and headspace solid-phase microextraction gas chromatography mass spectrometry, *Anal. Bioanal. Chem.*, **387**, 2547–2558.

Castro Mejías, R., Natera Marín, R., and García Moreno, M.d.V. (2003). Optimisation of headspace solid-phase microextraction for the analysis of volatile phenols in wine, *J. Chromatogr. A*, **995**, 11–20.

Chatonnet, P. and Dubourdieu D. (1998). Identification of substances responsible for the sawdust aroma in oak wood, *J. Sci. Food Agric.*, **76**(2), 179–188.

Chatonnet, P. and Boidron, J.N. (1988) Dosages de phénols volatils dans les vins par chromatographie en phase gazeuse, *Sci. Aliments*, **8**, 479–488.

Chatonnet, P. and Dubourdieu, D. (1996). Odeur de "planche" dans les bois de chêne: les responsables identifiés, *Rev. Oenolog. Tech. Vitivinicoles Œnologiques*, **82**, 17–19.

Chatonnet, P., Boidron, J.N., and Pons, M. (1990). Elevage des vins rouges en fûts de chêne: evolution de certains composés volatils et de leur impact aromatique, *Sci. Alim.*, **10**, 565–587.

Chatonnet, P., Dubourdieu, D., and Boidron, J.N. (1995). The influence of Brettanomyces/Dekkera sp. yeasts and lactic acid bacteria on the ethylphenol content of red wines, *Am. J. Enol. Vitic.*, **46**, 463–468.

Chatonnet, P., Dubourdieu, D., Boidron, J.N., and Lavigne, V. (1993). Synthesis of volatile phenols by Saccharomyces cerevisiae in Wines, *J. Sci. Food Agric.*, **62**, 191–202.

Chatonnet, P., Dubourdieu, D., Boidron, J.N., and Pons, M. (1992). The origin of ethylphenols in wines, *J. Sci. Food Agric.*, **60**, 165–178.

Chatonnet, P. (1993). Fenoli volatili: influenze organolettiche e metodi di prevenzione, *Vignevini*, **20**(7–8), 26–34.

Chatonnet, P., Viala, C., and Dubourdieu, D. (1997). Influence of polyphenol components of red wines on the microbial synthesis of volatile phenols, *Am. J. Enol. Vitic.*, **48**, 463–468.

Ciolfi, G., Garofolo, A., and Di Stefano, R. (1995). Identification of some o-aminophenones as secondary metabolites of *Saccharomyces cerevisiae*, *Vitis*, **34**(3), 195–196.

Coghe, S., Benoot, K., Delvaux, F., Vanderhaegen, B., and Delvaux, F.R. (2004). Ferulic acid release and 4-vinylguaiacol formation during brewing and fermentation: indications for feruloyl esterase activity in Saccharomyces cerevisiae, *J. Agric. Food Chem.*, **52**(3), 602–608.

Couto, J.A., Campos, F.M., Figueiredo, A.R., and Hogg, T.A. (2006). Ability of lactic acid bacteria to produce volatile phenols, *Am. J. Enol. Vitic.*, **57**, 166–171.

Cozzolino, D., Smyth, H.E., Lattey, K.A., Cynkar, W., Janik, L., Dambergs, R.G., Francis, I.L., and Gishen, M. (2006). Combining mass spectrometry based electronic nose, visible-near infrared spectroscopy and chemometrics to assess the sensory properties of Australian Riesling wines, *Anal. Chim. Acta*, **563**(1–2), 319–324.

Culleré, L., Cacho, J., and Ferreira, V. (2004). Analysis for wine C5–C8 aldehydes though the determination of their *O*-(2,3,4,5,6-pentafluorobenzyl) oximes formed directly in the solid phase cartridge, *Anal. Chim. Acta*, **524**, 201–206.

Davis, C.R., Wibowo, D., Eschenbruch R., Lee T.H., and Fleet, G.H. (1985). Practical implications of malolactic fermentation: A review, *Am. J. Enol. Vitic.*, **36**, 290–301.

De la Calle, G., Magnaghi, S., Reichenbächer, M., and Danzer, K. (1996). Systematic optimization of the analysis of wine bouquet components by solid phase microextraction, *J. High Res. Chromatogr.*, **19**(5), 257–262.

de Revel, G. and Bertrand, A. (1994). Dicarbonyl compounds and their reduction products in wine. Identification of wine aldehydes, In *Trends in Flavour Research*, H. Maarse and D.G. van der Heij (Eds.), Elsevier Science BV, p. 353.

De Revel, G. and Bertrand, A. (1993). A method for the detection of carbonyl compounds in wine: glyoxal and methylglyoxal, *J. Sci. Food Agric.*, **61**, 267–272.

De Revel, G. and Bertrand, A. (1994). Dicarbonyl compounds and their reduction products in wine. Identification of wine aldehydes, In *Trends in Flavour Research*, H. Maarse and D.G. van der Heij (Eds.), Elsevier Science B.V., pp. 353–361.

Demyttenaere, J.C.R., Dagher, C., Sandra, P., Kallithraka, S., Verhé, R., and De Kimpe, N. (2003). Flavour analysis of Greek white wine by solid-phase microextraction-capillary gas chromatography-mass spectrometry *J. Chromatogr. A*, **985**, 233–246.

Di Stefano, R. and Ciolfi, G. (1982). Produzione di acetaldeide da parte di stipiti di lieviti di specie diverse, *Riv. Vitic. Enol.*, **35**, 474–480.

Di Stefano, R. (1991). Proposition d'une methode de preparation de l'echantillon pour la determination des terpenes libres et glycosides des raisins et des vins, *Bull. O.I.V.*, **64**(721–722), 219–223.

Dias, L, Dias, S., Sancho, T., Stender, H., Querol, A., and Malfeito-Ferreira, M. (2003). Identification of yeasts isolated from wine: related environments and capable of producing 4-ethylphenol, *Food Microbiol.*, 567–574.

Díez, J., Domínguez, C., Guillén, D.A., Veas, R., and Barroso, C.G. (2004). Optimisation of stir bar sorptive extraction for the analysis of volatile phenols in wines, *J. Chromatogr. A*, **1025**, 263–267.

Dominguez, C., Guillén, D.A., and Barroso, C.G. (2002). Determination of volatile phenols in fino sherry wines, *Anal. Chim. Acta*, **458**, 95–102.

Drawert, F. and Rapp, A. (1968). Gaschromatographische Untersuchung pflanzicher Aromen. I. Anreicherung, Trennung und Identifizierung von flüchtigen Aromastoffen in Traubenmosten und Weinen, *Chromatographia*, **1**, 446–458.

Dubois, P. and Brulè, G. (1970). Étude des phénols volatiles des vins rouges, *C.R. Acad. Sci., Sér. D.*, 1797–1798.

Dubois, P. (1983). Volatile Phenols in Wines, In *Flavour of distilled beverages, origin and development*, J.R. Piggot (Ed.), Chichester: Ellis Horwood.

Edlin, D.A.N., Narbad, A., Dickinson, J.R., and Lloyd, D. (1995). The biotransformation of simple phenolic compounds by Brettanomyces anomalus, *FEMS-Microbiol. Lett.*, **125**, 311–316.

Edlin, D.A.N., Narbad, A., Gasson, M. J., Dickinson, J. R., and Lloyd, D. (1998). Purification and characterization of hydroxycinnamate decarboxylase from Brettanomyces anomalus, *Enzyme Microbiol Technol.*, **22**, 232–239.

Etievant, P. (1991). *Wine*, In *Volatile compounds in foods and beverages*, H. Maarse (Ed.), M. Dekker Inc. New York.

Fan, W., Tsai, I-M., and Qian, M.C. (2007). Analysis of 2-aminoacetophenone by direct-immersion solid-phase microextraction and gas chromatography–mass spectrometry and its sensory impact in Chardonnay and Pinot gris wines, *Food Chem.*, **105**, 1144–1150.

Fariña, L., Boido, E., Carrau, F., and Dellacassa, E. (2007). Determination of volatile phenols in red wines by dispersive liquid–liquid microextraction and gas chromatography-mass spectrometry detection, *J. Chromatogr. A*, **1157**, 46–50.

Favretto, D., Grandis, G., Allegri, G., and Traldi, P. (1998). An investigation of the aroma fraction of some Italian wines by solid phase micro extraction gas chromatography/mass spectrometry and membrane inlet mass spectrometry, *Rapid Commun. Mass Spectrom.*, **12**, 1595–1600.

Fedrizzi, B., Magno, F., Moser, S., Nicolini, G., and Versini, G. (2007a). Concurrent quantification of light and heavy sulphur volatiles in wine by headspace solid-phase microextraction coupled with gas chromatography/ mass spectrometry, *Rapid Commun. Mass Spectrom.*, **21**(5), 707–714.

Fedrizzi, B., Versini, G., Lavagnini, I., Nicolini, G., and Magno, F. (2007b). Gas Chromatography-Mass Spectrometry determination of 3-mercapto-hexan-1-ol and 3-mercaptohexyl acetate in wine. A comparison between Solid Phase Extraction and Headspace Solid Phase Microextraction methods, *Anal. Chim. Acta*, **596**(2), 291–297.

Ferreira, A.C.S. and de Pinho, P.G. (2003). Analytical method for determination of some aroma compounds on white wines by solid phase microextraction and gas chromatography, *J. Food Sci.*, **68**(9), 2817–2820.

Ferreira, V., Culleré, L., López, R., and Cacho, J. (2004). Determination of important odor-active aldehydes of wine through gas chromatography-mass spectrometry of their *O*-(2,3,4,5,6-pentafluorobenzyl)oximes formed directly in the solid phase extraction cartridge used for selective isolation, *J. Chromatogr A*, **1028**, 339–345.

Ferreira, V., López, R., Escudero, A., and Cacho, J.F. (1998). Quantitative determination of trace and ultratrace flavour active compounds in red wines through gas chromatographic-ion trap mass spectrometric analysis of microextracts, *J. Chromatogr. A*, **806**, 349–354.

Ferreira, V., Rapp, J A., Cacho, F., Hastrich, H., and Yavas, I. (1993). Fast and quantitative determination of wine flavor compounds using microextraction with Freon 113, *J. Agric. Food Chem.*, **41**(9), 1413–1420.

Ferreira, V., Sharman, M., Cacho, J.F., and Dennis, J. (1996). New and efficient microextraction/solid-phase extraction method for the gas chromatographic analysis of wine volatiles, *J. Chromatogr. A*, **731**(1–2), 247–259.

Ferreira, V., Jarauta, I., Ortega, L., and Cacho, J. (2004). Simple strategy for the optimization of solid-phase extraction procedures through the use of solid–liquid distribution coefficients. Application to the determination of aliphatic lactones in wine, *J. Chromatogr. A*, **1025**, 147–156.

Flamini R., Tonus T., and Dalla Vedova A. (2002b). A GC–MS method for determining acetaldehyde in wines, *Riv. Vitic. Enol.*, (**2/3**), 15–21.

Flamini, R. and Dalla Vedova, A. (2003). Glyoxal/glycolaldehyde: a redox system involved in malolactic fermentation of wine, *J. Agric. Food Chem.*, **51**, 2300–2303.

Flamini, R., Dalla Vedova, A., and Panighel, A. (2005a). Study of carbonyl compounds in some Italian marc distillate (grappa) samples by synthesis of *O*-(2,3,4,5,6-pentafluorobenzyl)-hydroxylamine derivatives, *Riv. Vitic. Enol.*, (1), 51–63.

Flamini, R., Dalla Vedova, A., Panighel, A., Perchiazzi, N., and Ongarato, S. (2005b). Monitoring of the principal carbonyl compounds involved in malolactic fermentation of wine by synthesis of *O*-(2,3,4,5,6-pentafluorobenzyl) hydroxylamine derivatives and solid-phase-micro-extraction positive-ion-chemical-ionization mass spectrometry analysis,. *J. Mass Spectrom.*, **40**(12), 1558–1564.

Flamini, R., De Luca, G., and Di Stefano, R. (2002a). Changes in carbonyl compounds in Chardonnay and Cabernet Sauvignon wines as a consequence of malolactic fermentation, *Vitis*, **41**(2), 107–112.

Flanzy, C. (1998). *Œnologie—fondements scientifiques et technologiques*, TEC & DOC, Paris.

Francioli, S., Guerra, M., López-Tamames, E., Guadayoi, J.M., and Caixach, J. (1999). Aroma of sparkling wines by headspace/solid phase microextraction and gas chromatography/mass spectrometry, *Am. J. Enol. Vitic.*, **50**(4), 404–408.

Froudiere, L. and Larue, F. (1988). Condition de survie de Brettanomyces (Dekkera) dans le moût de raisin et le vin, *Connaisance de la Vigne et du Vin*, **2**, 296–303.

Fugelsang, K. (1998). Brettanomyces: Dr Jekyll ou Mr. Hyde des vins? *Biofutur*, **182**, 22–23.

Fulcrand, H., Cameira-dos-Santos, P.J., Sarni-Manchado, P., Chenyer, V., and Favre-Bonvin, J. (1996). Structure of new anthocyanin derived wine pigments, *J. Chem. Soc.-Perkin Trans.*, **1**, 735–739.

Genovese, A., Dimaggio, R., Lisanti, M.T., Piombino, P., and Moio, L. (2005). Aroma components of red wines by different extraction methods and gas chromatography-SIM/mass spectrometry analysis, *Annali di Chimica*, **95**(6), 383–394.

Gianotti, S. and Di Stefano, R. (1991). Metodo per la determinazione dei composti volatili di fermentazione, *Annal. Ist. Sper. Enol. Asti*, XXII.

Grando, M.S., Versini, G., Nicolini, G., and Mattivi, F. (1993). Selective use of wine yeast strains having different volatile phenols production, *Vitis*, **32**, 43–50.

Guillou, I., Bertrand, A., de Revel, G., and Barbe, J.C. (1997). Occurrence of hydroxypropanedial in certain musts and wines, *J. Agric. Food Chem.*, **45**(9), 3382–3386.

Guillou, I. (1997). Study of low molecular weight substances combining sulfur dioxide in white wines stemming from botrytised harvest. Revelation of the existence and importance of the role of hydroxypropanedial, *Bull. O.I.V.*, **70**, 791.

Gunata, Y.Z., Bayonnove, C.L., Baumes, R.L., and Cordonnier, R.E. (1985). The aroma of grapes I. Extraction and determination of free and glicosidically bound fractions of some grape aroma components, *J. Chromatogr.*, **331**, 83–90.

Hardy, P.J. (1969). Extraction and concentration of volatiles from dilute aqueous-alcoholic solution using trichlorofluoromethane, *J. Agric. Food Chem.*, **17**(3), 656–658.

Hayasaka, Y. and Asenstorfer, R.E. (2002). Screening for potential pigments derived from anthocyanins in red wine using Nanoelectrospray Tandem Mass Spectrometry, *J. Agric. Food Chem.*, **50**, 756–761.

Hayasaka, Y. and Bartowsky, E.J. (1999). Analysis of diacetyl in wine using solid-phase microextraction combined with gas chromatography-mass spectrometry, *J. Agric. Food Chem.*, **47**, 612–617.

Hoenicke, K., Borchert, O., Grüning, K., and Simat, T.J. (2002a). "Untypical aging off-flavor" in wine: synthesis of potential degradation compounds of indole-3-acetic acid and kynurenine and their evaluation as precursors of 2-aminoacetophenone, *J. Agric. Food Chem.*, **50**, 4303–4309.

Hoenicke, K., Simat, T.J., Steinhart, H., Christoph, N., Geßner, M., and Köhler, H.J. (2002b). "Untypical aging off-flavor" in wine: formation of 2-aminoacetophenone and evaluation of its influencing factors, *Anal. Chim. Acta*, **458**, 29–37.

Howard, K.L., Mike, J.H., and Riesen, R. (2005). Validation of a solid-phase microextraction method for headspace analysis of wine aroma compounds, *Am. J. Enol. Vitic.*, **56**(1), 37–45.

Kinton, V.R., Collins, R.J., Kolahgar, B., and Goodner, K.L. (2003). Fast analysis of beverages using mass spectral based chemical sensor, *Gerstel App Note*, **4**, 1–10.

Lapolla, A., Flamini, R., Aricò, C.N., Rugiu, C., Reitano, R., Ragazzi, E., Seraglia, R., Dalla Vedova, A., Lupo, A., and Traldi, P. (2006). The fate of glyoxal and methylglyoxal in peritoneal dialysis, *J. Mass Spectrosc*, **41**(3), 405–408.

Lapolla, A., Flamini, R., Dalla Vedova, A., Senesi, A., Reitano, R., Fedele, D., Basso, E., Seraglia, R., and Traldi, P. (2003). Glyoxal and methylglyoxal levels in diabetic patients: quantitative determination by a new GC/MS method, *Clin. Chem. Lab. Med.*, **41**(9), 1166–1173.

Lestremau, F., Andersson, F.A.T., and Desauziers, V. (2004). Investigation of artefact formation during analysis of volatile sulphur compounds using solid phase microextraction (SPME), *Chromatographia*, **59**(9–10), 607–613.

López, R., Aznar, M., Cacho, J., and Ferreira, V. (2002). Determination of minor trace volatile compounds in wine by solid-phase extraction and gas chromatography with mass spectrometric detection, *J. Chromatogr. A*, **966**, 167–177.

Marais, J.A. (1986). A reproducible capillary gas chromatographic technique for the determination of specific terpenes in grape juice and wine, *S. Afr. J. Enol. Vitic.*, **7**(2), 21–25.

Marengo, E., Aceto, M., and Maurino, V. (2001). Classification of Nebbiolo-based wines from Piedmont (Italy) by means of solid-phase microextraction-gas chromatography-mass spectrometry of volatile compounds, *J. Chromatogr. A*, **943**, 123–137.

Martí, M.P., Mestres, M., Sala, C., Busto, O., and Guasch, J. (2003). Solid-phase microextraction and gas chromatography olfactometry analysis of successively diluted samples. A new approach of the aroma extract dilution analysis applied to the characterization of wine aroma, *J. Agric. Food Chem.*, **51**(27), 7861–7865.

Martineau, B., Acree, T.E., and Henick-Kling T. (1995). Effect of wine type on threshold for diacetyl, *Food Res. Int.*, **28**, 139–143.

Martorell, N., Martì, M.P., Mestres, M., Busto, O., and Guasch, J. (2002). Determination of 4-ethylguaiacol and 4-ethylphenol in red wines using headspace-solid-phase microextraction-gas chromatography, *J. Chromatogr. A*, **975**, 349–354.

Mateo-Vivaracho, L., Cacho, J., and Ferreira, V. (2007). Quantitative determination of wine polyfunctional mercaptans at nanogram per liter by gas chromatography-negative chemical ionization mass spectrometric analysis of their pentafluorobenzyl derivatives, *J. Chromatogr. A*, **1146**(2), 242–250.

Mateo-Vivaracho, L, Ferreira, V., and Cacho, J. (2006). Automated analysis of 2-methyl-3-furanthiol and 3-mercaptohexyl acetate at ng L–1 level by headspace solid-phase microextracion with on-fibre derivatisation and gas chromatography—negative chemical ionization mass spectrometric determination, *J. Chromatogr. A*, **1121**, 1–9.

Mestres, M., Busto, O., and Guasch, J. (2000). Analysis of organic sulfur compounds in wine aroma, *J. Chromatogr. A*, **881**(1–2), 569–581.

Monje, M.C., Chr. Privat, Gastine, V., and Nepveu, F. (2002). Determination of ethylphenol compound in wine by headspace solid-phase microextraction in conjunction with gas chromatography and flame ionization detection, *Anal. Chim. Acta*, **458**, 111–117.

Montedoro, G. and Bertuccioli, M. (1986). The flavour of wines, vermouth and fortified wines, In *Food flavours*, I.D., Morton and A.J., MacLeod (Eds.), Elsevier, Amsterdam.

Murat, M.L., Tominaga, T., Saucier, C., Glories, Y., and Dubourdieu, D. (2003). Effect of anthocyanins on stability of a key odorous compound, 3-mercaptohexan-1-ol, in Bordeaux rose wine, *Am. J. Enol. Vitic.*, **54**(2), 135–138.

Murray, R.A. (2001). Limitations to the use of solid-phase microextraction for quantitation of mixtures of volatile organic sulfur compounds, *Anal. Chem.*, **73**(7), 1446–1649.

Narziss, L., Miedaner, H., and Nitzsche, F. (1990). Ein Beitrag zur Bildung von 4-Vinyl-Guajakol bei der Herstellung von bayerischem Weizenbier, *Monatsschrift fuer Brauwissenschaft*, **43**(3), 96.

Nascimento, R.F., Marques, J.C., Lima Neto, B.S., De Keukeleire, D., and Franco, D.W. (1997). Qualitative and quantitative high-performance liquid chromatographic analysis of aldehydes in Brazilian sugar cane spirits and other distilled alcoholic beverages, *J. Chromatogr. A*, **782**, 13–23.

Nongonierma, A., Cayot, P., Le Quere, J.L., Springhett, M., and Voilley, A. (2006). Mechanisms of extraction of aroma compounds from foods, using adsorbents. Effect of various parameters. *Food Rev. Int.*, **22**(1), 51–94.

Ochiai, N., Sasamoto, K., Daishima, S., Heiden, A.C., and Hoffmann A. (2003). Determination of stale-flavour carbonyl compounds in beer by stir bar sorptive extraction with in-situ derivatization and thermal desorption-gas chromatography-mass spectrometry, *J. Chromatogr. A.*, **986**, 101–110.

Ojala, M., Kotiaho, T., Siirilae, J., and Sihvonen, M.L. (1994). Analysis of aldehydes and ketones from beer as *O*-(2,3,4,5,6-pentafluorobenzyl) hydroxylamine derivatives, *Talanta*, **41**(8), 1297–1309.

Pizarro, C., Pérez-del-Notario, N., and Gonzalez-Saiz, J.M. (2007). Determination of Brett character responsible compounds in wines by using multiple headspace solid-phase microextraction, *J. Chromatogr. A*, **1143**, 176–181.

Pollnitz, A.P., Pardon, K.H., and Sefton, M.A. (2000). Quantitative analysis of 4-ethylphenol and 4-ethylguaiacol in red wine, *J. Chromatogr. A*, **874**, 101–109.

Pozo-Bayón, M.A., Pueyo, E., Martín-Álvarez, P.J., and Polo, M.C. (2001). Polydimethylsiloxane soli-phase microextraction-gas chromatography methjod for the analysis of volatile compounds in wines. Its application to the characterization of varietal wines, *J. Chromatogr. A*, **922**, 267–275.

Pretorius, L.S. (2000). Tailoring wine yeast for the new millennium: novel approaches to the ancient art of winemaking, *Yeast*; **16**, 675–729.

Rapp, A. (1988). Wine aroma substances from gas chromatographic analysis, In *Modern methods of plant analysis*, H.F. Linskens and J.F. Jackson (Eds), Springer Verlag, Berlin.

Rapp, A. and Versini, G. (1991). Influence of nitrogen compounds in grape on aroma compounds of wines, *in Proceed. Intern. Symp. on Nitrogen in Grapes and Wines*, Seattle, Washington.

Rapp, A. and Versini, G. (1996). Methylanthranilate ("foxy taint") concentrations of hybrid and *Vitis vinifera* wines, *Vitis*, **35**(4), 215–216.

Rapp, A., Güntert, M., and Almy, J. (1985). Identification and significance of several sulfur-containing compounds in wine, *Am. J. Enol. Vitic.*, **36**(3), 219–221.

Rapp, A., Hastrich, H., Engel, L., and Knipser, W. (1978). Possibilities of characterizing wine quality and wine varieties by means of capillary chromatography, In *Flavor of food and beverages*, G. Charalambous (Ed.), Academic Press, New York.

Rapp, A. and Versini, G. (1996). Flüchtige phenolische Verbindungen in Wein, *Deutsche Lebensmittel-Rundschau*, **92**(2), 42–48.

Rapp, A., Versini, G., and Ullemeyer, H. (1993). 2-Aminoacetophenone: Causal component of "untypical aging flavour" ("naphthalene note", "hybrid note") of wine, *Vitis*, **32**, 61–62.

Rapp, A., Yavas, I., and Hastrich, H. (1994). Einfache und schnelle Anreicherung ("Kaltronmethode") von Aromastoffen des Weines und deren quantitative Bestimmung mittels Kapillargaschromatographie, *Deutsche Lebensm.-Rdsch.*, **90**, 171–174.

Rauhut, D. (1996). Qualitätsmindernde schwefelhaltige Stoffe im Wein— Vorkommen, Bildung, Beseitigung PhD Thesis, University of Giessen, Germany.

Ribéreau-Gayon, P., Glories, Y., Maujean, A., and Dubourdieu, D. (1998). *Traité d'Oenologie 2. Chimie du vin—stabilisation et traitements*, Dunod, Paris.

Rocha, S.M., Coutinho, P., Barros, A., Delgadillo, I., and Coimbra, M.A. (2006). Rapid tool for distinction of wines based on the global volatile signature, *J. Chromatogr. A*, **1114**, 188–197.

Rocha, S.M., Rodrigues, F., Coutinho, P., Delgadillo, I., and Coimbra, M.A. (2004). Volatile composition of Baga red wine. Assessment of the identification of the would-be impact odourants, *Anal. Chim. Acta*, **513**, 257–262.

Rodrigues, N., Gonçalves, G., Pereira-da-Silva, S., Malfeito-Ferreira, M., and Loureiro, V. (2001). Development and use of a new medium to detect yeast of the genera Dekkera/Brettanomyces sp., *J. Appl. Microbiol.*, **90**, 588–599.

Sanchéz-Palomo, E., Díaz-Maroto, M.C., and Pérez-Coello, S. (2005). Rapid determination of volatile compounds in grapes by HS-SPME coupled with GC–MS, *Talanta*, **66**, 1152–1157.

Sauvageot, F. and Vivier, P. (1997). Effects of malolactic fermentation on sensory properties of four burgundy wines, *Am. J. Enol. Vitic.*, **48**(2), 187–192.

Schneider, R., Kotseridis, Y., Ray, J., Augier, C., and Baumes, R. (2003). Quantitative determination of sulfur-containing wine odorants at sub parts per billion levels. Development and application of a stable isotope dilution assay, *J. Agric. Food Chem.*, **51**(11), 3243–3248.

Schmarr, H-G., Ganß, S., Sang, W., and Potouridis, T. (2007). Analysis of 2-aminoacetophenone in wine using a stable isotope dilution assay and multidimensional gas chromatography–mass spectrometry, *J. Chrom. A*, **1150**, 78–84.

Schreier, P. (1979). Flavor composition of wines: a review, *CRC Critical Rev. Food Sci. Nutrition*, **12**(1), 59–111.

Schwarz, M., Wabnitz, T.C., and Winterhalter, P. (2003). Pathway leading to the formation of anthocyanin-vinylphenol adducts and related pigments in red wine, *J. Agric. Food Chem.*, **51**, 3682–3687.

Segurel, M.A., Razungles, A.J., Riou, C., Trigueiro, M.G.L., and Baumes, R. (2005). Ability of possible DMS precursors to release DMS during wine aging and in the conditions of heat-alkaline treatment, *J. Agric. Food Chem.*, **53**(7), 2637–2645.

Siebert, T.E., Smyth, H.E., Capone, D.L., Neuwöhner, C., Pardon, K.H., Skouroumounis, G.K., Herderich, M.J., Sefton, M.A., and Pollnitz, A.P. (2005). Stable isotope dilution analysis of wine fermentation products by HS–SPME–GC–MS, *Anal. Bioanal. Chem.*, **381**, 937–947.

Simpson, R.F. (1979). Aroma composition of bottle aged white wine, *Vitis*, **18**, 148–154.

Suárez, R., Suárez-Lepe, J.A., Morata, A., and Calderon, F. (2007). The production of ethylphenols in wine by yeasts of the genera Brettanomyces and Dekkera: A review, *Food Chem.*, **102**, 10–21.

Swiegers, J.H. and Pretorius, I.S. (2007). Modulation of volatils sulfur compounds by wine yeast, *Appl. Microbiol. Biotech.*, **74**(5), 954–960.

Swiegers, J.H., Capone, D.L., Pardon, K.H., Elsey, G.M., Sefton, M.A., Francis, I.L., and Pretorius, I.S. (2007). Engineering volatile thiol release in Saccharomyces cerevisiae for improved wine aroma, *Yeast*, **24**(7), 561–574.

Tominaga, T., Murat, M.L., and Dubourdieu, D. (1998). Development of a method for analyzing the volatile thiols involved in the characteristic aroma of wines made from *Vitis vinifera* L. Cv. Sauvignon Blanc, *J. Agric. Food Chem.*, **46**(3), 1044–1048.

van Wyk, C.J. and Rogers, I.M. (2000). A "phenolic" off-odour in white table wines: causes and methods to diminish its occurrence, *S. Afr. J. Enol. Vitic.*, **21**, 52–57.

Vanderlinde, R., Bertrand, A., and Segur, M.C. (1992). Dosage des aldehydes dans les eaux-de-vie, In *Proceeding of 1er Symposium Scientifique International du Cognac, «Elaboration et connaissance des spiritueux»*, Cognac 11–15 Mai; Lavoiser TEC&DOC, Paris, 506–511.

Vas, G., Koteleky, K., Farkas, M., Dobó, A., and Vekey, K. (1998). Fast screening method for wine headspace compounds using solid-phase microextraction (SPME) and capillary GC technique, *Am. J. Enol. Vitic.*, **49**(1), 100–104.

Versini, G., Dalla Serra, A., Falcetti, M., and Sferlazzo, G. (1992). Rôle du clone, du millésime et de l'époque de la récolte sur le potentiel aromatique du raisin de Chardonnay, *Rev. Oenolog.*, **18**, 19–23.

Versini, G., Dellacassa, E., Carlin, S., Fedrizzi, B., and Magno, F. (2008). Analysis of Aroma Compounds in Wine, In *Hyphenated Techniques in Grape & Wine Chemistry*, Flamini, (Ed.), John Wiley & Sons, Ltd.

Versini, G., Rapp, A., Scienza, A., Dalla Serra, A., and Dell'Eva, M. (1988). *Evidence of some glycosidically bound new monoterpenes and norisoprenoids in grapes*, In *Proc. Bioflavour '87*. P. Schreier (Ed.), W. de Gruyter, Berlin.

Versini, G. (1985). Sull'aroma del vino Traminer Aromatico o Gewürztraminer, *Vignevini*, **12**(1–2), 57–65.

Vesely, P., Lusk, L., Basarova, G., Seabrooks, J., and Ryder, D. (2003). Analysis of aldehydes in beer using solid-phase microextraction with on-fiber derivatization and gas chromatography/mass spectrometry, *J. Agric. Food Chem.*, **51**, 6941–6944.

Vianna, E. and Ebeler, S. (2001). Monitoring ester formation in grape juice fermentations using solid phase microextraction coupled with gas chromatography-mass spectrometry, *J. Agric. Food Chem.*, **49**(2), 589–595.

Vidal, J.P., Estreguil, S., and Cantagrel, R. (1993). Quantitative analysis of Cognac carbonyl compounds at the PPB level by GC–MS of their *O*-(pentafluorobenzyl amine) derivatives, *Chromatographia*, **36**, 183–186.

Vidal, J.P., Mazerolles, G., Estreguil, S., and Cantagrel, R. (1992). Analyse quantitative de la fraction carbonylée volatile des eaux-de-vie de Cognac. In *Proceeding of 1er Symposium Scientifique International du Cognac, «Elaboration et connaissance des spiritueux»*, Cognac 11–15 May; Lavoiser TEC&DOC, Paris, 529–537.

Webb, A.D. (1967). Wine flavor: volatile aroma compounds of wines, Schultz, H.W., Day, E.A., and Libbey, L.M. (Eds.) In *The Chemistry and Physiology of Flavours*, The AVI Publishing Company Inc., Westport Co.

Williams, P.J. (1982). Use of C_{18} reversed-phase liquid chromatography for the isolation of monoterpene glycosides and nor-isoprenoid precursors from grape juice and wines, *J. Chromatogr. A*, **235**(2), 471–480.

Wright, J.M. and Parle, J.N. (1973). Brettanomyces in the New Zealand wine industry. *New Zealand J. Agric. Res.*, **17**, 273–278.

Zulema, P., Palma, M., and Barroso, C. (2004). Determionation of terpenoids in wines by solid phase extraction and gas chromatography, *Anal. Chim. Acta*, **513**, 209–214.

6

GRAPE AND WINE POLYPHENOLS

6.1 INTRODUCTION

Polyphenols are the principal wine compounds associated with beneficial health effects. Grape seed procyanidins and proanthocyanidins
are active ingredients used in medicinal products for the treatment of
circulatory disorders (capillary fragility, microangiopathy of the retina)
with antioxidant plasma activity, reduce platelet aggregation, decrease
the susceptibility of healthy cells to toxic and carcinogenic agents, and
have antioxidant activity toward human low-density lipoprotein.
Quercetin, the principal flavonol in grape, blocks aggregation of
human platelets and seems to inhibit carcinogens and cancer cell
growth in human tumors. Several studies evidenced the anticancer,
cardioprotection, anti-inflammatory, antioxidant, and platelet aggregation inhibition activity of resveratrol (Flamini, 2003 and references
cited therein). The principal nonanthocyanic polyphenols of grape are
the flavan-3-ols (+)-catechin and (−)-epicatechin; principal flavonols
are kaempferol, quercetin, and myricetin glycosides (mainly as gluco-
sides and glucoronides), and recently isorhamnetin, laricitrin, and
syringetin were identified (as were structures in Fig. 6.1).

Mass Spectrometry in Grape and Wine Chemistry, by Riccardo Flamini
and Pietro Traldi
Copyright © 2010 John Wiley & Sons, Inc.

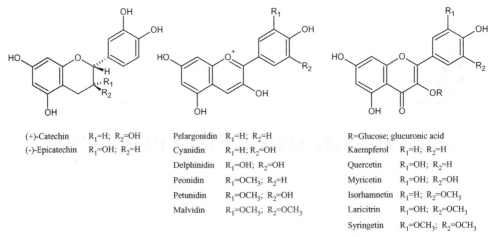

| (+)-Catechin | R₁=H; R₂=OH |
| (-)-Epicatechin | R₁=OH; R₂=H |

(+)-Catechin R_1=H; R_2=OH
(-)-Epicatechin R_1=OH; R_2=H

Pelargonidin R_1=H; R_2=H
Cyanidin R_1=H; R_2=OH
Delphinidin R_1=OH; R_2=OH
Peonidin R_1=OCH$_3$; R_2=H
Petunidin R_1=OCH$_3$; R_2=OH
Malvidin R_1=OCH$_3$; R_2=OCH$_3$

R=Glucose; glucuronic acid
Kaempferol R_1=H; R_2=H
Quercetin R_1=OH; R_2=H
Myricetin R_1=OH; R_2=OH
Isorhamnetin R_1=H; R_2=OCH$_3$
Laricitrin R_1=OH; R_2=OCH$_3$
Syringetin R_1=OCH$_3$; R_2=OCH$_3$

Figure 6.1. Principal flavan-3-ols, anthocyanidins, and flavonols of grape. The glucose residue of anthocyanidins can be linked to an acetyl, *p*-coumaroyl, or caffeoyl (for malvidin, Mv) group.

Anthocyanins confer color to red grapes and wines. The anthocyanin profiles of grape varieties are studied for chemotaxonomic purposes and allow to distinguish, e.g., between *Vitis vinifera* and hybrid grape varieties, the latter being characterized by the peculiar presence of 3,5-*O*-diglucoside anthocyanins. Moreover, grape anthocyanins are natural colorants used in the food and pharmaceutical industries (Hong and Wrolstad, 1990). In the mouth, the formation of a complex between tannins and the saliva proteins induces the sensorial characteristic of astringency to the wine.

Vitis vinifera red grapes are characterized from the anthocyanins delphinidin (Dp), cyanidin (Cy), petunidin (Pt), peonidin (Pn), and malvidin (Mv) present in 3-*O*-monoglucoside, 3-*O*-acetylmonoglucoside, and 3-*O*-(6-*O*-*p*-coumaroyl)monoglucoside forms, the Mv 3-*O*-(6-*O*-caffeoyl)monoglucoside also can be present (Fig. 6.1). In the non-*Vitis vinifera* (hybrid) grapes, anthocyanins containing a second glucose molecule linked to the C5 hydroxyl group are also often present (diglucosides). Recently, pelargonidin (Pg) 3-*O*-monoglucoside was reported (Wang et al., 2003).

Changes in the color of red wines that occur during aging are due to the anthocyanins undergoing chemical reactions and polymerization with the other wine compounds. More than 100 structures belong to the pigment families of anthocyanins, pyranoanthocyanins, direct flavanol-anthocyanin condensation products, and acetaldehyde-mediated

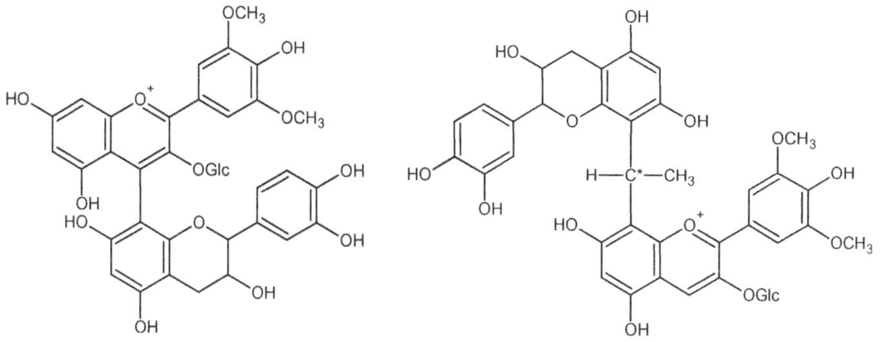

Figure 6.2. Compounds formed in wines during aging: (a) structure with direct linkage between anthocyanin and flavan-3-ol proposed by Somers (1971) and (b) the anthocyanin-flavan-3-ol structure with an ethyl bridge proposed by Timberlake and Bridle (1976).

flavanol-anthocyanin condensation products (anthocyanin can be linked either directly or by an ethyl bridge to a flavan-3-ol), were identified (Alcade-Eon et al., 2006). The principal structures are shown in Figs. 6.2 and 6.3.

The B- and A-type procyanidins and proanthocyanidins (condensed or nonhydrolyzable tannins, Fig. 6.4) are polymers of flavan-3-ols present in the skin and seeds of the grape berry. In winemaking, they are transferred to the wine, and the sensorial characteristics of astringency and bitterness of wine are linked to the galloylation degree (DG) and degree of polymerization (PD) of flavan-3-ols (Cheynier and Rigaud, 1986; Vidal et al., 2003).

Structural characterization of anthocyanins and polyphenols in grape extracts and wine by liquid chromatography (LC) coupled with ultraviolet–visible (UV–vis) methods requires hydrolysis or thiolysis of the sample (de Freitas et al., 1998). Liquid chromatography mass spectrometry (LC/MS) coupled with multiple mass spectrometry (MS/MS and MS^n) resulted in the more suitable tool to study the structures formed in wine during aging (Alcade-Eon et al., 2004; 2006) and to characterize high molecular weight (MW) compounds, such as procyanidins, proanthocyanidins, prodelphinidins, and tannins (Niessen and Tinke, 1995; de Hoffmann, 1996; Abian, 1999; Flamini et al., 2007). In general, these methods require minor sample purification and MS/MS allows characterization of both the aglycone and sugar moiety.

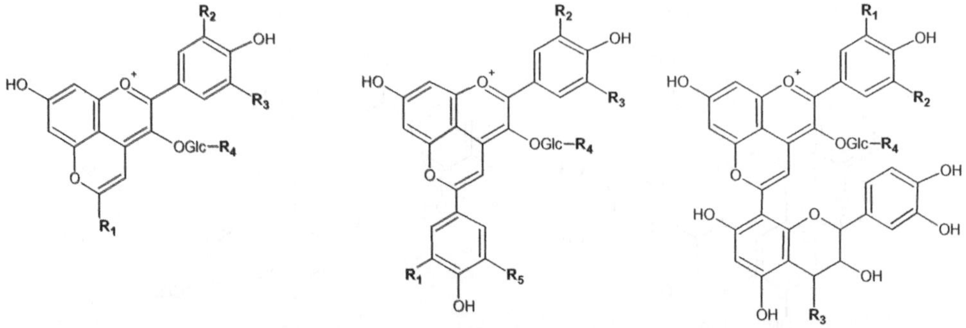

R_1=H; R_2=OCH$_3$; R_3=OCH$_3$; R_4=H (vitisin B)
R_1=H; R_2=OCH$_3$; R_3=OCH$_3$; R_4=acetyl
R_1=H; R_2=OCH$_3$; R_3=OCH$_3$; R_4=coumaroyl
R_1=CH$_3$; R_2=OCH$_3$; R_3=OCH$_3$; R_4=H
R_1=OH; R_2=OCH$_3$; R_3=OCH$_3$; R_4=H
R_1=COOH; R_2=OCH$_3$; R_3=OH; R_4=H
R_1=COOH; R_2=OCH$_3$; R_3=H; R_4=H
R_1=COOH; R_2=OCH$_3$; R_3=OCH$_3$; R_4=H
R_1=COOH; R_2=OCH$_3$; R_3=OCH$_3$; R_4=acetyl
R_1=COOH; R_2=OCH$_3$; R_3=OCH$_3$; R_4=coumaroyl
R_1=COOH; R_2=OCH$_3$; R_3=H; R_4=coumaroyl
R_1=COOH; R_2=OH; R_3=OH; R_4=H

R_1=H; R_2=OCH$_3$; R_3=OCH$_3$; R_4=H; R_5=H (pigment A)
R_1=H; R_2=OCH$_3$; R_3=OCH$_3$; R_4=H; R_5=OH
R_1=H; R_2=OCH$_3$; R_3=OCH$_3$; R_4=acetyl; R_5=H
R_1=H; R_2=OCH$_3$; R_3=OCH$_3$; R_4=coumaroyl; R_5=H
R_1=H; R_2=OCH$_3$; R_3=OCH$_3$; R_4=acetyl; R_5=OCH$_3$
R_1=H; R_2=OCH$_3$; R_3=OCH$_3$; R_4=coumaroyl; R_5=OCH$_3$
R_1=OCH$_3$; R_2=OCH$_3$; R_3=OCH$_3$; R_4=H; R_5=OCH$_3$
R_1=H; R_2=OCH$_3$; R_3=OH; R_4=H; R_5=H
R_1=H; R_2=OCH$_3$; R_3=OH; R_4=acetyl; R_5=H
R_1=H; R_2=OCH$_3$; R_3=OH; R_4=coumaroyl; R_5=H
R_1=H; R_2=OCH$_3$; R_3=H; R_4=coumaroyl; R_5=H
R_1=H; R_2=OCH$_3$; R_3=H; R_4=H; R_5=H
R_1=H; R_2=OCH$_3$; R_3=H; R_4=H; R_5=OH
R_1=H; R_2=OCH$_3$; R_3=H; R_4=coumaroyl; R_5=OH
R_1=H; R_2=OCH$_3$; R_3=OCH$_3$; R_4=coumaroyl; R_5=OH
R_1=H; R_2=OCH$_3$; R_3=H; R_4=H; R_5=OCH$_3$
R_1=H; R_2=OCH$_3$; R_3=OCH$_3$; R_4=H; R_5=OCH$_3$

R_1=OCH$_3$; R_2=OCH$_3$; R_3=(epi)catechin; R_4=H
R_1=OCH$_3$; R_2=H; R_3=H; R_4=H
R_1=OCH$_3$; R_2=OCH$_3$; R_3=H; R_4=H
R_1=OCH$_3$; R_2=OCH$_3$; R_3=H; R_4=coumaroyl

Figure 6.3. Structures of C4 substituted anthocyanins identified in aged red wines formed by reaction with pyruvic acid, vinylphenol, vinylcatechol, vinylguaiacol, vinyl(epi)catechin (Fulcrand et al., 1998; Hayasaka and Asenstorfer, 2002; Alcade-Eon et al., 2004; Gomez-Ariza et al., 2006.)

6.2   THE LC/MS OF NON-ANTHOCYANIC POLYPHENOLS OF GRAPE

Lee et al. (2005) proposed a method for analysis of flavonols in grape by performing berry extraction with acidified methanol (0.01% of 12 N HCl). After filtration, the solvent is removed under vacuum and the residue is dissolved in a 0.1 M citric acid buffer with pH 3.5. First, polyphenols are fractionated on a reverse-phase C$_{18}$ cartridge (e.g., Sep-Pak 5 g), then on a Sephadex LH-20 3-g cartridge (a cross-linked dextran-based stationary phase used for gel permeation, normal-phase partition, and adsorption chromatography). Four fractions finally are recovered by ethyl acetate and methanol, as shown in the flow diagram Fig. 6.5.

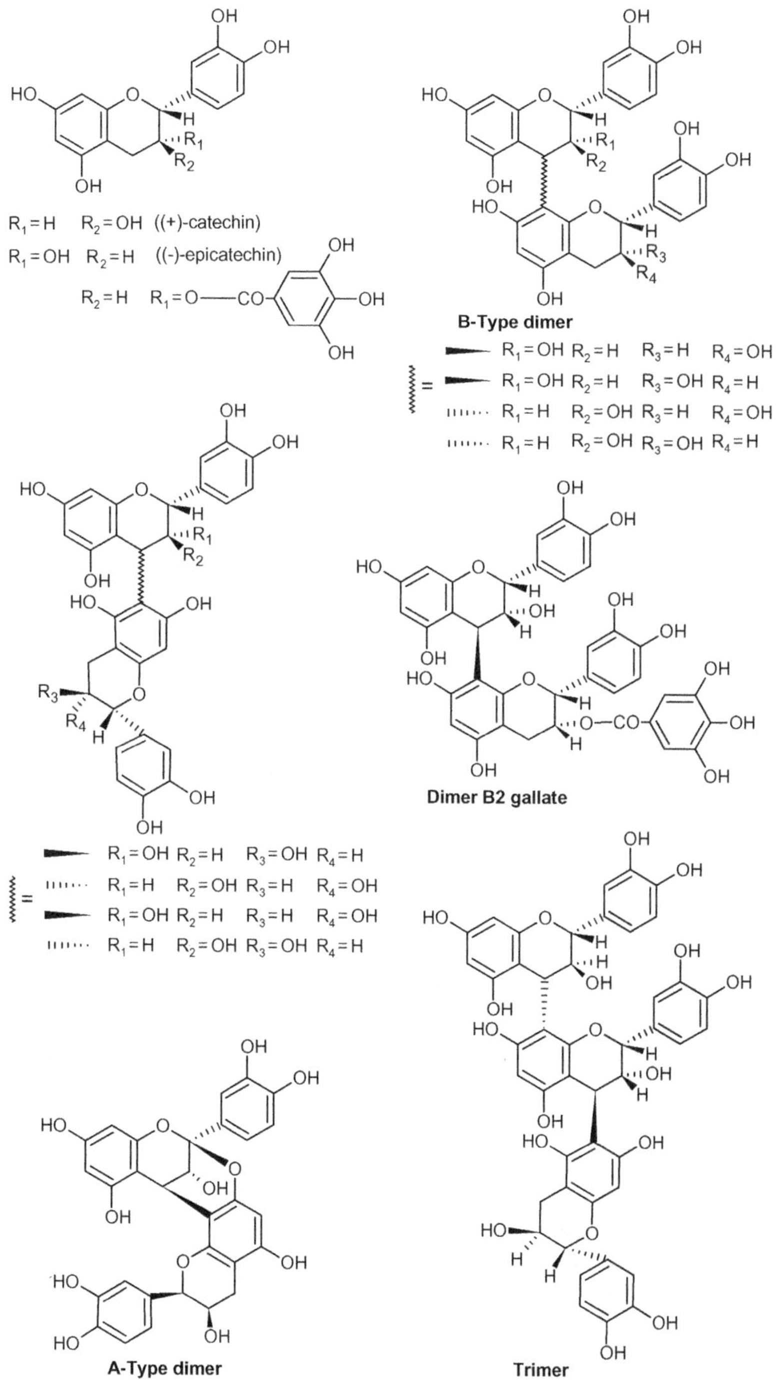

R₁=H   R₂=OH ((+)-catechin)

R₁=OH R₂=H   ((-)-epicatechin)

R₂=H   R₁=O—CO-

B-Type dimer

R₁=OH R₂=H   R₃=H   R₄=OH
R₁=OH R₂=H   R₃=OH R₄=H
R₁=H   R₂=OH R₃=H   R₄=OH
R₁=H   R₂=OH R₃=OH R₄=H

R₁=OH R₂=H   R₃=OH R₄=H
R₁=H   R₂=OH R₃=H   R₄=OH
R₁=OH R₂=H   R₃=H   R₄=OH
R₁=H   R₂=OH R₃=OH R₄=H

Dimer B2 gallate

A-Type dimer

Trimer

Figure 6.4. The B- and A-type flavan-3-ol dimers and trimers present in grape seeds.

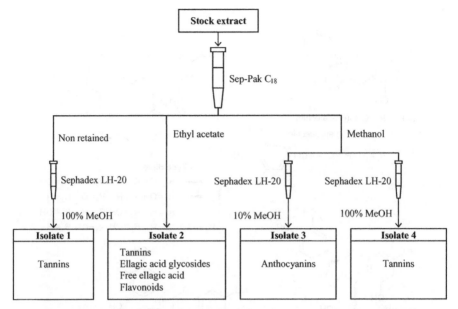

Figure 6.5. Fractionation of polyphenols in grape extract (Lee et al., 2005).

The methanolic fraction is evaporated to dryness, dissolved in a pH 3.5 buffer, and partitioned on a Sephadex LH-20 cartridge in two stages. Ethyl acetate of isolate 2 is evaporated, then the residue is dissolved in the pH 3.5 buffer. Fractions are then characterized by liquid chromatography–electrospray ionization mass spectrometry (LC/ESI–MS) analysis, collision-induced dissociation (CID), and MS/MS to confirm the compounds identification. The compounds identified in isolate 2 of muscadine grapes (cv. *Albemarle*) are reported in Table 6.1 with the characteristic fragments originated from MS/MS experiments.

Recently, a study on flavonols in different *V. vinifera* red grape varieties revealed, in addition to myricetin and quercetin 3-glucosides and 3-glucuronides and to kaempferol and isorhamnetin 3-glucosides, the presence of laricitrin and syringetin 3-glucosides. In addition, minority flavonols such as kaempferol and laricitrin 3-galactosides, kaempferol-3-glucuronide, and quercetin and syringetin 3-(6-acetyl)glucoside were identified (Castillo-Muñoz et al., 2007). Compounds identified in a *Petit Verdot* grape skins extract are reported in Table 6.2. Extraction of grape skin was performed by a methanol(MeOH)–H_2O–formic acid 50:48.5:1.5 (v/v/v) solution. Flavonols in the extract were separated from anthocyanic compounds by solid-phase extraction (SPE) using a commercial cartridge composed of reverse-phase and cationic-exchange

TABLE 6.1. The LC/ESI-MS Analysis of Isolate 2 in Fig. 6.5 of *Albemarle* (Muscadine) Grape Extract and Characteristic Fragments Originated from MS/MS Experiments[a]

RT	UV (nm)	Compound	MW	ESI	BP (m/z); ID	MS² (m/z)	MS³ (m/z)
58–60	261, 280sh	Ellagitannins	800	(−)	799; [M−H]⁻	781, 763, 745, 735, 495, 481, 451, 317, 301, 273	763, 745, 735, 719, 479, 461, 301, 275, 247
				(+)	818; [M+NH₄]⁺	801, 783, 447, 429, 385, 357, 337, 303, 277, 259, 231	429, 411, 385, 357, 303, 277
			814	(−)	813; [M−H]⁻	781, 763, 753, 735, 301	763, 745, 419, 317, 301, 273, 229
				(+)	832; [M+NH₄]⁺	797, 779, 461, 447, 443, 397, 335, 317, 303, 277, 259, 241	427, 411, 385, 357, 335, 303, 277
86.0	352	Myricetin rhamnoside	464	(−)	463; [M−H]⁻	359, 337, 317	287, 271, 179, 151
				(+)	465; [M+H]⁺	447, 429, 361, 319	301, 290, 283, 273, 263, 255, 245, 165, 163, 137
90.5	360	Ellagic acid xyloside	434	(−)	433; [M−H]⁻	301	257, 229
				(+)	435; [M+H]⁺	303	285, 275, 257, 247, 229, 165, 153, 137
91.3	361	Ellagic acid rhamnoside	448	(−)	447; [M−H]⁻	300, 301	272, 257, 244, 229
				(+)	449; [M+H]⁺	303	285, 275, 259
92.3	366	Ellagic acid	302	(−)	301 [M−H]⁻	301, 284, 257, 229, 185	
				(+)	nd		
94.2	351	Quercetin rhamnoside	448	(−)	447; [M−H]⁻	301	283, 271, 255, 179, 169, 151, 121, 107
				(+)	449; [M+H]⁺	431, 413, 303	303, 285, 275, 257, 247, 229, 165, 153, 137
97.5	344	Kaempferol rhamnoside	432	(−)	431; [M−H]⁻	327, 299, 285, 256	267, 257, 255, 241, 229, 213, 197, 163
				(+)	433; [M+H]⁺	415, 397, 375, 287	287, 269, 241, 231, 213, 197, 183, 165, 153

[a]Analytical conditions: C₁₈ 80 Å (150 × 2 mm; 4 μm) column; binary solvent composed of (A) 0.5% formic acid containing 5-mM ammonium formate and (B) 0.5% formic acid in methanol; gradient program from 5 to 30% of B in 5 min, from 30 to 65% of B in 70 min, from 65 to 95% of B in 30 min, 95% B isocratic for 20 min (flow 0.15 mL/min). Mass spectrometry conditions: both positive and negative ion mode; sheath gas N₂ 60 units/min; auxiliary gas N₂ 5 units/min; spray voltage 3.3 kV; capillary temperature 250°C; capillary voltage 1.5 V; tube lens offset 0 V (Lee et al., 2005). Not detected = nd.

TABLE 6.2. The LC Retention Times (RT), UV–vis, and Mass Spectra Data of Flavonols Identified in Petit Verdot Grape Skins[a]

Flavonol	HPLC[b] RT (min)	λ_{max} (nm)	[M–H]+ and Product Ion (m/z)
Myricetin-3-glucuronide	13.9	257(sh), 261, 301(sh), 353	495, 319
Myricetin-3-glucoside	14.5	257(sh), 262, 298(sh), 355	481, 319
Quercetin-3-glucuronide	18.0	257, 265(sh), 299(sh), 354	479, 303
Rutin[c]		256, 264(sh), 300(sh), 354	611, 303
Quercetin-3-glucoside	18.8	256, 265(sh), 295(sh), 354	465, 303
Laricitrin-3-glucoside	19.9	256, 265(sh), 301(sh), 357	495, 333
Kaempferol-3-glucoside	22.6	265, 298(sh), 320(sh), 348	449, 287
Isorhamnetin-3-glucoside	24.3	255, 265(sh), 297(sh), 354	479, 317
Syringetin-3-glucoside	24.9	255, 265(sh), 300(sh), 357	509, 347
Laricitrin-3-galactoside	19.4	256, 265(sh), 302(sh), 357	495, 333
Kaempferol-3-galactoside	21.1	266, 292(sh), 320(sh), 348	449, 287
Kaempferol-3-glucuronide	21.9	265, 290(sh), 320(sh), 348	463, 287
Quercetin-3-(6-acetyl)glucoside	22.9	257, 265(sh), 295(sh), 352	517, 303
Syringetin-3-(6-acetyl)glucoside	30.4	255, 265(sh), 298(sh), 358	551, 347

[a]The LC/ESI–MS conditions: C_{18} column (4.6×250 mm; 5 μm) at 40 °C; solvents water–acetonitrile–formic acid 87:3:10 v/v/v (A) and 40:50:10 v/v/v (B); elution gradient from 6 to 30% of B in 15 min, then to 50% of B in 30 min, to 60% of B in 35 min, 60% B isocratic for 38 min, return to 6% B in 46 min. Positive-ion mode detection, dry gas N_2 (11 mL/min), drying temperature 350 °C, nebulizer 65 psi, capillary –2500 V, capillary exit offset 70 V, skimmer 1: 20 V, skimmer 2: 6 V (Castillo-Muñoz et al., 2007).
[b]High-performance liquid chromatography = HPLC.
[c]Quercetin-3-O-(6″-rhamnosyl)glucoside (Castillo-Muñoz et al., 2009).

materials previously conditioned with methanol and washed with water. After the sample loading, the cartridge is washed with 0.1 M HCl and water, and the flavonol fraction containing neutral or acidic polyphenols is eluted with methanol. Anthocyanins were removed from the stationary phase by washing with an aqueous solution containing 2% ammonia and 80% methanol.

The LC/MS analysis of resveratrol (3,5,4'-trihydroxystilbene) and piceatannol (3,4,3',5'-tetrahydroxy-*trans*-stilbene) in grape is usually performed operating in the negative-ion mode (ESI source voltage 4500V, entrance capillary voltage 4V, entrance capillary temperature 280 °C), using a C_{18} column with a binary solvent composed of $H_2O/0.1\%$ formic acid and MeOH (elution gradient program: 33% MeOH for 40 min, 33 → 100% MeOH in 15 min, 100% MeOH for 5 min at a flow rate 0.6 mL/min) (De Rosso et al., 2009).

A study of the MS^n fragmentation of resveratrol and piceatannol was performed by deuterium exchange experiments and accurate mass measurements (Stella et al., 2008). The product ion spectrum of the [M–H]⁻ ion of *trans*-resveratrol at *m/z* 227, is reported in Fig. 6.6, and that of piceatannol at *m/z* 243 are shown in Fig. 6.7. Fragmentation patterns of the [M–H]⁻ ion of two compounds are reported in Fig. 6.8. Fragmentations were confirmed with deuterium labeling experiments by dissolving the standard compounds in deuterated methanol: The

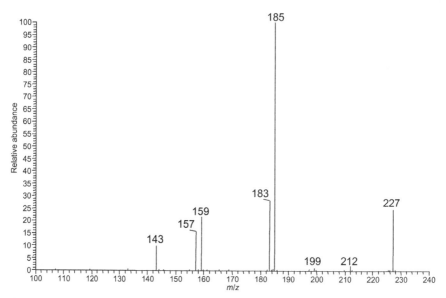

Figure 6.6. Product negative-ion spectrum of direct infusion ESI-generated [M–H]⁻ species of *trans*-resveratrol at a flow rate of 10 µL/min. The ESI conditions: source voltage 4500V, entrance capillary voltage –4V, entrance capillary temperature 280 °C, sheat gas flow rate 40 (arbitrary units), scan range *m/z* 70–700; collisional supplementary radio frequency voltage to the ion trap end-caps 2V; ion trap collision gas He pressure 1.1×10^{-5} Torr. (Reprinted from Rapid Communications in Mass Spectrometry 22, Stella et al., Collisionally induced fragmentation of [M–H]⁻ species of resveratrol and piceatannol investigated by deuterium labeling and accurate mass measurements, p. 3868, Copyright © 2008, with permission from John Wiley & Sons, Ltd.)

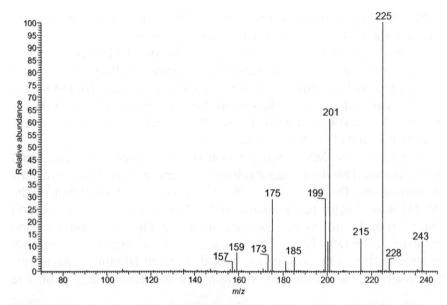

Figure 6.7. Product ion spectrum of direct infusion ESI generated [M–H]⁻ species of piceatannol. The ESI conditions are the same reported in the caption of Fig. 6.6. (Reprinted from Rapid Communications in Mass Spectrometry 22, Stella et al., Collisionally induced fragmentation of [M–H]⁻ species of resveratrol and piceatannol investigated by deuterium labeling and accurate mass measurements, p. 3870, Copyright © 2008, with permission from John Wiley & Sons, Ltd.)

deprotonated molecules of *trans*-resveratrol and piceatannol were shifted at m/z 229 and 246, respectively, proving the occurrence of OH hydrogen exchanges. The MS/MS spectrum of the ion at m/z 229 shows ions at m/z 187, 186, and 185. The species at m/z 187 corresponds to the ion at m/z 185 of Scheme 1a containing two D atoms. Substitution of a D for H leads to the fragment ion at m/z 186 corresponding to C_2HDO loss, while the ion at m/z 185 could correspond to a deuterated fragment ion at m/z 183, or to the loss of C_2D_2O. The presence of ions at m/z 161 and 159 confirmed the mechanisms of formation of the ions at m/z 159 and 157 shown in Schemes 1b and 1a, respectively. Similar results were obtained in the MS/MS spectrum of the ion at m/z 246 of piceatannol. In particular, ions at m/z 228, 227, and 226, corresponding to losses of H_2O, HDO, and D_2O, respectively, confirmed the primary water loss mechanisms proposed (fragment ion at m/z 225 in Fig. 6.9). Fragmentations were definitively confirmed by accurate mass measurements.

Extraction of proanthocyanidins (PAs) from grape seeds can be accomplished by grinding the dry seeds until a homogeneous powder

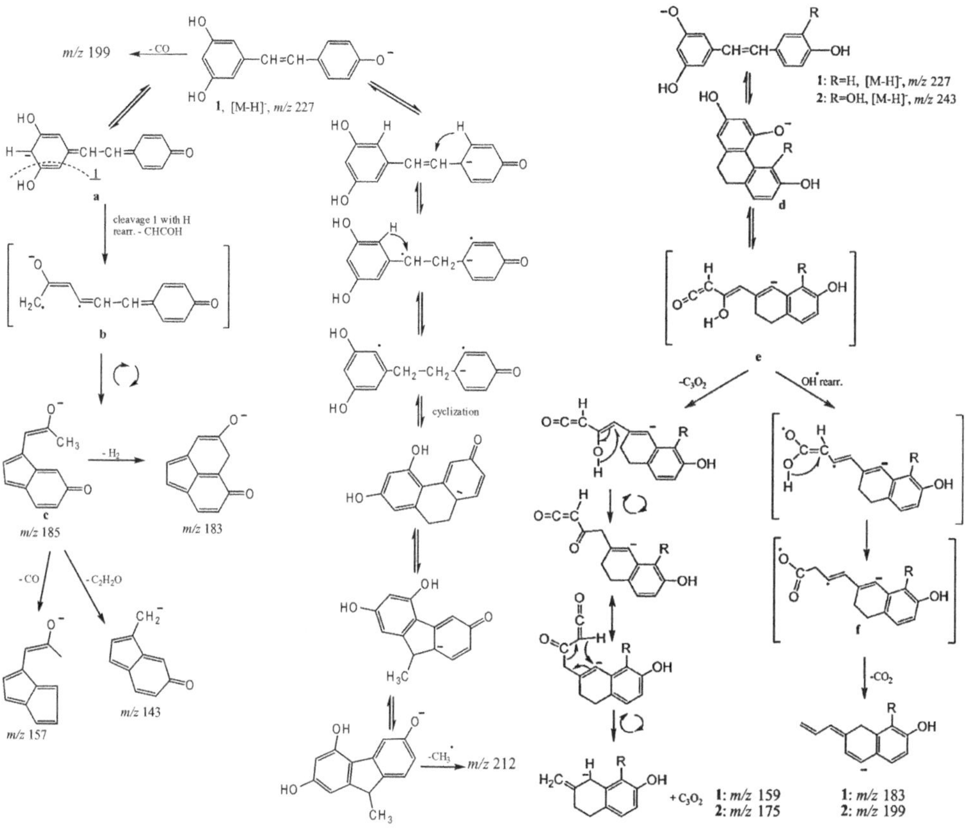

Scheme 1a   **Scheme 1b**

Figure 6.8. Collisionally induced fragmentation patterns of [M–H]$^-$ ions of *trans*-resveratrol at *m/z* 227 (compound **1**) considering that the deprotonation reaction occurred on the phenol moiety (Scheme 1a), and of [M–H]$^-$ ions of *trans*-resveratrol (R = H) and piceatannol at *m/z* 243 (R = OH, compound **2**) considering that the deprotonation reaction occurred on the resorcinol moiety (Scheme 1b). (Reprinted from Rapid Communications in Mass Spectrometry 22, Stella et al., Collisionally induced fragmentation of [M–H]$^-$ species of resveratrol and piceatannol investigated by deuterium labelling and accurate mass measurements, p. 3869, Copyright © 2008, with permission from John Wiley & Sons, Ltd.)

is obtained and then performing three consecutive extractions with an aqueous 75% methanol solution lay stirring at room temperature for 15 min each with ultrasound. Methanol is removed by concentration of the extract under vacuum at 30 °C, the aqueous residue is washed with hexane in order to eliminate lipophilic substances, and fractionated on a Sephadex LH-20 column. The first fractions are eluted with aqueous

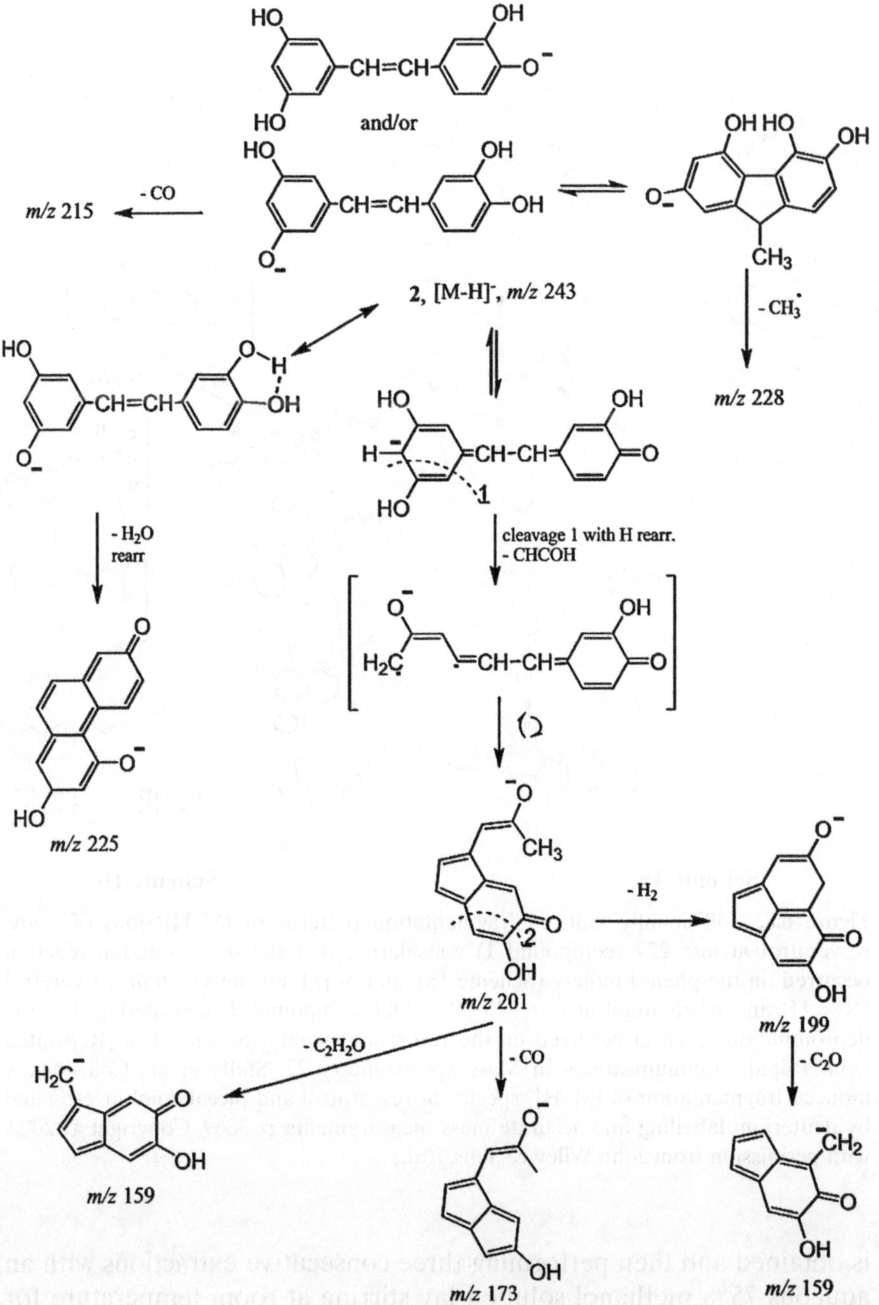

Figure 6.9. Collisionally induced fragmentation pattern of [M–H]⁻ ions of piceatannol considering that the deprotonation reaction occurred on the catechol moiety. (Reprinted from Rapid Communications in Mass Spectrometry 22, Stella et al., Collisionally induced fragmentation of [M–H]⁻ species of resveratrol and piceatannol investigated by deuterium labeling and accurate mass measurements, p. 3871, Copyright © 2008, with permission from John Wiley & Sons, Ltd.)

90% ethanol, the others with acetone–water solutions (Gonzales-Manzano et al., 2006; Gabetta et al., 2000). Fractions of 500 mL with a composition similar to those reported in Table 6.3 are obtained.

Another method for extraction of tannins from grape and purification of extract was proposed by Vidal et al. (2003). For grape seed extraction, 360 g of seeds are frozen in liquid nitrogen and ground with a blender. The powder is extracted twice with 1.5 L of an acetone/water 60:40 (v/v) solution and the extracts are pooled. After centrifugation, the supernatant is concentrated under vacuum and lipophilic compounds are removed by washing with hexane (250 mL). Purification of the aqueous acetone extract is performed by chromatography on a methacrylic size-exclusion resin Toyopearl TSK HW-50 (F) (18–35 cm) column. Two fractions are eluted from the column: the first is with ethanol(EtOH)/H_2O/TFA 55:45:0.02 (v/v/v) (three bed volumes), the other is with acetone/H_2O 30:70 (v/v) (one-bed volume). The two solutions are pooled, concentrated under vacuum, and freeze dried. Further purification is performed using a DVB–PS resin by dissolution of the residue in water (6 g in 200 mL) and fractionation on a 25 × 50-cm column. After washing with water and ether to eliminate the flavan-3-ol monomers, PAs with a polymerization degree of 3 units (DP3) are recovered with MeOH. The fraction containing PAs DP10 is recovered from the column with acetone/H_2O 60:40 (v/v).

TABLE 6.3. Composition of Fractions Obtained by Separation on Sephadex LH-20[a]

Fraction	Main Constituents
1	(+)-Catechin; (–)-epicatechin
2	(–)-Epicatechin-3-*O*-gallate; dimer; dimer gallate
3	Dimer; dimer gallate; trimer
4	Dimer gallate; dimer digallate; trimer; trimer gallate; tetramer
5	Trimer; trimer gallate; tetramer; tetramer gallate; pentamer
6	Trimer gallate; trimer digallate; tetramer; tetramer gallate; tetramer digallate; pentamer; pentamer gallate
7	Tetramer gallate; tetramer digallate; pentamer; pentamer gallate; pentamer digallate; hexamer
8	Pentamer gallate; tetramer trigallate pentamer digallate; hexamer; hexamer gallate
9	Pentamer digallate; pentamer trigallate; hexamer gallate; hexamer digallate
10	Pentamer digallate; pentamer trigallate; hexamer gallate; hexamer digallate; heptamer; heptamer gallate

[a]A 160-g resin, 50 × 4.5 cm i.d. column. Elution with (a) 90% ethanol and (b) 20%, (c) 40% and (d) 70% acetone aqueous solutions (flow rate 16 mL/min) (Gabetta et al., 2000).

To extract tannins from skins, low MW phenolics (mainly anthocyanins) are removed previously by immerging skins in a 12% (v/v) ethanol solution for 72 h at 4 °C. The solution is discarded, the skins are ground in MeOH, and the solution is kept immersed for 2 h at 4 °C. After filtration, solid parts are again extracted overnight at 4 °C with acetone/H_2O 60:40 (v/v) and the two extracts (methanolic and aqueous acetone) are concentrated under vacuum and fractionated separately. Fractionation of PAs is performed by chromatography on a Toyopearl TSK HW-50(F) column. After sample passage, sugars and phenolic acids are removed by washing of the column with EtOH/H_2O/TFA 55:45:0.02 (v/v/v) followed by acetone/H_2O 30:70 (v/v). The fraction containing PAs DP 12–20 is recovered from the column with acetone/H_2O 60:40 (v/v).

Purification of PAs also can be performed by SPE using a C_{18} (6–20 mL) cartridge. Seeds (250 mg) or grape juice (50 mL) extract is suspended in 20 mL of water. The solution is passed through the cartridge previously conditioned by passing 5-mL MeOH followed by water. After the sample passage, the cartridge is rinsed with 40 mL of water and the PAs are eluted with 6–10 mL of acetone/water/acetic acid 70:29.5:0.5 (v/v/v) solution (Lazarus et al., 1999; Núñez et al., 2006).

To study the composition of PAs can be useful to perform thiolysis of the extract: the tannin powder is dissolved in methanol (1 mg/mL) and introduced into a glass vial with an equal volume of a 5% toluene-α-thiol methanolic solution containing 0.2 M HCl. The reaction is carried out at 90 °C for 2 min and the thiolyzed solution is analyzed (Fulcrand et al., 1999; Vidal et al., 2003).

The LC/ESI–MS analysis of PAs is usually performed by reverse-phase chromatography, chromatograms relative to analysis of a grape seed extract are reported in Fig. 6.10.

Normal-phase LC of PAs using silica columns (e.g., 250 × 4.6 mm; 5 μm at 37 °C) provides satisfactory separation of oligomers based on their MW (Lazarus et al., 1999). Due to the weak acidic nature of PAs, most LC/MS applications are performed in negative-ion mode. Experimental conditions of two normal-phase LC/ESI–MS methods are summarized in Table 6.4.

The PAs in extracts can be characterized by direct infusion ESI–MS. Dissolution in methanol/acetonitrile (1:1) showed the highest intensity of ions operating in the negative mode, including multiply charged ions (Hayasaka et al., 2003). Negative mode shows simpler mass spectra due to the absence of intense adduct ion species and to the production of more multiply charged ions than the positive-ion mode. For PAs with DP3 and DP9, mass spectra similar to those shown in Fig. 6.11a and b are recorded. The [M–H]$^-$ and [M–2H]$^{2-}$

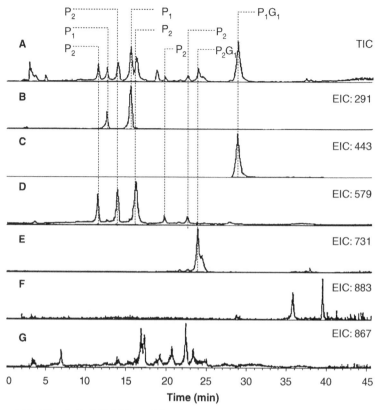

Figure 6.10. The LC–ESI–MS positive-ion mode chromatograms of a grape seeds extract analysis recorded in the range m/z 200–1000. (A) Total ion chromatogram (TIC); extracted ion chromatograms: (B) P_1 (catechin or epicatechin m/z 291), (C) P_1G_1 (catechin/epicatechin gallate, m/z 443), (D) P_2 (catechin–epicatechin dimer, m/z 579, (E) P_2G_1 (m/z 731), (F) P_2G_2 (m/z 883), (G) P_3 (m/z 867). Analytical conditions: column C_{18} (250 × 4.6 mm, 5 μm) at 25 °C; binary solvent composed of (A) 0.1% formic acid (v/v) and (B) 0.1% formic acid (v/v) in acetonitrile; gradient program: from 10 to 20% of B in 20 min, 20% B isocratic for 10 min, from 20 to 50% of B in 10 min, 50% B isocratic for 10 min (flow rate 1.0 mL/min, 1/4 of eluent split into mass spectrometer). The ESI needle voltage 3.5 kV; drying gas N_2 (8 L/min); interface capillary temperature 325 °C; nebulized gas He 40 psi. (Reprinted from Rapid Communications in Mass Spectrometry 19, Wu et al., Determination of proanthocyanidins in fresh grapes and grape products using liquid chromatography with mass spectrometric detection, p. 2065, Copyright © 2005, with permission from John Wiley & Sons, Ltd.)

species of PAs are reported in Table 6.5. Abundant [M–H]⁻ singly charged ions separated by 288 Da are observed in the m/z 289–2017 and 441–1881 ranges. These ions correspond to the molecular masses of procyanidins (PCs) with DP 1–7 and procyanidin monogallates (PC1Gs) of DP 1–6, respectively. The PAs with DP9 show the

TABLE 6.4. Methods for LC/ESI–MS Analysis of Proanthocyanidins with Normal-Phase Columns

METHOD 1 (Lazarus et al., 1999)

Silica column $250 \times 4.6\,mm$; $5\,\mu m$

Mobile phase: (A) dichloromethane, (B) methanol, (C) HAc/H_2O 1:1 (v/v).

Elution linear gradient of B into A with a constant 4% C: start 14% B in A, $14 \rightarrow 28.4\%$ B in 30 min, $28.4 \rightarrow 50\%$ B in 30 min; $50 \rightarrow 86\%$ B in 5 min; isocratic 5 min (flow rate 1 mL/min)

LC/ESI–MS conditions: negative-ion mode, buffering reagent 0.75 M NH_4OH in the eluent stream at a flow rate $40\,\mu L/min$, capillary voltage 3 kV, fragmentor voltage 75 V, nebulizing pressure 25 psig, drying gas temperature 350 °C

METHOD 2 (Núñez et al., 2006)

Silica column $250 \times 2.0\,mm$; $5\,\mu m$

Mobile phase: (A) dichloromethane/methanol/H_2O/HAc 82:14:2:2 (v/v/v/v), (B) MeOH/H_2O/HAc 96:2:2 (v/v/v)

Elution linear gradient of B into A: from 0 to 18% B in 30 min, $18 \rightarrow 31\%$ B in 15 min, $31 \rightarrow 88\%$ B in 5 min (flow rate 0.2 mL/min)

LC/ESI–MS conditions: negative mode, ionization reagent ammonium acetate 10 mM in the eluant stream at flow rate of $30\,\mu L/min$, capillary voltage 3.2 kV, cone voltage 30 V, source temperature 150 °C, desolvation gas temperature 300 °C

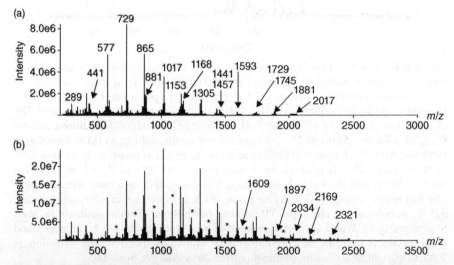

Figure 6.11. The ESI mass spectra of proanthocyanidins with a degree of polymerization (DP) 3 (a) and DP9 (b) obtained by signal accumulation of 20 consecutive scans. m/z values: major singly charged ions of PAs. The asterisk (*) symbolizes doubly charged ions of PAs monogallates. Analytical conditions: ESI needle, orifice, and ring potentials –4500, –60, and –350 V, respectively; curtain gas N_2; nebulizer gas air. (Reprinted from Rapid Communications in Mass Spectrometry 17, Hayasaka et al., Characterization of proanthocyanidins in grape seeds using electrospray mass spectrometry, p. 11, Copyright © 2003, with permission from John Wiley & Sons, Ltd.)

additional larger [M–H]⁻ ions derived from PC1G with DP7, PC2Gs (procyanidin digallates) with DP6 and DP7, and from PC3Gs (trigallates) with DP4 and DP5 (Fig. 6.11b; Table 6.4). The ESI mass spectrum in Fig. 6.12a shows the ions resulting from the product ion spectra obtained from m/z 865 (PC with DP3, Fig. 6.12b) and m/z 1017 (PC1G with DP3, Fig. 6.12c). Their intensity increases with the orifice potential. The fragmentation pathways observed for the PA [M–H]⁻ and [M–3H]⁻ ions (m/z 577, 575, 729, 727, and 441) could be due to the cleavage of the interflavanic bond, retro-Diels–Alder (RDA) fission on the C ring followed by the elimination of water with formation of [M–H-152]⁻ (m/z 713, 425, 865, and 577) and [M–H-152-H_2O]⁻ (m/z 695, 407, 847, and 559) ions, and [M–H-126]⁻ (m/z 739 and 451) ions by elimination of the phloroglucinol molecule. The ion at m/z 881 corresponds to the dimer of epicatechin–gallate or to the epicatechin–epicatechin–epigallocatechin trimer (isobaric compounds). Doubly charged ions show a series of abundant ions separated by 144 Da from m/z 652.4 to 1948.8 (signals marked with an asterisk in Fig. 6.11b), which correspond to the [M–2H]²⁻ ions of PC1Gs with DP 4–13. Two different fragmentation patterns of trimeric species were observed by increasing the orifice voltage. From the ions at m/z 863 (A-type) two ions at m/z 575 and 573 form and fragmentation of the ions at m/z 711 are observed by RDA. As a consequence of the 152-Da neutral loss corresponding to 3,4-dihydroxy-α-hydroxystyrene, two fragments are observed at m/z 285 and 289, which are generated by cleavage of the A-type interflavanic linkage. The fragmentation schemes are reported in Fig. 6.13 (Cheynier et al., 1997).

Analysis of PCs and PAs also can be performed in positive-ion mode and compounds are identified on the m/z values of their protonated molecules. The [M+H]⁺ ions of dimers, trimers, and tetramers show the signals at m/z 579, 867, 1155, their mono- and digalloyl derivatives signals at m/z 731, 1019, 1307, 883, 1171, 1459, trigalloyl derivatives of trimers and tetramers at m/z 1323 and 1611. Also the [M+H]⁺ ion signals of flavan-3-ol pentamers, hexamers and heptamers at m/z 1443, 1731, 2019, their monogalloyl derivatives at m/z 1595, 1883, 2171, pentamers and hexamers digalloyl derivatives at m/z 1747 and 2035, and pentamers and hexamers trigalloyl derivatives at m/z 1899 and 2187, are observed (Gabetta et al., 2000).

The positive-ion mode fragmentation patterns proposed for trimeric procyanidins studied by isolation and ion trap fragmentation of the most intense MS spectra signals, fragmentation of the principal ions of the MS² spectra, and acquisition of MS³ spectra, are shown in Fig. 6.14 (Pati et al., 2006). In Figs. 6.15 and 6.16, the schemes of positive

TABLE 6.5. The [M–H]⁻ and [M–2H]²⁻ Ions of Proanthocyanidins with Degree of Polymerization (DP) 3 and 9

	Procyanidins (PCs)			Monogallates (PC1Gs)			Digallates (PC2Gs)		
	Sdp3	Sdp9		Sdp3	Sdp9		Sdp3	Sdp9	
DP	[M–H]⁻	[M–H]⁻	[M–2H]²⁻	[M–H]⁻	[M–H]⁻	[M–2H]²⁻	[M–H]⁻	[M–H]⁻	[M–2H]²⁻
1	289.2	289.4		441.2	441.4				
2	577.4	577.4		729.4	729.4		881.4	881.4	
3	865.4	865.4		1017.6	1017.4		1169.8	1169.8	
4	1153.6	1153.4		1305.8	1305.8	652.4	1457.6	1457.4	
5	1441.8	1441.6	720.4	1593.4	1593.4	796.4	1745.4	1745.2	872.6
6	1729.8	1729.2	na[a]	1881.8	1881.6	940.8		2034.0	na[a]
7	2017.2	2017.2	1009.2		2169.8	1084.2		2322.2	1161.0
8			na[a]			1228.6			na[a]
9			1296.6			1373.2			1448.6
10			na[a]			1516.8			na[a]
11			1584.0			1661.0			1737.0
12						1805.4			na[a]
13						1948.8			2025.2

[a]Not assigned = na (Hayasaka et al., 2003).

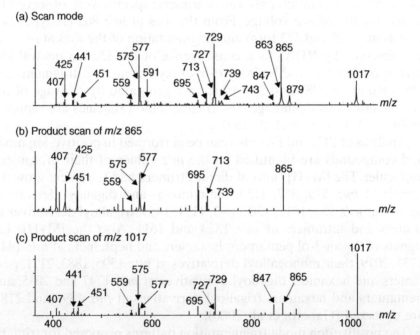

Figure 6.12. The MS/MS of proanthocyanidins using N₂ as the collision gas (2 units); collision energy potential 40–60V. (Reprinted from Rapid Communications in Mass Spectrometry 17, Hayasaka et al., Characterization of proanthocyanidins in grape seeds using electrospray mass spectrometry, p. 13, Copyright © 2003, with permission from John Wiley & Sons, Ltd.)

Trigallates (PC3Gs)			Tetragallates (PC4Gs)			Pentagallates (PC5Gs)		
Sdp3	Sdp9		Sdp3	Sdp9		Sdp3	Sdp9	
$[M–H]^-$	$[M–H]^-$	$[M–2H]^{2-}$	$[M–H]^-$	$[M–H]^-$	$[M–2H]^{2-}$	$[M–H]^-$	$[M–H]^-$	$[M–2H]^{2-}$
	1609.2	804.4						
	1897.8	948.8						1100.2
		1092.6		1024.8				1243.8
		1236.8		na^a				1388.4
		1380.6		1312.6				1532.2
		1524.6		na^a				1676.4
		1669.2		1601.2				1821.8
		1813.2		na^a				
				1889.2				

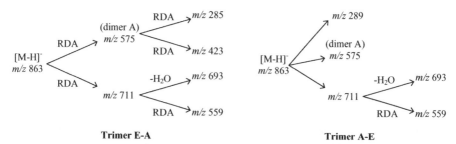

Figure 6.13. Schemes of A-type procyanidin trimers fragmentation observed by LC/ESI–MS negative-ion mode.

fragmentation patterns for monomer catechin (Fig. 6.15) and a B-type trimer (Fig. 6.16) are reported (Li and Deinzer, 2007). Table 6.6 reports the positive-ion ESI tandem mass product ions of flavan-3-ol monomers and PA dimers, trimers, and oligomers. Figures 6.17a and b show the positive-ion mode mass spectra of a grape seed extract, and the ESI–MS^2 full-scan spectra of PCs DP 2–5.

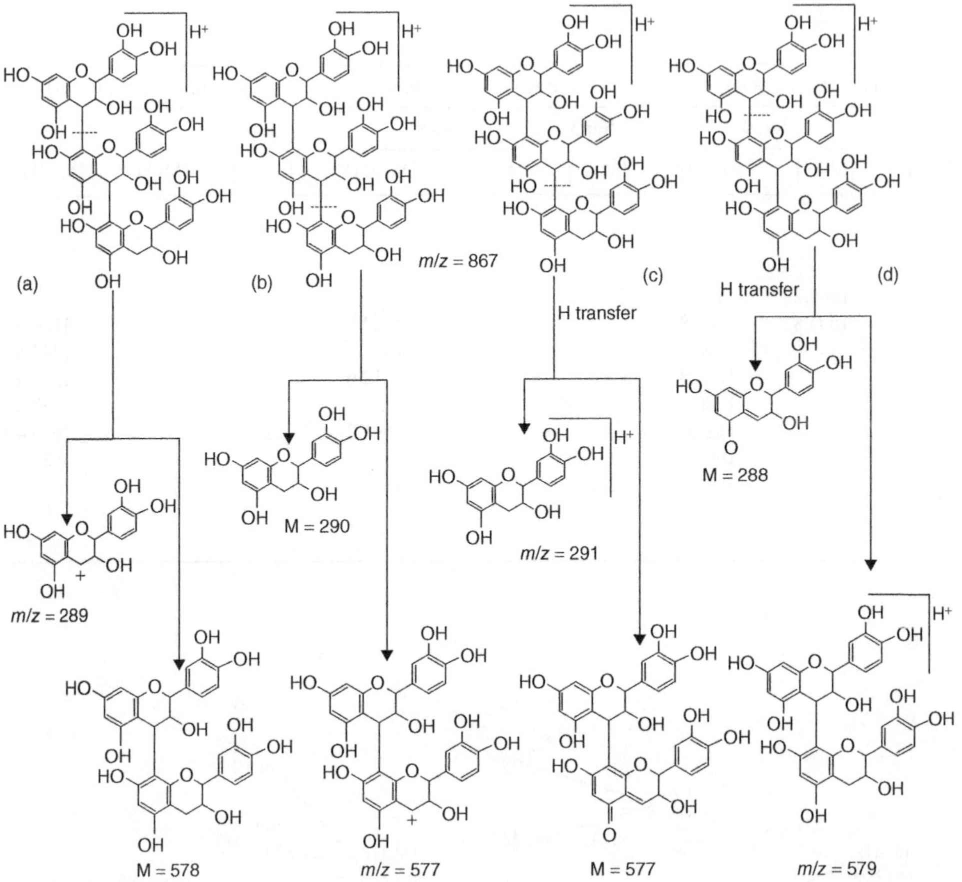

Figure 6.14. Positive-ion mode fragmentation patterns of trimeric procyanidins. The ESI–MS conditions: spray voltage 4.5 kV; sheath gas nitrogen 0.9 L/min; capillary voltage 35 V; capillary temperature 200°C; tube lens offset voltage 15 V. (Reprinted from Pati et al., 2006, Simultaneous separation and identification of oligomeric procyanidins and anthocyanidins-derived pigments in raw red wine by HPLC-UV-ESI-MSn, *Journal of Mass Spectrometry*, 41, p. 869, with permission from John Wiley & Sons, Ltd.)

6.3 THE LC/MS OF NON-ANTHOCYANIC POLYPHENOLS OF WINE

The polyphenols in wine reported in Table 6.7 can be determined by LC/ESI–MS using a C_{18} column (e.g., 250×4.6 mm, 3μm) and a method like: binary solvent composed of (A) aqueous formic acid 0.5% (v/v) and (B) formic acid–acetonitrile–H_2O 5:400:595 (v/v/v) with a gradient program from 0 to 20% B in 15 min, 20% B isocratic for 10 min,

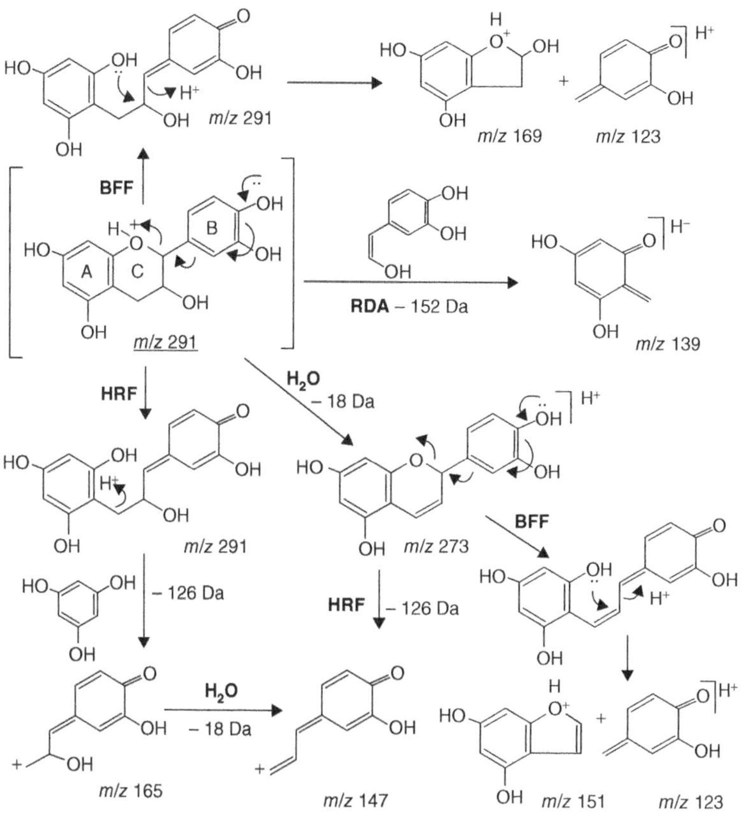

Figure 6.15. Positive fragmentation pathways of the monomer catechin: retro-Diels–Alder fission (RDA), heterocyclic ring fission (HRF), benzofuran-forming fission (BFF), and loss of water molecule. (Reprinted from Li and Deinzer, 2007, Tandem Mass Spectrometry for Sequencing Proanthocyanidins, Analytical Chemistry, 79, p. 1740, with permission from American Chemical Society.)

$20 \rightarrow 70\%$ B in 45 min, 70% B isocratic for 5 min, $70 \rightarrow 100\%$ B in 10 min, 100% B isocratic for 5 min at flow rate 0.7 mL/min (Bravo et al., 2006).

The ESI is effective also in the analysis of flavan-3-ols operating in the positive-ion mode. A better sensitivity can be achieved but, since most acid phenols in wine are not detectable in this mode, it is preferable to work in the negative mode. An example of a chromatogram relative to a flavan-3-ols wine analysis performed in the negative-ion mode is reported in Fig. 6.18. By operating with a cone voltage of 60 V, these compounds show high formation of the [M–H]⁻ ion. A reduction of the molecular species intensity is observed by increasing the cone voltage up to 120 V; the most abundant fragments originate from losses

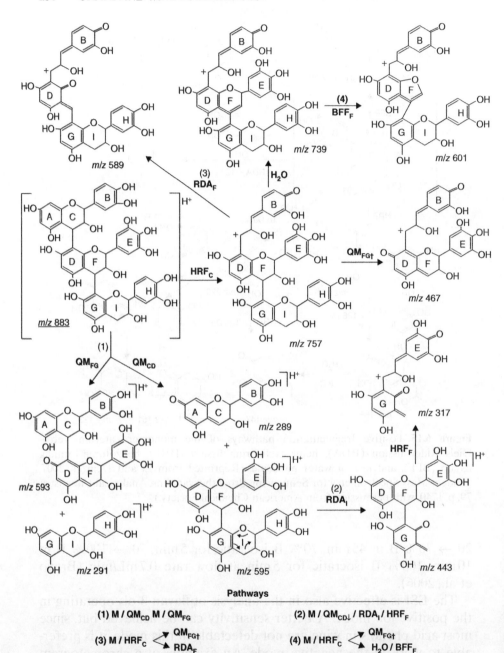

Figure 6.16. Positive fragmentation pathways of *m/z* 883 B-Type trimer: RDA, HRF, BFF, quinone methide fission (QM), and loss of water molecule. The QM$_{CD↓}$ ion derived from the QM fisson of ring-C/ring-D linkage bond by the loss of upper unit; QM$_{FG↑}$ the ion derived from the QM fisson of ring-F/ring-G linkage bond by the loss of lower unit. (Reprinted from Li and Deinzer, 2007, Tandem Mass Spectrometry for Sequencing Proanthocyanidins, Analitical Chemistry, 79, p. 1744, with permission from American Chemical Society.)

TABLE 6.6. Positive-Ion ESI Tandem Mass Product Ions of Flavan-3-ol monomers and PAs Dimers and Oligomers[a]

Compound	[M+H]+	HRF_C	RDA_C	BFF_C	BFF_C/H2O	HRF_C/H2O
C	291	165(126)	139(152)	169(122), 123	151(140), 123	147(144)
EC	291	165(126)	139(152)	169(122), 123	151(140), 123	147(144)
GC	307	181(126)	139(168)	169(138), 139	151(156), 139	163(144)

Compound	[M+H]+	QM_CD↑	QM_CD↑	HRF_C	RDA_C	BFF_C	RDA_F	HRF_C		
								RDA_F	H2O/BFF_F	RDA_F HRF_C
(E)GC-(4,8)-(E)C	595	305		469(126)	427(168)	443(152)	317(152)	317(152)	329(140)	317(126)
(E)GC-(4,6)-(E)C	595			469(126)	427(168)	443(152)	317(152)	317(152)	329(140)	317(126)
(E)C-(4,8)-(E)C	579		291	453(126)	427(152)	427(152)	301(152)	301(152)	313(140)	301(126)
(E)C-(4,8)-(E)C	579	289		453(126)	427(152)	427(152)	301(152)	301(152)	313(140)	301(126)
(E)C-(4,8)-(E)C	579		291	453(126)	427(152)	427(152)	301(152)	301(152)	313(140)	301(126)
(E)C-(4,8)-(E)C	579	289		453(126)	427(152)	427(152)	301(152)	301(152)	313(140)	301(126)

TABLE 6.6. (*Continued*)

Compound	[M+H]+	QM_CD↑/ QM_CD↓	QM_FG↑/ QM_FG↓	QM_CD↓/RDA_↑/HRF_F	HRF_C/QM_FG↑	HRF_C/RDA_F	HRF_C/H2O/BFF_F
(E)C-(E)GC-(E)C	883	-/595	593/-	595(288)/443(152)/ 317(126)	757(126)/ 467(290)	757(126)/ 589(168)	757(126)/ 601(156)
(E)GC-(E)GC-(E)C	899	-/595	609/291	595(304)/443(152)/ 317(126)	773(126)/ 483(290)	773(126)/ 605(168)	773(126)/ 617(156)
(E)C-(E)C-(E)C	867	-/579	577/291	579(288)/427(152)/ 301(126)	741(126)/ 451(290)	741(126)/ 589(152)	741(126)/ 601(140)
(E)C-(E)C-(E)C	867	-/579	577/-	579(288)/427(152)/ 301(126)	741(126)/ 451(290)	741(126)/ 589(152)	741(126)/ 601(140)
(E)C-(E)C-(E)C	867	-/579	577/-	579(288)/427(152)/ 301(126)	741(126)/ 451(290)	741(126)/ 589(152)	741(126)/ 601(140)

Compound	[M+H]+	Diagnostic Ions
(E)C-(E)C-(E)C-(E)C	1155	867, 865, 579
(E)C-(E)C-GC-(E)C	1171	883, 881, 595
(E)GC-(E)GC-(E)C-(E)C	1187	899, 897, 595
(E)GC-(E)GC-(E)GC-(E)C	1203	913, 899, 595
(E)C-(E)C-(E)C-(E)C-(E)C	1443	1155, 1153, 867, 579
(E)GC-(E)C-(E)C-(E)C-(E)C	1459	1171, 1169, 867, 579
(E)GC-(E)GC-(E)C-(E)C-(E)C	1475	1187, 1185, 883, 579
(E)C-(E)C-(E)C-(E)C-(E)C-(E)C	1731	1441, 1143, 1155, 867, 579

[a]Neutral losses are shown in parentheses. C, EC, and GC: catechin, epicatechin, and gallocatechin, respectively. (E)C and (E)GC: (epi)catechin and (epi) gallocatechin, respectively. (E) indicates either catechin–epicatechin or gallocatechin–epigallocatechin. RDA, retro-Diels-Alder fission; HRF, heterocyclic ring fission; QM, quinone methide fission; BFF, benzofuran forming fission; QM_CD↓, the ion derived from the QM fission of ring-C/ring-D linkage bond by the loss of upper unit; QM_CD↑, the ion derived from the QM fisson of ring-C/ring-D linkage bond by the loss of lower unit; QM_FG↓, the ion derived from the QM fisson of ring-F/ring-G linkage bond by the loss of upper unit; QM_FG↑, the ion derived from the QM fission of ring-F/ring-G linkage bond by the loss of lower unit (Li and Deinzer, 2007).

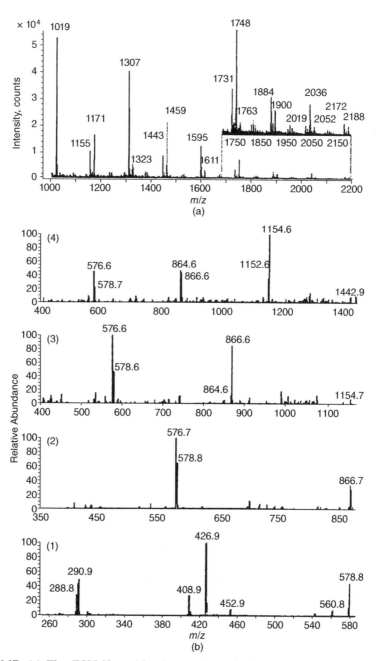

Figure 6.17. (a) The ESI/MS positive-ion mode analysis of a grape seeds extract: $[M+H]^+$ ions of proanthocyanidins from P_3G_1 to P_7G_1 (needle voltage 3.5 kV; drying gas N_2; interface capillary temperature 325 °C; nebulizer gas He 40 psi). (Reprinted from Rapid Communications in Mass Spectrometry 19, Wu et al., Determination of proanthocyanidins in fresh grapes and grape products using liquid chromatography with mass spectrometric detection, p. 2065, Copyright © 2005, with permission from John Wiley & Sons, Ltd.) **(b)** The ESI–MS² full-scan spectra of (1) dimeric (*m/z* 579), (2) trimeric (*m/z* 867), (3) tetrameric (*m/z* 1155), and (4) pentameric (*m/z* 1449) pro-cyanidins. (Reprinted from *Journal of Mass Spectrometry*, 41, Pati et al., Simultaneous separation and identification of oligomeric procyanidins and anthocyanin-derived pigments in raw red wine by HPLC-UV-ESI-MSn, p. 868, Copyright © 2006, with permission from John Wiley & Sons, Ltd.)

TABLE 6.7. Ions of Wine Phenols Produced by Negative-Ion ESI[a]

MW	Compound	Main Ions Observed (m/z)	
		Fragm. 60V	Fragm. 120V
154	Protocatechuic acid	153(109)	109(153)
138	Protocatechuic aldehyde	137	137(108)
138	p-Hydroxybenzoic acid	137(93)	93(137)
122	p-Hydroxybenzoic aldehyde	121	121(92)
168	Vanillic acid	167	108(167, 123, 91)
152	Vanillin	151(136)	136(151)
194	Ferulic acid	193(134, 149)	134(193)
182	Syringic aldehyde	181(166)	181(166, 151)
164	p-Coumaric acid	147(103)	147(103)
180	Caffeic acid	179(135)	135(179)
178	Esculetin	177	177(133)
170	Gallic acid	169(125)	125(169)

[a]Pérez-Margariño et al., 1999. In parentheses: Ions with lower abundance.

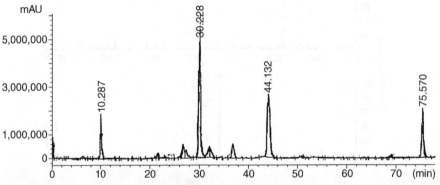

Figure 6.18. The LC/ESI–MS total ion current (TIC) chromatogram of a wine sample analysis performed in negative-ion mode. Analytical conditions: column C_{18} (250 × 4.6 mm, 3 μm); binary solvent composed of (A) 4.5% formic acid in water and (B) solvent A/acetonitrile 90:10; gradient program: from 0 to 50% of B in 25 min, from 50 to 80% of B in 35 min, 80% B isocratic for 20 min (flow rate 0.7 mL/min). Retention time at 10.287-min gallic acid, 30.228-min catechin, 44.132-min epicatechin, 75.57-min epicatechin gallate. (Reprinted from *Journal of Chromatography A*, 847, Pérez-Magariño et al., Various applications of liquid chromatography–mass spectrometry to the analysis of phenolic compounds, p. 80, Copyright © 1999, with permission from Elsevier.)

of carboxyl $[M–H-45]^-$, hydroxyl $[M–H-17]^-$, and/or formaldehyde $[M–H-30]^-$ (Table 6.8).

The sample for analysis can be prepared by liquid–liquid extraction of 50-mL wine using diethyl ether (3 × 5 mL) and ethyl acetate (3 × 15 mL) after previous concentration to 15 mL under vacuum at

TABLE 6.8. Fragment Ions of Flavan-3-ols Generated in Both Negative- and Positive-Ion Mode[a]

		Main Ions Observed (m/z)		
		API[+]	API[-]	
MW	Compound	Fragm. 60 V	Fragm. 60 V	Fragm. 120 V
290	(+)-Catechin	291(139)	289	289(245)
290	(−)-Epicatechin	291(139, 150)	289	289(245)
306	(−)-Epigallocatechin	307(139)	305	305
442	(−)-Epicatechin-3-O-gallate	443(123, 273)	441	441(289, 169)
458	Epigallocatechin-3-O-gallate	459(139, 289)	457	169(457)

[a]Pérez-Margariño et al., 1999. In parentheses: Ions with lower abundance.

30 °C in order to eliminate ethanol. The organic phases are combined, the resulting solution is dried over Na_2SO_4, and the solvent is removed under vacuum, then the residue is dissolved in 2 mL of methanol/water (1:1) and the solution is filtrated 0.45 μm before analysis (Monagas et al., 2005). The compounds identified by LC/ESI–MS in four different red wines are reported in Table 6.9 with maximum absorption wavelengths in the UV–vis spectra. Several classes of non-anthocyanic wine phenols were determined with this method: flavan-3-ols, flavonols, hydroxycinnamoyltartaric acids, cis- and trans-resveratrol, piceid (resveratrol glucoside), dimeric (B1, B3, B4, and B5), and trimeric (C1, T2, and T3) procyanidins, phenolic acids.

Two different sample preparation methods by size-exclusion and reverse-phase chromatography were proposed for analysis of PCs and PAs in wine. In the former, a volume of 5 mL of dealcoholized wine is passed through a Fractogel Toyopearl TSK gel HW-50 (F) (12 × 120 mm) column. The stationary phase is washed with 25 mL of water and the simple polyphenols are eluted with 50 mL of an ethanol/water/trifluoroacetic acid 55:45:0.005 (v/v/v) solution. The polymeric fraction is recovered with 50 mL of acetone/H_2O 60:40 (v/v). Figure 6.19 shows the LC/ESI–MS extracted ion chromatograms of dimers and trimers in a wine (Fulcrand et al., 1999).

Sample preparation by reverse-phase chromatography can be performed using a C_{18} SPE cartridge: 30 mL of dealcoholized wine are loaded onto the cartridge, after rinsing with 40 mL of water PAs are recovered with 10 mL of acetone/water/acetic acid 70:29.5:0.5 (v/v/v) (Lazarus et al., 1999). A method for fractionation of polyphenols in wine by reverse-phase chromatography is reported in the flow diagram in Fig. 6.20 (Sun et al., 2006).

TABLE 6.9. Non-anthocyanin Phenolic Compounds Identified by LC/ESI–MS in Wines from Different Vitis Vinifera Varieties (Tempranillo, Garciano, Cabernet Sauvignon, Merlot) with their Principal MS Fragments and the Maximum Absorption Wavelengths of UV–vis Spectra[a]

| RT (min) | Compound | (m/z) | | λ_{max} (nm) |
		[M–H]⁻	Fragments	
7.6	Gallic acid	169	125	272
14.0	Protocatechuic acid	153	109	294, 260
15.7	Dihydroxyphenylethanol	153		280
16.7	*trans*-Caffeyltartaric acid	311	179	330, 298(s)
19.2	2,3-Dihydroxy-1-(4-hydroxy-3-methoxyphenyl)-propan-1-one	211		310, 280
19.5	Methyl gallate	183	169, 125	272
20.5	Tyrosol	137		275
22.3	(epi)Gallocatechin-(epi)catechin	593	425	276
22.7	Procyanidin B3	577	425, 289	280
23.0	Procyanidin B1	577	425, 289	280
24.2	*trans*-Coumaroyltartaric acid	295	163	313
27.3	(+)-Catechin	289		279
28.0	Procyanidin T2	865	713, 577, 289	280
28.3	*trans*-Feruryltartaric acid	325	193	329, 301(s)
28.7	Hexose ester of vanillic acid	329	167	nd
29.0	Procyanidin T3	865	713, 577, 289	280
29.7	Vanillic acid	167		289, 262
30.5	Procyanidin B4	577	425, 289	283
31.3	*trans*-Caffeic acid	179	135	323
33.0	Hexose ester of *trans*-*p*-coumaric acid (1)	325	163, 145	311
33.6	Procyanidin B2	577	425, 289	280
34.6	Syringic acid	197		277
37.0	Hexose ester of *trans*-*p*-coumaric acid (2)	325	163, 145	312
38.7	(−)-Epicatechin	289		279
40.2	Trimeric procyanidin	865	713, 577, 289	280
41.3	Ethyl gallate	197	169, 125	273
42.2	Procyanidin C1	865	713, 577, 289	282
43.2	*trans*-*p*-Coumaric acid	163	119	309
43.5	Trimeric procyanidin	865	713, 577, 289	280
43.8	Procyanidin dimer gallate	729	577	278
44.0	Procyanidin B5	577	425, 289	280
48.5	Myricetin-3-*O*-glucuronide	493	317	349, 300(s), 261
50.1	Myricetin-3-*O*-glucoside	479	317	349, 300(s), 261
51.3	Epicatechin-3-*O*-gallate	441	289, 169	277

TABLE 6.9. (*Continued*)

| RT (min) | Compound | (*m/z*) | | λ_{max} (nm) |
		[M–H]⁻	Fragments	
53.2	*trans*-Resveratrol-3- *O*-glucoside	389	227	306(s), 319
55.2	Ellagic acid	301		368
57.2	Quercetin-3-*O*-galactoside	463	301	354, 300(s), 256
57.7	Quercetin-3-*O*-glucuronide	477	301	354, 300(s), 256
58.2	Astilbin	449	303	288
58.7	Quercetin-3-*O*-glucoside	463	301	354, 300(s), 256
60.6	Tryptophol	160		279
62.4	Kaempferol-3-*O*-glucoside	447	285	346, 300(s), 265
67.6	Myricetin	317		371, 300(s), 254
68.9	*cis*-Resveratrol-3- *O*-glucoside	389	227	285
71.9	*trans*-Resveratrol	227		306, 319(s)
92.7	Quercetin	301		369, 300(s), 255
95.7	*cis*-Resveratrol	227		284

[a]The ESI/MS conditions: negative-ion mode, drying gas N_2 flow 10 L/min, temperature 350 °C; nebulizer pressure 55 psi; capillary voltage 4000 V; fragmentation program: *m/z* 0–200 at 100 V, *m/z* 200–3000 at 200 V. Analytical conditions: C_{18} column (300 × 3.9 mm; 4 μm); binary solvent composed of (A) H_2O/acetic acid 98:2 (v/v) and (B) H_2O/MeCN/acetic acid 78:20:2 (v/v/v); gradient program: from 0 to 80% of B in 55 min, from 80 to 90% of B in 2 min, 90% B isocratic for 3 min, from 90 to 95% of B in 10 min, from 95 to 100% of B in 10 min (flow rate 0.7 mL/min) (Monagas et al., 2005). (s) shoulder.

Fractions 1–4 can be characterized by direct-infusion ESI in the negative ionization mode. Polyphenols in fractions 8–10 are more complex and this approach does not provide any useful information. Therefore MS^n is necessary. Compounds identified in fractions 8 and 9 of a red wine are reported in Table 6.10, the analytical conditions used are reported below.

6.4 LIQUID-PHASE MASS SPECTROMETRY OF GRAPE ANTHOCYANINS

Analysis of the grape anthocyanins is usually performed recording the reverse-phase LC/UV (520 nm) profile. Figure 6.21 shows the chromatogram relative to analysis of a hybrid grape extract recorded using the chromatographic conditions described below. Compounds identified are reported in Table 6.11. If just a UV detector is used, the peak assignment is based mainly on the compounds elution sequence from the column. The coupling with MS operating in positive ion mode provides more confident structural data.

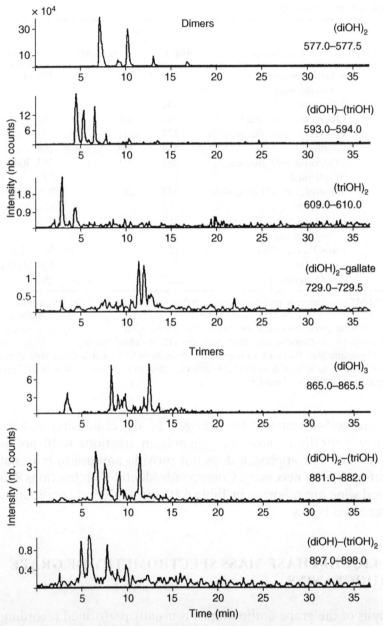

Figure 6.19. Extracted ion chromatograms recorded in negative mode of procyanidins and proanthocyanidins dimers and trimers from LC/ESI–MS analysis of a wine. Analytical conditions: C_{18} (125 × 2 mm, 3 μm) narrow-bore column; ion spray voltage –4000 V, orifice voltage –60 V. Binary solvent composed of (A) aqueous 2% formic acid and (B) acetonitrile/H_2O/formic acid 80:18:2 (v/v/v). Gradient program: from 5 to 30% of B in 20 min, 30 → 50% B in 10 min (flow rate 200 μL/min, flow rate in ESI source 50 μL/min). (Reproduced from *Journal of Agricultural and Food Chemistry*, 1999, 47, p. 1026, Fulcrand et al., with permission of American Chemical Society.)

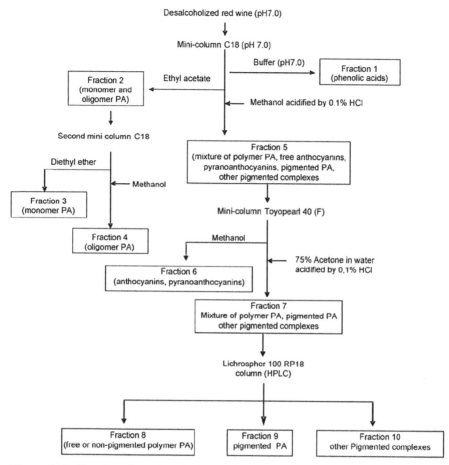

Figure 6.20. Fractionation of polyphenols in red wine (PA = proanthocyanidins). (Reprinted from *Journal of Chromatography A*, 1128, Sun et al., Fractionation of red wine polyphenols by solid phase extraction and liquid chromatography, p. 29, Copyright © 2006, with permission from Elsevier.)

Non-acidified methanol is the more suitable solvent to extract anthocyanins from grape reducing risks of hydrolysis of acetylated compounds: 20 berry skins are extracted with 50 mL of methanol for 12 h at room temperature (Revilla et al., 1998). Alternatively, the use of a methanol/water/formic acid 50:48.5:1.5 (v/v/v) mixture was reported as well (Gao et al., 1997). The volume of extract is reduced to about one-half under vacuum at 30 °C, adjusted to 100 mL with water, and 10 mL of this solution is diluted to 50 mL with water in order to further reduce the MeOH content. The resulting solution is purified by a SPE C_{18} cartridge (e.g., 1 g) previously activated by passage of

TABLE 6.10. The LC/ESI-MS and MS" Analysis of Fractions 8 and 9 Prepared as Showed in Fig. 6.20[a]

Column Elution Order	Polyphenols	(m/z)						
		[M–H]–	[M+H]+	[M]+	[M–2H]2–	MS2	MS3	MS4
1	PC1PD1	593				**441**, 289	315, 289, 153	
2	PC3	865				847, **577**, 289	451, 289	
3	PC4	**1153**				863, **577**	451, 289	
4	PC6-4G or PC6PD1-2G				1168	**1017**, 729	575	
5	PC7PD1-6G or PC7PD2-4G or PC7PD3-2G				1616	1084, 1641		
6	PC2PD1		**883**			847, 731, **579**	427, 289	
7	PC3	865	**867**			**577**, 291	289	
8	PC4	1153	1155			863, **577**	451, 289	
9	PC5		**1443**			1425, 1291, 1155, **865**, 577	847, 577	
10	PC2-Mv-3-glu			1069		781	619	601, 331
11	Mv-3-glu-PC3	1357				1067, **779**	617	437

[a]Bold ions were subjected to next stage of fragmentation. PC*n* = *n* units of (epi)catechin; *n*G = number of galloyl units. Analytical conditions: C18 column (250 × 4 mm, 5 μm) at 30°C. Binary solvent (A) acetonitrile/water/formic acid 2.5/97.4/0.1 (v/v/v) and (B) acetonitrile/water/formic acid 30/69.9/0.1 (v/v/v). Gradient program for fraction 8: from 0 to 30% of B in 20 min, 30 → 50% B in 20 min, 50 → 100% B in 20 min. Gradient program for fraction 9: from 0 to 15% of B in 20 min, 15 → 40% B in 20 min, 40 → 100% B in 40 min (flow-rate = 0.7 mL/min). Skimmer and capillary voltages for negative-ion mode –40 and +40 and –4000 V, respectively (+40 and –4000 V for positive); nebulizer gas N2 10 psi; drying gas N2 10 L/min; dry temperature 250°C (Sun et al., 2006).

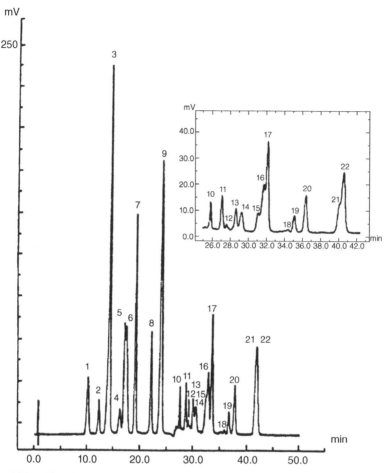

Figure 6.21. The LC/UV anthocyanin profile of a hybrid grape (*Clinton*) skin extract recorded at 520 nm. Analytical conditions: column C_{18} (250 × 4.6 mm, 5 μm), binary solvent composed of (A) water/formic acid 90:10 (v/v) and (B) methanol/water/formic acid 50:40:10 (v/v/v); gradient program from 15 to 45% of B in 15 min, 45 → 70% of B in 30 min, 70 → 90% of B in 10 min, 90 → 99% of B in 5 min, 99 → 15% of B in 5 min (flow rate 1 mL/min). (Reprinted from *American Journal of Enology and Viticulture*, 51, Favretto and Flamini, Copyright © 2000.)

3-mL methanol followed by 5 mL of water. After the sample passage, the cartridge is washed with 5 mL of water in order to remove sugars and more polar compounds, non-anthocyanic phenols are eluted with 3 mL of ethyl acetate, and anthocyanins are recovered with 3 mL of methanol.

Direct-injection ESI–MS/MS provides the structural characterization of anthocyanins in the extract and semiquantitative data too

TABLE 6.11. Retention Times (RT) and Molecular Weights (MW) of Anthocyanins Identified in the Chromatogram[a]

Peak	RT (min)	Anthocyanin	MW
1	10.50	Delphinidin-3,5-O-diglucoside (1)	627
2	12.50	Cyanidin-3,5-O-diglucoside (2)	611
3	14.22	Petunidin-3,5-O-diglucoside (3A)+	641
		Delphinidin-3-O-monoglucoside (3B)	465
4	16.08	Peonidin-3,5-O-diglucoside (4)	625
5	16.67	Malvidin-3,5-O-diglucoside (5)	655
6	17.14	Cyanidin-3-O-monoglucoside (6)	449
7	18.59	Petunidin-3-O-monoglucoside (7)	479
8	21.22	Peonidin-3-O-monoglucoside (8)	463
9	22.88	Malvidin-3-O-monoglucoside (9)	493
10	25.78	Delphinidin-3-O-acetylmonoglucoside (10)	507
11	27.01	Delphinidin-3-(6-O-p-coumaroyl),5-O-diglucoside (11)	773
12	27.52	Cyanidin-3-(6-O-p-coumaroyl),5-O-diglucoside (12A)+	757
		Cyanidin-3-O-acetylmonoglucoside (12B)	491
13	28.53	Petunidin-3-(6-O-p-coumaroyl),5-O-diglucoside (13)	787
14	29.15	Petunidin-3-O-acetylmonoglucoside (14A)+	521
		Malvidin-3-(6-O-p-coumaroyl),5-O-diglucoside (14B)	801
15	30.98	Peonidin-3-O-acetylmonoglucoside (15)	505
16	31.58	Malvidin-3-O-acetylmonoglucoside (16)	535
17	31.97	Delphinidin-3-(6-O-p-coumaroyl)monoglucoside (17)	611
18	34.37	Malvidin-3-(6-O-caffeoyl)monoglucoside (18)	655
19	35.02	Cyanidin-3-(6-O-p-coumaroyl)monoglucoside (19)	595
20	36.29	Petunidin-3-(6-O-p-coumaroyl)monoglucoside (20)	625
21	40.11	Peonidin-3-(6-O-p-coumaroyl)monoglucoside (21)	609
22	40.50	Malvidin-3-(6-O-p-coumaroyl)monoglucoside (22)	639

[a]Figure 6.21 Favretto and Flamini, 2000.

with a short time and low solvent consuming analysis. An ESI-direct injection positive ion mass spectrum of the extract analyzed in Fig. 6.21 is reported in Fig. 6.22: all anthocyanins show the evident signal of an M^+ ion. Characterization of compounds is achieved by MS/MS and collision induced dissociation (CID) experiments applying a supplementary radio frequency field to the endcaps of the ion trap (1–15 V) in order to make the selected ions collide with He. The fragments recorded are reported in Table 6.12. A list of other monomer anthocyanins identified in extracts of different grape varieties by LC/ESI–MS/MS is reported in Table 6.13.

In general, MS^n is highly effective in differentiation of isobaric compounds. The fragment ions $[M-162]^+$, $[M-324]^+$ (two consecutive losses of sugar residue), $[M-204]^+$, $[M-308]^+$, $[M-324]$, and $[M-470]^+$ (consecutive losses of acylated glucose and the sugar residues) allows characterization

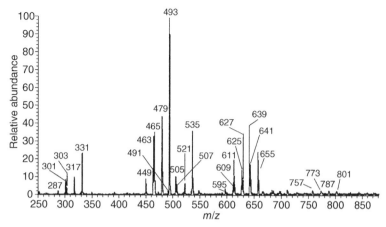

Figure 6.22. Direct-injection ESI/MS anthocyanin profile of *Clinton* grape skin extract. Analytical conditions: source voltage 4.2-kV positive-ion mode, capillary voltage 3.14 V, capillary temperature 220 °C, sheath gas flow rate 40 (arbitrary units), flow rate injection into the ESI source 3 μL/min. (Reprinted from *American Journal of Enology and Viticulture*, 51, Favretto and Flamini, Copyright © 2000.)

of both monoglucoside and diglucoside anthocyanins. Of course, the collision energy applied affects the relative abundance of diagnostic fragments. In the case of Mv-3,5-*O*-diglucoside and Mv-3-*O*-(6-caffeoyl) monoglucoside, differentiation between two compounds by MS^n experiments is not possible due to the identical molecular mass and aglycone of molecules. They were distinguished by dissolving the extract in a deuterated solvent (water or methanol), in agreement with a different number of exchangeable acidic protons present in the molecules different mass shifts were observed (Fig. 6.23).

The ESI/MS semiquantitative data of anthocyanins in the extract were achieved by calculating the calibration curves of Mv-3-*O*-glucoside (recording the intensity of the M^+ signal at *m/z* 493) as the standard for monoglucosides, and Mv-3,5-*O*-diglucoside (M^+ at *m/z* 655) for diglucosides, since both compounds are commercially available. A standard solution of Mv-3-*O*-glucoside 40 ppm in water/acetonitrile 95:5 (v/v) was used to optimize the ESI parameters and to maximize the signals (experimental conditions: spray voltage 4.5 kV, sheath gas nitrogen 0.9 L/min, capillary voltage 35 V, capillary temperature 200 °C, tube lens offset voltage 15 V) (Favretto and Flamini, 2000; Pati et al., 2006).

Also, oligomeric anthocyanins reported in Table 6.14 and Mv-4-vinyl-polycatechins were identified in grape marc and skins (Asenstorfer et al., 2001; Vidal et al., 2004). Identification of two dimeric anthocyanins by direct infusion ESI–MS/MS is shown in Fig. 6.24.

TABLE 6.12. Fragmentation of M⁺ Ions of Anthocyanins Identified in *Clinton* Grape Skin Extract by Direct-ESI and MS^n [a]

Anthocyanin	m/z						
	M^+	$[M-C_6H_{10}O_5]^+$	$[M-C_8H_{12}O_6]^+$	$[M-C_{15}H_{16}O_7]^+$	$[M-2(C_6H_{10}O_5)]^+$	$[M-C_{15}H_{16}O_7-C_6H_{10}O_5]^+$	$[M-C_{15}H_{16}O_8]^+$
Malvidin-3-O-monoglucoside	493	331					
Petunidin-3-O-monoglucoside	479	317					
Delphinidin-3-O-monoglucoside	465	303					
Peonidin-3-O-monoglucoside	463	301					
Cyanidin-3-O-monoglucoside	449	287					
Malvidin-3-O-acetylmonoglucoside	535		331				
Petunidin-3-O-acetylmonoglucoside	521		317				
Delphinidin-3-O-acetylmonoglucoside	507		303				
Peonidin-3-O-acetylmonoglucoside	505		301				
Cyanidin-3-O-acetylmonoglucoside	491		287				
Malvidin-3-(6-O-p-coumaroyl)monoglucoside	639			331			
Petunidin-3-(6-O-p-coumaroyl)monoglucoside	625			317			
Delphinidin-3-(6-O-p-coumaroyl)monoglucoside	611			303			
Peonidin-3-(6-O-p-coumaroyl)monoglucoside	609			301			
Cyanidin-3-(6-O-p-coumaroyl)monoglucoside	595			287			
Malvidin-3,5-O-diglucoside	655	493			331		
Petunidin-3,5-O-diglucoside	641	479			317		
Delphinidin-3,5-O-diglucoside	627	465			303		
Peonidin-3,5-O-diglucoside	625	463			301		
Cyanidin-3,5-O-diglucoside	611	449			287		
Malvidin-3-(6-O-p-coumaroyl),5-O-diglucoside	801	639		493		331	
Petunidin-3-(6-O-p-coumaroyl),5-O-diglucoside	787	625		479		317	
Delphinidin-3-(6-O-p-coumaroyl),5-O-diglucoside	773	611		465		303	
Cyanidin-3-(6-O-p-coumaroyl),5-O-diglucoside	757	595		449		287	
Malvidin-3-(6-O-caffeoyl)monoglucoside	655						331

[a] Favretto and Flamini, 2000.

TABLE 6.13. Monomer Anthocyanins Identified by LC/ESI–MS/MS in Skin Extract or Juice of Different Grape Cultivars[a]

Anthocyanin	m/z (M⁺)	Cultivar
Cy-3-O-pentoside	419	*Casavecchia*
Pg-3-O-glucoside	433	*Concord, Salvador, Rubired*
Cy-3-O-(6-O-acetyl)pentoside	461	*Casavecchia*
Cy-3-O-(6-O-p-coumaryl)pentoside	565	*Casavecchia*
Dp-3-O-glucoside-pyruvic acid	533	*Isabelle*
Dp-3-O-(6-O-p-coumaryl)glucosidepyruvic acid	679	*Isabelle*
Pn-3-O-glucoside-acetaldehyde	487	*Isabelle, Pallagrello*
Mv-3-O-glucoside-acetaldehyde	517	*Isabelle*
Pt-3-O-(6-O-p-caffeoyl)-5-O-diglucoside	803	*Isabelle, Casavecchia*
Dp-3-O-(6-O-acetyl)-5-O-diglucoside	669	*Isabelle*
Dp-3-O-(6-O-feruloyl)-5-O-diglucoside	803	*Isabelle, Casavecchia*
Pn-3-O-(6-O-p-coumaryl)-5-O-diglucoside	771	*Concord, Salvador, Isabelle, Casavecchia*

[a]Mazzuca et al., 2005; Wang, et al., 2003. Pelargonidin = Pg; Dp = delphinidin; Cy = cyanidin; Pt = petunidin; Pn = peonidin; Mv = malvidin.

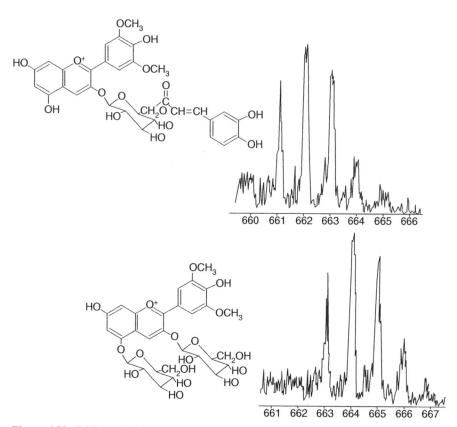

Figure 6.23. Differentiation of two isobaric compounds with the same aglycone moiety by deuterium-exchange experiment: positive ESI mass spectra of Mv-3-O-(6-O-caffeoyl)monoglucoside (above) and Mv-3,5-O-diglucoside (below) after dissolution of the *Clinton* grape skin extract residue in D₂O. (Reprinted from *American Journal of Enology and Viticulture*, 51, Favretto and Flamini, Copyright © 2000.)

TABLE 6.14. Oligomeric Anthocyanins Identified in Shiraz Grape Skins[a]

m/z	Assignment	m/z	Assignment	m/z	Assignment
287(F)	Cy	795(F)	MvDp+G	1315(F)	MvMvMv+2G
301(F)	Pn	809(F)	MvPt+G	1417(F)	MvMvCy+G·pCG
303(F)	Dp	823(F)	MvMv+G	1431(F)	MvMvPn+G·pCG
317(F)	Pt	941(M)	MvCy+2G	1433(F)	MvMvDp+G·pCG
331(F)	Mv	955(M)	MvPn+2G	1433(M)	MvMvCy+3G
449(M)	Cy+G	957(M)	MvDp+2G	1447(F)	MvMvPt+G·pCG
463(M)	Pn+G	971(M)	MvPt+2G	1447(M)	MvMvPn+3G
465(M)	Dp+G	985(M)	MvMv+2G	1449(M)	MvMvDp+3G
479(M)	Pt+G	1087(M)	MvCy+G·pCG	1461(F)	MvMvMv+G·pCG
493(M)	Mv+G	1101(M)	MvPn+G·pCG	1463(M)	MvMvPt+3G
617(F)	MvCy	1103(M)	MvDp+G·pCG	1477(M)	MvMvMv+3G
631(F)	MvPn	1117(M)	MvPt+G·pCG	1579(M)	MvMvCy+2G·pCG
633(F)	MvDp	1131(M)	MvMv+G·pCG	1593(M)	MvMvPn+2G·pCG
647(F)	MvPt	1271(F)	MvMvCy+2G	1595(M)	MvMvDp+2G·pCG
661(F)	MvMv	1285(F)	MvMvPn+2G	1609(M)	MvMvPt+2G·pCG
779(F)	MvCy+G	1287(F)	MvMvDp+2G	1623(M)	MvMvMv+2G·pCG
793(F)	MvPn+G	1301(F)	MvMvPt+2G		

[a]Fragment ion = F; M = molecular ion; Dp = delphinidin; Cy = cyanidin; Pt = petunidin; Pn = peonidin; Mv = malvidin; G = glucose, pCG = p-coumaroyl glucoside (Vidal et al., 2004).

6.5   THE LC/MS OF ANTHOCYANIN DERIVATIVES IN WINE

The LC/MS analysis of anthocyanins and their derivatives in wine can be performed by direct injection of the sample without a prior sample preparation. Several analytical methods with different chromatographic conditions were proposed by this approach (Table 6.15).

In method D, the TFA percentage of solvent is kept low to limit formation of ionic pairs that may decrease MS sensibility. Table 6.16 reports the compounds identified in three different wines (*Graciano*, *Tempranillo*, and *Cabernet Sauvignon*) using Method A in Table 6.15, and Table 6.17 reports the compounds detected in a *Primitivo* wine by using Method D.

Purification of the wine sample prior analysis can be performed by SPE: 5 mL of wine are diluted 1:4 with water and the solution is passed through a C_{18} cartridge previously activated by passages of methanol followed by water. After sample loading, the cartridge is washed with 6 mL of 0.3% formic acid aqueous solution and with 4 mL of water, then anthocyanins are recovered with 5 mL of methanol. The solution is evaporated to dryness and the residue is redissolved in the LC mobile phase (Kosir et al., 2004).

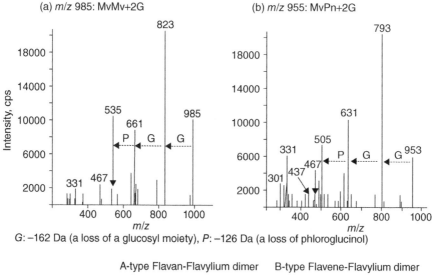

G: −162 Da (a loss of a glucosyl moiety), P: −126 Da (a loss of phloroglucinol)

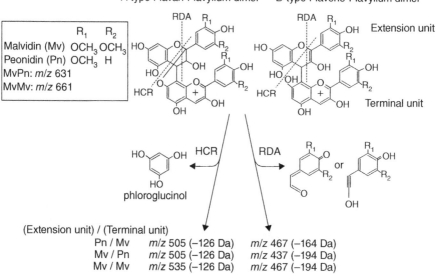

HCR: heterocyclic ring fission, RDA: retro-Diels–Alder fission

Figure 6.24. Above: direct-infusion ESI–MS/MS product ion spectra of anthocyanin dimers identified in grape skins extract composed of Mv-glucoside and: (a) Mv-3-glucoside (*m/z* 985, MvMv+2G), (b) Pn-3-glucoside (*m/z* 955, MvPn+2G) (ESI needle, orifice, and ring potentials at 5000, 150, and 250V, respectively; collision gas N₂; collision energy 30–60V). Below: fragmentation scheme of two anthocyanin dimers. (Reproduced from *Journal of Agricultural and Food Chemistry*, 2004, 52, p. 7148, Vidal et al., with permission of American Chemical Society.)

TABLE 6.15. Four Different Methods Used for LC/ESI–MS Analysis of Anthocyanins and Anthocyanin Derivatives in Wine by Direct Injection of the Sample into the Column

METHOD A (Monagas et al., 2003)
Column C_{18} (150 × 3.9 mm, 4 μm) at room temperature
Solvent: (A) H_2O/formic acid 90:10 (v/v), (B) H_2O/MeOH/formic acid 45:45:10 (v/v/v)
Elution gradient program: from 15 to 80% of B in 30 min, 80% B isocratic for 13 min (flow rate 0.8 mL/min)
METHOD B (Kosir et al., 2004)
Column C_{18} (250 × 4.6 mm, 5 μm) at 30 °C
Solvent: (A) 0.3% perchloric acid in water, (B) 96% ethanol
Elution gradient program: from 71.5 to 54.5% of B in 32 min, 54.5 → 31.5% B in 13 min, 31.5 → 100% B in 2 min, 100% B isocratic for 3 min (flow rate 0.8 mL/min)
METHOD C (Salas et al., 2004)
Column C_{18} (250 × 2.0 mm, 5 μm) at 30 °C
Solvent: (A) H_2O/formic acid 95:5 (v/v), (B) acetonitrile/solvent A 80:20 (v/v)
Elution gradient program: 2% B isocratic for 7 min, from 2 to 20% of B in 15 min, 20 → 30% B in 8 min, 30 → 40% B in 10 min, 40 → 50% B in 5 min, 50 → 80% B in 5 min (flow rate 0.25 mL/min)
METHOD D (Pati et al., 2006)
Column C_{18} (150 × 2.0 mm, 5 μm) at room temperature
Solvent: (A) H_2O/acetonitrile 95:5 (v/v) containing 0.1% (v/v) TFA, (B) water/acetonitrile 10:90 (v/v) containing 0.1% (v/v) TFA
Elution gradient program: 2% B isocratic for 2 min, from 2 to 10% B in 6 min, 10 → 13% B in 22 min, 13 → 20% B in 20 min, 20 → 30% B in 25 min (flow rate 0.2 mL/min)

TABLE 6.16. Anthocyanins and Their Derivatives Identified by LC/ESI–MS in *Graciano*, *Tempranillo*, and *Cabernet Sauvignon* Wines by Performing Analysis with Method A Reported in Table 6.15[a]

		(m/z)		
RT (min)	Compound	[M]+	Fragments	λ_{max} (nm)
4.5	Mv-3-O-glucoside-(epi)catechin	781		530
6.9	Dp-3-O-glucoside	465	303	524
8.6	Cy-3-O-glucoside	449	287	515
9.8	Pt-3-O-glucoside	479	317	526
11.2	Pn-3-O-glucoside	463	301	516
12.0	Mv-3-O-glucoside	493	331	520
13.4	Pn-3-O-glucoside pyruvate	531	369	509
14.0	Dp-3-(6-O-acetylglucoside)	507	303	533
14.3	Mv-3-O-glucoside pyruvate	561	399	513
15.2	Mv-3-(6-O-acetylglucoside) pyruvate	603	399	518
15.8	Cy-3-(6-O-acetylglucoside)	491	287	516

TABLE 6.16. (*Continued*)

		(m/z)		
RT (min)	Compound	[M]$^+$	Fragments	λ_{max} (nm)
16.1	Mv-3-*O*-glucoside-8-ethyl-(epi) catechin	809		543
16.2	Pt-3-(6-*O*-acetylglucoside)	521	317	532
18.0	Mv-3-(6-*O*-*p*-coumaroylglucoside) pyruvate	707	399	513
18.7	Pn-3-(6-*O*-acetylglucoside)	505	301	520
19.0	Dp-3-(6-*O*-*p*-coumaroylglucoside)	611	303	532
19.3	Mv-3-(6-*O*-acetylglucoside)	535	331	530
20.1	Pn-3-(6-*O*-caffeoylglucoside)	625	301	524
20.6	Mv-3-(6-*O*-caffeoylglucoside)	655	331	536
21.1	Cy-3-(6-*O*-*p*-coumaroylglucoside)	595	287	527
21.2	Mv-3-(6-*O*-*p*-coumaroylglucoside) cis isomer	639	331	537
21.9	Pt-3-(6-*O*-*p*-coumaroylglucoside)	625	317	532
22.2	Mv-3-*O*-glucoside-4-vinyl-catechin	805		503
22.3	Mv-3-(6-*O*-*p*-coumaroylglucoside)-8-ethyl-(epi)catechin	955		540
23.6	Mv-3-(6-*O*-acetylglucoside)-4-vinyl-catechin	847		508
24.1	Pn-3-(6-*O*-*p*-coumaroylglucoside)	609	301	524
24.4	Mv-3-(6-*O*-*p*-coumaroylglucoside) trans isomer	639	331	535
25.3	Mv-3-*O*-glucoside-4-vinylcathecol	625	463	514
26.0	Mv-3-(6-*O*-acetylglucoside)-4-vinyl-epicatechin	847		514
26.3	Mv-3-*O*-glucoside-4-vinyl-epicatechin	805		508
27.8	Mv-3-*O*-glucoside-4-vinylphenol	609	447	504
28.6	Mv-3-*O*-glucoside-4-vinylguaiacol	639	447	504
29.7	Mv-3-(6-*O*-acetylglucoside)-4-vinylphenol	651	477	509
34.7	Mv-3-(6-*O*-*p*-coumaroylglucoside)-4-vinylphenol	755	447	504

[a]The ESI–MS parameters: positive-ion mode; drying gas N$_2$; temperature 350°C; nebulizer pressure 380 Pa (55 psi); capillary voltage 4 kV; fragmentator voltage: 100 V from 0 to 17 min, 120 V from 17 to 55 min (Monagas et al., 2003).

Isolation of oligomeric pigments from the wine and fractionation of extract can be performed by cation-exchange chromatography in the presence of a bisulfite buffer. The procedure is described in the flow diagram in Fig. 6.25.

Prior fractionation of the wine sample was concentrated under vacuum. The methanolic solution from the C$_{18}$ 50-g column (**b**) was

TABLE 6.17. Anthocyanin Derivatives Identified in a *Primitivo* Wine by LC/ESI–MS Analysis Using Method D Reported in Table 6.15 with their Characteristic Fragment Ions Produced by MS/MS and MS^3 Using as Precursor Ions the Most Intense *m/z* Signal in the Mass Spectrum^a

RT (min)	Compound	[M]$^+$	MS^2	MS^3
			(*m/z*)	
16.8	Pn-3-*O*-glucoside-(epi)catechin	751	589	571, 437, 463
18.3	Mv-3-*O*-glucoside-(epi)catechin	781	619	601, 493, 467
20.0	Dp-3-*O*-glucoside	465	303	
20.3	Mv-3-*O*-glucoside-di(epi)catechin	1069	907, 781, 619	
22.3	Cy-3-*O*-glucoside	449	287	
23.4	Mv-3-*O*-glucoside-(epi)catechin	781	619	601, 493, 467
23.5	Mv-3-*O*-glucoside-di(epi)catechin	1069	907, 781, 619	
24.0	Pt-3-*O*-glucoside	479	317	
26.7	Pn-3-*O*-glucoside	463	301	
28.3	Mv-3-*O*-glucoside	493	331	
29.5	Mv-3-*O*-glucoside pyruvate	561	399	
33.5	Mv-3-*O*-glucoside acetaldehyde	517	355	
38.4	Mv-3-*O*-glucoside-8-ethyl-(epi)catechin	809	647, 519, 357	
40.7	Mv-3-*O*-glucoside-8-ethyl-(epi)catechin	809	647, 519, 357	
43.5	Mv-3-*O*-glucoside-4-vinyl-di(epi)catechin	1093	931 803	641
44.4	Mv-3-*O*-glucoside-8-ethyl-(epi)catechin	809	647, 519, 357	
45.3	Mv-3-*O*-glucoside-4-vinyl-di(epi)catechin	1093	931 803	641
46.4	Mv-3-*O*-glucoside-8-ethyl-(epi)catechin	809	647, 519, 357	
47.3	Mv-3-(6-*O*-*p*-coumaroylglucoside)-(epi)catechin	927	619	601, 493, 467
48.7	Mv-3-(6-*O*-*p*-coumaroylglucoside) pyruvate	707	399	
48.7	Pn-3-(6-*O*-acetylglucoside)	505	301	
50.4	Mv-3-(6-*O*-acetylglucoside)	535	331	
52.1	Dp-3-(6-*O*-*p*-coumaroylglucoside)	611	303	
54.3	Mv-3-(6-*O*-acetylglucoside)-4-vinyl-(epi)catechin	847	643	491
56.1	Mv-3-(6-*O*-caffeoylglucoside)	655	331	
56.8	Cy-3-(6-*O*-*p*-coumaroylglucoside)	595	287	
58.3	Pt-3-(6-*O*-*p*-coumaroylglucoside)	625	317	
58.5	Mv-3-(6-*O*-*p*-coumaroylglucoside)	639	331	

TABLE 6.17. (*Continued*)

RT (min)	Compound	$[M]^+$	MS^2	MS^3
			(*m/z*)	
59.4	Mv-3-(6-*O*-*p*-coumaroylglucoside)-4-vinyl-di(epi)catechin	1239	931 641	641
59.7	Mv-3-*O*-glucoside-4-vinyl-(epi)catechin	805	643	491
60.5	Pn-3-(6-*O*-*p*-coumaroylglucoside)-8-ethyl-(epi)catechin	925	635, 617, 327	
61.5	Mv-3-(6-*O*-*p*-coumaroylglucoside)-8-ethyl-(epi)catechin	955	665, 357	
62.8	Pn-3-(6-*O*-*p*-coumaroylglucoside)	609	301	
63.7	Mv-3-(6-*O*-*p*-coumaroylglucoside)	639	331	
63.9	Mv-3-*O*-glucoside-4-vinylcathecol	625	463	
65.5	Mv-3-*O*-glucoside-4-vinyl-(epi)catechin	805	643	491
65.7	Mv-3-(6-*O*-*p*-coumaroylglucoside)-4-vinyl-(epi)catechin	951	643	491
67.5	Mv-3-(6-*O*-*p*-coumaroylglucoside)-4-vinyl-(epi)catechin	951	643	491
67.9	Mv-3-*O*-glucoside-4-vinylphenol	609	447	
69.8	Mv-3-*O*-glucoside-4-vinylguaiacol	639	477	
71.2	Mv-3-(6-*O*-*p*-coumaroylglucoside)-8-ethyl-(epi)catechin	955	665, 357	

[a]The MS conditions: positive-ion mode; spray voltage 4.5 kV; sheath gas N_2 0.9 L/min; capillary voltage 35 V; capillary temperature 200 °C; tube lens offset voltage, 15 V (Pati et al., 2006).

concentrated under vacuum, the water volume was adjusted to 200 mL with water, and ~50 mL of the resulting solution was loaded onto a sulfoxyethyl cellulose 40 × 200-mm column. Vitisin is present in the neutral–anionic fraction (1) recovered with 10% methanol (v/v). For elution of fraction (2), non-acidified methanol was used in order to avoid hydrolysis of the pigments. To remove NaCl, the extract was passed through a C_{18} column and the stationary phase was washed with water. Fraction (3) contains, for the most part, anthocyanins existing primarily as their anionic bisulfite addition products. The pigments retained on the column were eluted with a 2 M NaCl in a 50% methanol solution (fraction 4). This fraction was further purified by preparative thin-layer chromatography (TLC) using a silica gel plate and 70% (v/v) aqueous propanol as eluent. The separated red band (Rf 0.8) was extracted with a 10% (v/v) aqueous methanol solution.

The MS analysis can be performed either by ESI-direct injection of the sample or LC separation. Table 6.18 reports the anthocyanidin C4 derivatives identified by ESI/MS of fraction 4 (point m in Fig. 6.25) of a red wine and a grape marc extract.

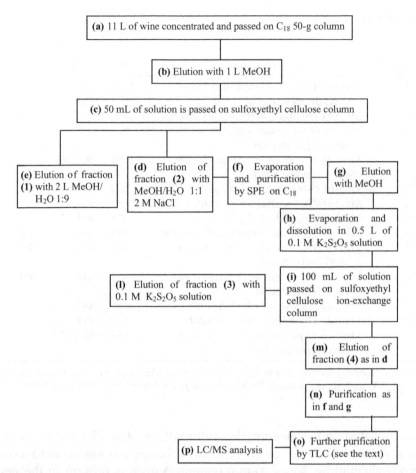

(a) 11 L of wine concentrated and passed on C_{18} 50-g column

(b) Elution with 1 L MeOH

(c) 50 mL of solution is passed on sulfoxyethyl cellulose column

(e) Elution of fraction **(1)** with 2 L MeOH/ H_2O 1:9

(d) Elution of fraction **(2)** with MeOH/H_2O 1:1 2 M NaCl

(f) Evaporation and purification by SPE on C_{18}

(g) Elution with MeOH

(h) Evaporation and dissolution in 0.5 L of 0.1 M $K_2S_2O_5$ solution

(l) Elution of fraction **(3)** with 0.1 M $K_2S_2O_5$ solution

(i) 100 mL of solution passed on sulfoxyethyl cellulose ion-exchange column

(m) Elution of fraction **(4)** as in **d**

(n) Purification as in **f** and **g**

(p) LC/MS analysis

(o) Further purification by TLC (see the text)

Figure 6.25. A method for fractionation of oligomeric pigments in wine (Asenstorfer et al., 2001).

Another sample preparation method proposed for LC/MS analysis of pyranoanthocyanidins and anthocyanin derivatives in wine is reported in the flow diagram in Fig. 6.26.

Eluates 1–4 in Fig. 6.26 were polled in Fraction A and the solution was analyzed. The UV–vis chromatograms was recorded by connecting the LC/ESI–MS system to the probe of the mass spectrometer via the UV cell outlet. Figure 6.27 shows the chromatograms relative to analyses of Fraction A of a wine at different stages of aging. The great number of anthocyanins and derivatives identified in the chromatograms in Fig. 6.27 are reported in Table 6.19: simple anthocyanins, ethyl-bridge derivatives, pyranoanthocyanins, and pigments formed by anthocyanin-flavanol linkage. As seen from the table, some compounds

TABLE 6.18. The ESI/MS Data of Pigments Isolated in Fraction 4 in Fig. 6.25 of the *Shiraz* Grape Marc Extract and Wine[a]

Marc ([M]+ *m/z*)[b]	Wine ([M]+ *m/z*)[b]	Compound
609.4	609.4	Pigment A
nd	639.4	3″-*O*-Methyl-pigment A
651.4	651.4	(Acetyl)pigment A
707.2	707.2	(*p*-Coumaryl)vitisin A
nd	755.6	(*p*-Coumaryl)pigment A
805.4	805.4	Mv-3-glucose-4-vinyl-catechin
847.4	nd	Mv-3-(acetyl)glucose-4-vinyl-catechin
951.4	951.4	Mv-3-(*p*-coumaryl)glucose-4-vinyl-catechin
1093.4	1093.4	Mv-3-glucose-4-vinyl-dicatechin
1135.4	nd	Mv-3-(acetyl)glucose-4-vinyl-dicatechin
1239.6	nd	Mv-3-(*p*-coumaryl)glucose-4-vinyl-dicatechin
1381.6	nd	Mv-3-glucose-4-vinyl-tricatechin
1423.4	nd	Mv-3-(acetyl)glucose-4-vinyl-tricatechin
1527.6	nd	Mv-3-(*p*-coumaryl)glucose-4-vinyl-tricatechin
1669.4	nd	Mv-3-glucose-4-vinyl-tetracatechin

[a]Analytical conditions: ion source and orifice potentials 5.5 kV and 30 V, respectively, positive-ion mode. Curtain gas N_2 8 units; nebulizer gas air 10 units; injected solution 50% acetonitrile acidified with 2.5% acetic acid (rate of 5 μL/min) (Asenstorfer et al., 2001).
[b]Not detected = nd.

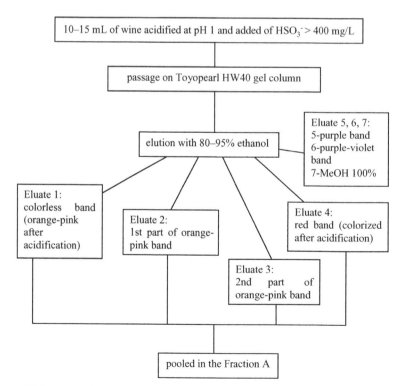

Figure 6.26. A sample preparation method for analysis of pyranoanthocyanidins and anthocyanin derivatives in wine (Alcade-Eon et al., 2004).

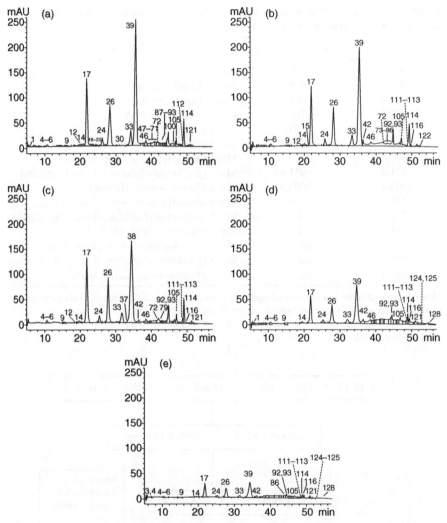

Figure 6.27. Chromatograms of a wine sample aged (a) 4 months, (b) 8 months; (c) 13 months; (d) 16 months; and (e) 23 months recorded at 520 nm. Compounds corresponding to the peaks are reported in Table 6.19. Chromatographic conditions: Column C_{18} (150 × 4.6 mm, 5 μm) at 35 °C; binary solvent (A) aqueous solution of TFA 0.1% and (B) acetonitrile; gradient elution program: 10% B isocratic for 5 min, from 10 to 15% of B in 15 min, 15% B isocratic for 5 min, from 15 to 18% of B in 5 min, from 18 to 35% of B in 20 min (flow rate 0.5 mL/min). (Reprinted from *Analytica Chimica Acta*, 563, Alcade-Eon et al., Changes in the detailed pigment composition of red wine during maturity and ageing. A comprehensive study, p. 240, Copyright © 2006, with permission from Elsevier.)

TABLE 6.19. Anthocyanins and Their Derivatives Identified in 4–23 Month Aged Wines[a]

RT (min)	Compound	$[M]^+$	MS^2	MS^3	λ_{max} (nm)	4	8	13	16	23
			m/z Fragment Ions			Aged Wine (months)				
21.7	Dp-3-glc	465	303	303	277, 342, 524	x	x	x	x	x
26.1	Cy-3-glc	449	287	287	279, 516	x	x	x	x	x
28.1	Pt-3-glc	479	317	317	277, 347, 525	x	x	x	x	x
34.1	Pn-3-glc	463	301	301	280, 517	x	x	x	x	x
35.5	Mv-3-glc	493	331	331	277, 348, 527	x	x	x	x	x
38.3	Dp-3-acetylglc	507	303	303	276, 346, 527	x	x	x	x	x
41.0	Cy-3-acetylglc	491	287	287	280, 523	x	x	x	x	x
41.6	Pt-3-acetylglc	521	317	317	270, 529	x	x	x	x	x
43.6	Pn-3-acetylglc	505	301	301	280, 522	x	x	x	x	x
44.3	Mv-3-acetylglc	535	331	331	278, 350, 530	x	x	x	x	x
43.1	Dp-3-*p*-coumglc *cis*	611	303	303	280, 301, 534	x	x	x	x	x
44.3	Dp-3-*p*-coumglc *trans*	611	303	303	282, 313, 531	x	x	x	x	x
45.1	Cy-3-*p*-coumglc *cis*	595	287	287	280, 301, 533	x	x	x	x	x
46.3	Cy-3-*p*-coumglc *trans*	595	287	287	284, 314, 524	x	x	x	x	x
45.3	Pt-3-*p*-coumglc *cis*	625	317	317	281, 301, 536	x	x	x	x	x
46.6	Pt-3-*p*-coumglc *trans*	625	317	317	282, 313, 532	x	x	x	x	x
47.5	Pn-3-*p*-coumglc *cis*	609	301	301	283, 300, 535	x	x	x	x	x
48.6	Pn-3-*p*-coumglc *trans*	609	301	301	283, 313, 526	x	x	x	x	x
47.5	Mv-3-*p*-coumglc *cis*	639	331	331	280, 301, 535	x	x	x	x	x
48.7	Mv-3-*p*-coumglc *trans*	639	331	331	282, 313, 532	x	x	x	x	x
41.1	Dp-3-cafglc *trans*	627	303	303	283, 331, 532	x	nd	nd	nd	nd
43.6	Pt-3-cafglc *trans*	641	317	317	283, 328, 531	x	x	nd	nd	nd
45.6	Pn-3-cafglc *trans*	625	301	301	283, 328, 525	x	nd	nd	nd	nd
44.8	Mv-3-cafglc *cis*	655	331	331		x	x	x	x	nd
45.7	Mv-3-cafglc *trans*	655	331	331	282, 328, 534	x	x	x	x	x
16.7	Dp-3,7-diglc	627	303	303	279, 523	x	nd	nd	nd	nd
20.4	Pt-3,5-diglc	641	317	317	275, 521	x	nd	nd	nd	nd

TABLE 6.19. (Continued)

		m/z				Aged Wine (months)				
			Fragment Ions							
RT (min)	Compound	[M]⁺	MS²	MS³	λ_max (nm)	4	8	13	16	23
24.6	Pt-3,7-diglc	641	317	317	275, 349, 522	x	x	x	x	x
28.7	Pn-3,7-diglc	625	301	301		x	x	nd	nd	nd
23.5	Mv-3,5-diglc	655	331	331	275, 524	x	x	nd	nd	nd
30.8	Mv-3,7-diglc	655	331	331	278, 350, 526	x	x	x	x	x
34.7	Dp-3-glc + L(+)lactic acid	537				x	x	x	x	x
37.5	Pt-3-glc + D(−)lactic acid	551	317			x	x	x	x	nd
39.2	Pt-3-glc + L(+)lactic acid	551	317	317	278, 526	x	x	nd	x	x
40.5	Pn-3-glc + D(−)lactic acid	535				x	x	nd	nd	nd
41.7	Pn-3-glc + L(+)lactic acid	535	301	301	281, 525	x	x	x	x	x
40.8	Mv-3-glc + D(−)lactic acid	565	331	331	278, 350, 530	x	x	x	x	x
42.1	Mv-3-glc + L(+)lactic acid	565	331	331	278, 348, 531	x	x	x	x	x
5.7	Dp-3-glc-GC	769	607	439	531	x	x	x	x	x
7.1	Cy-3-glc-GC	753	591	453	282, 524	x	x	x	x	x
7.2	Pt-3-glc-GC	783	621	453	279, 532	x	x	x	x	x
10.8	Pn-3-glc-GC	767	605	437		x	x	x	x	x
10.6	Mv-3-glc-GC	797	635	467	281, 531	x	x	x	x	x
22.3	Mv-3-glc-EGC	797	635	467		x	x	x	x	x
24.4	Mv-3-acetylglc-GC	839				nd	x	x	x	nd
30.9	Dp-3-p-coumglc-GC	915	607	439		nd	x	x	x	nd
35.1	Cy-3-p-coumglc-GC	899				nd	x	x	nd	nd
35.4	Pt-3-p-coumglc-GC	929				nd	x	x	x	nd
38.5	Pn-3-p-coumglc-GC	913	605	437		nd	x	x	x	x
38.3	Mv-3-p-coumglc-GC	943	635	467		nd	x	x	x	x
41.1	Mv-3-p-coumglc-EGC	943				nd	x	x	x	x
10.8	Dp-3-glc-C	753	591	439	282, 534	x	x	x	x	x
14.8	Dp-3-glc-EC	753	591	439		x	x	x	x	x

Compound	RT	[M]+			λmax					
Cy-3-glc-C	14.9	737	575	423	286, 526	x	x	x	x	x
Cy-3-glc-EC	18.0	737	575	423		x	x	x	x	x
Pt-3-glc-C	16.2	767	605	453	279, 532	x	x	x	x	x
Pt-3-glc-EC	21.6	767	605	453		x	x	x	x	x
Pn-3-glc-C	20.3	751	589	437	283, 524	x	x	x	x	x
Pn-3-glc-EC	24.3	751	589	437		x	x	x	x	x
Mv-3-glc-C	21.0	781	619	467	280, 532	x	x	x	x	x
Mv-3-glc-EC	29.9	781	619	467		x	x	x	x	x
Mv-3-acetylglc-C	35.9	823	619	467	279, 533	x	x	x	x	x
Dp-3-p-coumglc-C	39.0	899	591	439		nd	nd	nd	nd	nd
Dp-3-p-coumglc-EC	40.5	899				nd	nd	nd	nd	nd
Cy-3-p-coumglc-C	39.5	883				nd	x	nd	x	nd
Cy-3-p-coumglc-EC	41.4	883				nd	nd	nd	nd	nd
Pt-3-p-coumglc-C	41.0	913	605	453		nd	x	x	nd	x
Pt-3-p-coumglc-EC	42.6	913				nd	nd	nd	nd	nd
Pn-3-p-coumglc-C	41.8	897	589	437		nd	x	x	nd	x
Pn-3-p-coumglc-EC	43.8	897				nd	nd	nd	nd	nd
Mv-3-p-coumglc-C	43.4	927	619	467	290, 538	x	x	x	x	x
Mv-3-p-coumglc-EC	46.0	927				nd	nd	nd	nd	nd
Dp-3-glc-ethyl-C	35.8	781	329	329		x	nd	nd	nd	x
Dp-3-glc-ethyl-EC	36.7	781				x	nd	nd	nd	nd
Cy-3-glc-ethyl-C	39.5	765				x	nd	nd	nd	nd
Pt-3-glc-ethyl-C	39.6	795	343	343		x	nd	nd	nd	x
Pn-3-glc-ethyl-C	42.0	779	327	327		x	x	nd	nd	x
Pn-3-glc-ethyl-EC	43.1	779				x	nd	nd	nd	x
Mv-3-glc-ethyl-C	41.1	809	357	357		x	nd	x	x	x
Mv-3-glc-ethyl-C	42.2	809	357	357	282, 539	x	x	x	x	x
Mv-3-glc-ethyl-C	43.1	809	357	357	276, 537	x	x	x	x	x
Mv-3-acetylglc-ethyl-C	45.8	851	357	357		x	nd	x	nd	nd
Mv-3-p-coumglc-ethyl-C	47.4	955	357	357		x	x	x	x	nd
Dp-3-glc-ethyl-GC	34.7	797	329	329		x	nd	nd	nd	x
Cy-3-glc-ethyl-GC	38.6	781				x	nd	nd	nd	nd
Pt-3-glc-ethyl-GC	38.7	811	343	343		x	nd	x	x	x

TABLE 6.19. (*Continued*)

RT (min)	Compound	[M]⁺	MS²	MS³	λ_{max} (nm)	4	8	13	16	23
40.9	Pn-3-glc-ethyl-GC	795				x	nd	nd	nd	nd
40.6	Mv-3-glc-ethyl-GC	825	357	357		x	nd	x	x	x
41.1	Mv-3-glc-ethyl-GC	825	357	357	539	x	x	x	x	x
41.8	Mv-3-glc-ethyl-GC	825	357	357		x	nd	x	x	x
45.0	Mv-3-acetylglc-ethyl-GC	867				x	nd	nd	nd	nd
21.0	A-type vitisin of Dp-3-glc	533	371	371	297, 368, 507	nd	x	x	x	x
27.0	A-type vitisin of Cy-3-glc	517				nd	nd	nd	x	nd
28.7	A-type vitisin of Pt-3-glc	547	385	385	299, 371, 508	nd	x	x	x	x
35.0	A-type vitisin of Pn-3-glc	531	369	369	503	nd	nd	nd	x	x
36.0	Vitisin A	561	399	399	299, 372, 510	nd	x	x	x	x
40.7	A-type vitisin of Pt-3-*p*-coumglc	693				nd	nd	x	x	nd
43.8	A-type vitisin of Pn-3-*p*-coumglc	677	369	369	284, 508	nd	x	x	x	x
44.1	A-type vitisin of Mv-3-*p*-coumglc	707	399	399	271, 514	nd	x	x	x	x
24.4	B-type vitisin of Dp-3-glc	489	327	327		nd	x	x	x	x
33.5	B-type vitisin of Pt-3-glc	503	341	341	492	nd	x	x	x	x
38.5	B-type vitisin of Pn-3-glc	487	325	325		x	x	x	x	nd
39.5	Vitisin B	517	355	355	294, 358, 490	x	x	x	x	x
41.4	B-type vitisin of Pn-3-acetylglc	529	325	325		x	x	nd	nd	nd
42.4	B-type vitisin of Mv-3-acetylglc	559	355	355	298, 361, 494	x	x	x	x	nd
41.1	Acetone derivative of Pn-3-glc	501	339	339	475	nd	x	x	x	nd
42.1	Acetone derivative of Mv-3-glc	531	369	369	480	nd	x	x	x	x
45.5	Dp-3-glc 4-vinylphenol adduct	581	419	419	264, 412, 503	x	x	x	x	x
47.5	Cy-3-glc 4-vinylphenol adduct	565				nd	x	x	x	x
48.3	Pt-3-glc 4-vinylphenol adduct	595	433	433	264, 413, 502	x	x	x	x	x
50.5	Pn-3-glc 4-vinylphenol adduct	579	417	417	278, 406, 500	x	x	x	x	x
51.0	Mv-3-glc 4-vinylphenol adduct	609	447	447	263, 412, 504	x	x	x	x	x

m/z — Fragment Ions; Aged Wine (months)

53.2	Mv-3-acetylglc 4-vinylphenol adduct	651	447	447	298, 416, 505	x	x	x	x	x	x
49.9	Dp-3-p-coumglc 4-vinylphenol adduct	727	419	419		nd	nd	nd	nd	x	x
52.4	Pt-3-p-coumglc 4-vinylphenol adduct	741	433	433	314, 504	nd	nd	nd	nd	x	x
54.6	Pn-3-p-coumglc 4-vinylphenol adduct	725	417	417	314, 501	nd	nd	nd	nd	x	x
55.2	Mv-3-p-coumglc 4-vinylphenol adduct	755	447	447	264, 313, 416, 505	nd	x	x	x	x	x
43.5	Dp-3-glc 4-vinylcatechol adduct	597	435	435	509	x	x	x	x	x	x
46.5	Pt-3-glc 4-vinylcatechol adduct	611	449	449	510	x	x	x	x	x	x
48.6	Pn-3-glc 4-vinylcatechol adduct	595	433	433	506	x	x	x	x	x	x
49.2	Mv-3-glc 4-vinylcatechol adduct	625	463	463	510	x	x	x	x	x	x
50.9	Mv-3-acetylglc 4-vinylcatechol adduct	667	463	463	513	nd	nd	x	x	x	x
47.6	Dp-3-p-coumglc 4-vinylcatechol adduct	743	435	435		x	x	x	x	x	x
50.5	Pt-3-p-coumglc 4-vinylcatechol adduct	757	449	449		nd	x	x	x	x	x
53.2	Mv-3-p-coumglc 4-vinylcatechol adduct	771	463	463	312, 511	x	x	x	x	x	x
52.0	Mv-3-glc 4-vinylguaiacol adduct	639	477	477	511	x	x	x	x	x	x
54.0	Mv-3-acetylglc 4-vinylguaiacol adduct	681	477	477	514	nd	x	x	x	x	x
55.7	Mv-3-p-coumglc 4-vinylguaiacol adduct	785	477	477	514	nd	x	x	x	x	x
49.0	Mv-3-glc 4-vinylepi-catechin adduct	805				x	x	x	x	x	x

[a] Delphinidin = Dp; Cy = cyanidin; Pt = petunidin; Pn = peonidin; Mv = malvidin; glc = glucose; p-coumglc = p-coumaroylglucoside; cafglc = caffeoylglucoside; acetylglc = acetylglucoside; C = catechin; GC = gallocatechin; EC = epicatechin; ECG = epigallocatechin; (x) = detected; nd = not detected. Chromatographic conditions: Column C_{18} (150 × 4.6 mm, 5 μm) at 35 °C; binary solvent (A) aqueous solution TFA 0.1 % and (B) MeCN; gradient program: 10% B isocratic for 5 min, from 10 to 15% of B in 15 min, 15% B isocratic for 5 min, from 15 to 18% of B in 5 min, from 18 to 35% of B in 20 min (flow rate of 0.5 mL/min). The MS conditions: positive-ion mode; sheath and auxiliary gas mixture of N_2 and He; sheath gas flow 1.2 L/min; auxiliary gas flow 6 L/min; capillary voltage 4 V; capillary temperature 195 °C; normalized collision energy 45% (Alcade-Eon et al., 2006).

are already present in wine in the first stages of aging and disappear in time, others form as a consequence of long aging.

Anthocyanin-flavan-3-ols derivatives can be characterized by performing MS^n experiments. Fragmentation spectra of (epi)catechin-Pn-3-glu (M^+ at m/z 751), Mv-3-glu-8-ethyl-(epi)catechin (M^+ at m/z 809), and Mv-3-glu-8-vinyl(epi)catechin (M^+ at m/z 805), are reported in Fig. 6.28. Fragmentation schemes proposed for (epi)catechin-Mv-3-glu (M^+ at m/z 781) and A-type Mv-3-glu-(epi)catechin (M^+ at m/z 783), are shown in Fig. 6.29.

Recently, in *Tempranillo* aged wines several oligomeric pigments of the F-A-A$^+$ type were identified and characterized by ESI/MS^n. The compounds are reported in Table 6.20. Possible structures proposed for species M^+ at m/z 1273 are reported in Fig. 6.30 (Alcade-Eon et al., 2007).

6.6   THE MALDI–TOF OF GRAPE PROCYANIDINS

Matrix-assisted laser desorption ionization–time of flight (MALDI–TOF) MS has been used in the characterization of grape procyanidins (Yang and Chien, 2000; Krueger et al., 2000; Vivas et al., 2004). The MALDI–TOF (see Section 1.5) is widely used in the grape and wine proteins analysis (Flamini and De Rosso, 2006). An acidic solution containing an energy-absorbing molecule (matrix) is mixed with the analyte and highly focused laser pulses are directed to the mixture. Proteins are desorbed, ionized, and accelerated by a high electrical potential. The ions arrive at the detector in the order of their increasing m/z ratio. Because of the robustness, tolerance to salt- and detergent-related impurities and the ability to be automated by MALDI–TOF, this technique is regularly used to perform generation of a mass map of proteins after enzymatic digestion (Ashcroft, 2003). An α-cyano-4-hydroxycinnamic acid (CHCA) matrix is commonly used for analysis of peptides and small proteins; sinapinic acid (SA) is used for analysis of higher MW proteins (10–100 kDa). The advantages of MALDI–TOF that it gives are good mass accuracy (0.01%), sensitivity (proteins in femtomole range can be detect), and requires very little sample for analysis.

The LC/MS does not allow separation and identification of oligomers higher than pentamers because the number of diastereoisomers is large and their separation is not possible. By having positive-ion MALDI–TOF in the reflectron mode, determination of (+)-catechin, (−)-epicatechin oligomers, and their galloylated derivatives up to a heptamer in grape seed extracts as sodium adducts [M+Na]$^+$ is possible

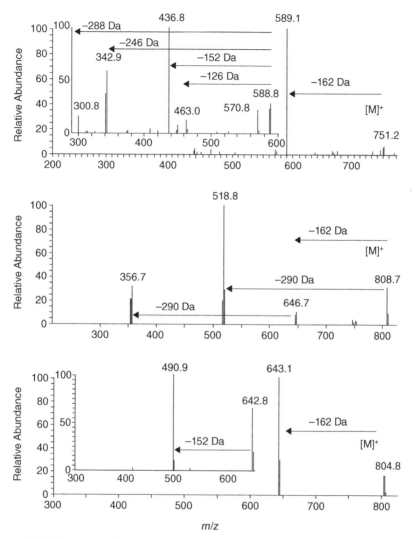

Figure 6.28. Fragmentation spectra of (epi)catechin-Pn-3-glu (M⁺ at *m/z* 751), Mv-3-glu-8-ethyl-(epi)catechin (M⁺ at *m/z* 809) and Mv-3-glu-8-vinyl(epi)catechin (M⁺ at *m/z* 805). (Reprinted from *Journal of Mass Spectrometry*, 41, Pati et al., Simultaneous separation and identification of oligomeric procyanidins and anthocyanin-derived pigments in raw red wine by HPLC-UV-ESI-MSn, p. 867, Copyright © 2006, with permission from John Wiley & Sons, Ltd.)

with resolution >3000. This resolution allows separation of individual ions for different isotope composition, for example, the ion at *m/z* 1177.46 was further resolved into a group of four peaks, as shown in the expanded view of the spectrum in Fig. 6.31 (Yang and Chien, 2000). The positive-ion reflectron mode also allows identification of a series

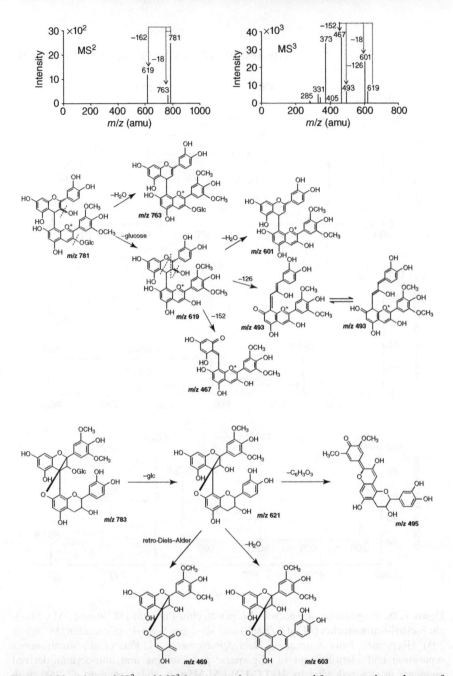

Figure 6.29. Above: MS² and MS³ fragmentation spectra and fragmentation scheme of (epi)catechin-Mv-3-glu (M⁺ at *m/z* 781) (MS conditions: positive-ion mode; source voltage 4.5 kV; capillary voltage 23.5 V; capillary temperature 250 °C; collision energy fragmentation 25% for MS², 30% for MS³, and 35% for MS⁴, Salas et al., 2004). (Reprinted from *Analytica Chimica Acta*, 513, Salas et al., Demonstration of the occurrence of flavanols–anthocyanin adducts in wine and in model solutions, p. 328, Copyright © 2004, with permission from Elsevier.) Below: fragmentation scheme proposed for A-type Mv-3-glu-(epi)catechin (M⁺ at *m/z* 783). (Reprinted from *Rapid Communications in Mass Spectrometry*, 21, Sun et al., High-performance liquid chromatography/electrospray ionization mass spectrometric characterization of new product formed by the reaction between flavanols and malvidin 3-glucoside in the presence of acetaldehyde, p. 2232, Copyright © 2007, with permission from John Wiley & Sons, Ltd.)

TABLE 6.20. Molecular and Fragment Ions of the Flavanol–Anthocyanin–Anthocyanin (F-A-A⁺) Trimers Identified in *Tempranillo* Aged Wines[a]

| | | | m/z | | |
| | | | Fragment Ions | | |
RT (min)	Proposed Identity	[M]⁺	MS²	MS³	MS⁴
7–9	(E)C-DpG-MvG	1245	1083 [M⁺-162]	921 [MS2⁺-162]	
			795 [M⁺-450]	903 [MS2⁺-180]	
			921 [M⁺-324]	657 [MS2⁺-426]	
			903 [M⁺-342]	633 [MS2⁺-450]	
				837 [MS2⁺-246]	
9–15	(E)C-CyG-MvG	1229	1067 [M⁺-162]	905 [MS2⁺-162]	
			904 [M⁺-324]	917 [MS2⁺-150]	
9–16	(E)C-PtG-MvG	1259	1097 [M⁺-162]	935 [MS2⁺-162]	629 [MS3⁺-306]
			971 [M⁺-288]	917 [MS2⁺-180]	
			935 [M⁺-324]	899 [MS2⁺-198]	
			809 [M⁺-450]	671 [MS2⁺-426]	
			1079 [M⁺-180]	747 [MS2⁺-350]	
				971 [MS2⁺-126]	
12–18	(E)C-PnG-MvG	1243	1081 [M⁺-162]	919 [MS2⁺-162]	
			1063 [M⁺-180]		
13–19	(E)C-MvG-MvG	1273	1111 [M⁺-162]	949 [MS2⁺-162]	
			949 [M⁺-324]	931 [MS2⁺-180]	
			661 [M⁺-612]	823 [MS2⁺-288]	
			931 [M⁺-342]	685 [MS2⁺-426]	
			823 [M⁺-450]	661 [MS2⁺-450]	
				737 [MS2⁺-374]	
				913 [MS2⁺-198]	
				535 [MS2⁺-576]	
				331 [MS2⁺-780]	
5–7	(E)GC-DpG-MvG	1261	1099 [M⁺-162]	937 [MS2⁺-162]	
7–9	(E)GC-PtG-MvG	1275	1113 [M⁺-162]	951 [MS2⁺-162]	
				647 [MS2⁺-466]	
9–14	(E)GC-MvG-MvG	1289	1127 [M⁺-162]	965 [MS2⁺-162]	
			965 [M⁺-324]	929 [MS2⁺-198]	
			661 [M⁺-628]	947 [MS2⁺-180]	
			823 [M⁺-466]	823 [MS2⁺-304]	
				865 [MS2⁺-262]	
				661 [MS2⁺-466]	
				535 [MS2⁺-592]	
				839 [MS2⁺-288]	
				467 [MS2⁺-660]	
				331 [MS2⁺-796]	

[a]Fragment ions in order of abundance. RT, LC retention time or range of time for the peaks eluting as a hump; MS2⁺ = major fragment ion obtained in the MS² analysis; MS3⁺ = major fragment ion obtained in the MS³ analysis; Dp = delphinidin; Cy = cyanidin; Pt = petunidin; Pn = peonidin; Mv = malvidin; G = glucose; (E)C = (epi)catechin; (E)GC = (epi)gallocatechin (Alcade-Eon et al., 2007).

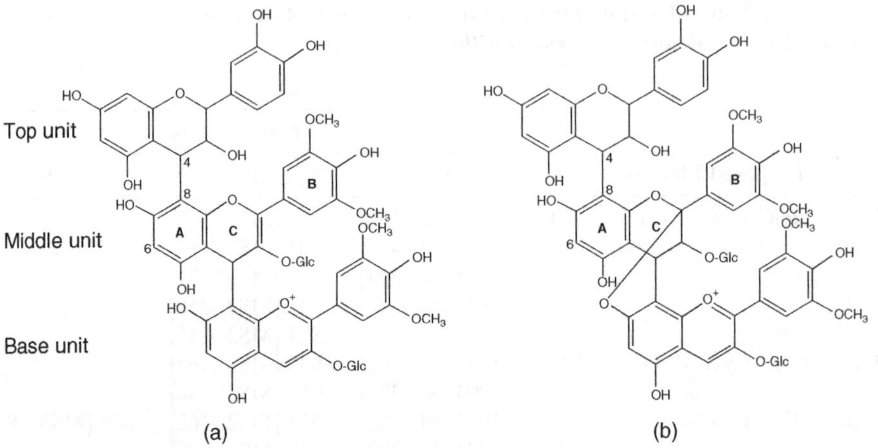

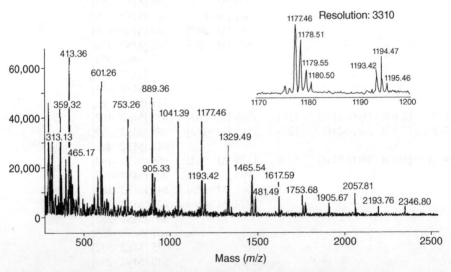

Figure 6.30. Possible structures proposed for M$^+$ at m/z 1273: (a) (E)C-MvG-Mv+G trimer in flavene–flavylium form (B-type linkage); (b) (E)C-MvG-Mv+G trimer in flavan–flavylium form (A-type linkage). (Reprinted from *Journal of Mass Spectrometry*, 42, Alcalde-Eon et al., Identification of dimeric anthocyanins and new oligomeric pigments in red wine by means of HPLC-DAD-ESI/MS^n, p. 744, Copyright © 2007, with permission from John Wiley & Sons, Ltd.)

Figure 6.31. Positive-ion MALDI–TOF reflectron mode mass spectrum of grape seed extract (matrix 2,5-dihydroxybenzoic acid). (Reprinted from *Journal of Agricultural and Food Chemistry*, 48, Yang and Chien, Characterization of grape procyanidins using high-performance liquid chromatography/mass spectrometry and matrix-assisted laser desorption time-of-flight mass spectrometry, p. 3993, Copyright © 2000, with permission from American Chemical Society.)

of compounds with MW 2 mass units lower than those of the above described compounds, corresponding to A-type polycatechins (structure in Fig. 6.4) (Krueger et al., 2000).

The MALDI–TOF–MS in positive-ion linear mode allows us to detect procyanidins oligomers up to nonamers as sodium adducts [M+Na]$^+$ even if they possess lower resolution (Fig. 6.32). The lower sensitivity of the reflectron mode for the large ions is reasonably due to their collisionally induced decomposition occurring in the flight path (Yang and Chien, 2000; Krueger et al., 2000). Procyanidin masses observed and calculated in both reflectron and liner modes are reported in Table 6.21. On the basis of the galloylated structures, an equation was developed to predict the mass distribution of polygalloyl polyflavan-3-ols (PGPF) in grape seed extracts: $290 + 288c + 152g + 23$, where 290 is the MW of the terminal catechin–epicatechin unit, c is the degree of polymerization, g is the number of galloyl esters, and 23 is the atomic weight of Na. This equation provides an easy description of the MS data (Krueger et al., 2000).

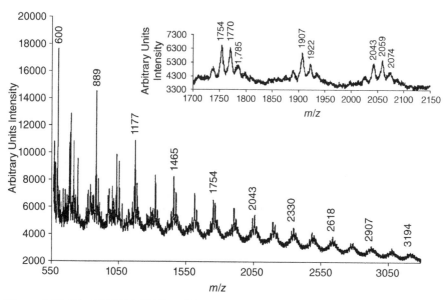

Figure 6.32. The MALDI–TOF positive linear mode mass spectrum of [M+Na]$^+$ procyanidin series from the dimer (*m/z* 600) to the undecamer (*m/z* 3194) (matrix *trans*-3-indoleacrylic acid). Above, the enlarged section of the spectrum with masses representing a polygalloyl polyflavan-3-ols (PGPF) series is shown. (Reprinted from *Journal of Agricultural and Food Chemistry*, 48, Krueger et al., Matrix-assisted laser desorption/ionization time-of-flight mass spectrometry of polygalloyl polyflavan-3-ols in grape seed extract, p. 1666, Copyright © 2000, with permission from American Chemical Society.)

TABLE 6.21. Masses Observed by MALDI–TOF and those Calculated by the Equation $290 + 288c + 152g + 23^a$

Polymer	n° Galloyl Ester	Calculated [M+Na]+	Observed [M+Na][+b]	
			Positive Linear	Positive Reflectron
Dimer	0	601	600	601
	1	753	752	753
	2	905	905	905
Trimer	0	889	889	889
	1	1041	1041	1041
	2	1193	1193	1193
	3	1345	1346	1345
Tetramer	0	1177	1177	1177
	1	1329	1329	1329
	2	1481	1482	1482
	3	1634	1634	1634
	4	1786	1785	1786
Pentamer	0	1466	1465	1466
	1	1618	1618	1618
	2	1770	1770	1770
	3	1922	1922	1922
	4	2074	2074	2074
	5	2226	n.d	n.d
Hexamer	0	1754	1754	1754
	1	1906	1907	1906
	2	2058	2059	2058
	3	2210	2211	n.d
	4	2362	2362	2362
	5	2514	2513	n.d
	6	2666	2667	n.d
Heptamer	0	2042	2043	2042
	1	2194	2195	2194
	2	2346	2346	2346
	3	2398	2499	2499
	4	2651	2651	n.d
	5	2803	2800	n.d
	6	2955	2954	n.d
	7	3107	n.d	n.d
Octamer	0	2330	2330	2330
	1	2483	2483	2483
	2	2635	2635	n.d
	3	2787	2787	n.d
	4	2939	2938	n.d
	5	3091	3090	n.d
	6	3243	n.d	n.d
Nanomer	0	2619	2618	2618
	1	2771	2770	n.d
	2	2923	2923	n.d
	3	3075	3075	n.d
	4	3227	n.d	n.d

TABLE 6.21. (*Continued*)

Polymer	n° Galloyl Ester	Calculated [M+Na]+	Observed [M+Na]+[b]	
			Positive Linear	Positive Reflectron
Decamer	0	2907	2907	n.d
	1	3059	3060	n.d
	2	3211	3212	n.d
	3	3363	n.d	n.d
Undecamer	0	3195	3194	n.d
	1	3347	3349	n.d

[a]290 MW of the terminal catechin unit, c degree of polymerization, g number of galloyl ester, 23 Na atomic mass (Krueger et al., 2000).
[b]not observed = n.d.

The dry grape seed extract is dissolved in acetone or methanol at 2 mg/mL, a 2,5-dihydroxybenzoic (DHB) acid matrix is prepared in tetrahydrofuran (THF) at 20 mg/mL, and the sample and matrix solutions are mixed at a 1:1 (v/v) ratio. Sodium apparently arises from the seeds themselves and only a minute amount of sodium is needed. The use of DHB and water-free solvents, such as anhydrous THF, acetone, or methanol for the sample and matrix preparation was reported to lead to the best analytical conditions in reflectron mode, providing the broadest mass range with the least background noise (Yang and Chien, 2000).

REFERENCES

Abian, J. (1999). The coupling of gas and liquid chromatography with mass spectrometry, *J. Mass Spectrom.*, **34**(3), 157–168.

Alcade-Eon, C., Escribano-Bailón, M.T., Santos-Buelga, C., and Rivas-Gonzalo, J.C. (2006). Changes in the detailed pigment composition of red wine during maturity and ageing. A comprehensive study, *Anal. Chim. Acta*, **563**(1–2), 238–254.

Alcade-Eon, C., Escribano-Bailón, M.T., Santos-Buelga, C., and Rivas-Gonzalo, J.C. (2004). Separation of pyranoanthocyanins from red wine by column chromatography, *Anal. Chim. Acta*, **513**(1), 305–318.

Alcalde-Eon, C., Escribano-Bailón, M.T., Santos-Buelga, C., and Rivas-Gonzalo, J.C. (2007). Identification of dimeric anthocyanins and new oligomeric pigments in red wine by means of HPLC-DAD-ESI/MS[n], *J. Mass Spectrom.*, **42**, 735–748.

Ashcroft, A.E. (2003). Protein and peptide identification: the role of mass spectrometry in proteomics, *Nat. Prod. Rep.*, **20**, 202–215.

Asenstorfer, R.E., Hayasaka, Y., and Jones, G.P. (2001). Isolation and structures of oligomeric wine pigments by bisulfite-mediated ion-exchange chromatography, *J. Agric. Food Chem.*, **49**(12), 5957–5963.

Bravo, M.N., Silva, S., Coelho, A.V., Vilas Boas, L., and Bronze, M.R. (2006). Analysis of phenolic compounds in Muscatel wines produced in Portugal, *Anal. Chim. Acta*, **563**(1–2), 84–92.

Castillo-Muñoz, N., Gómez-Alonso, S., García-Romero, E., and Hermosn-Gutiérrez, I. (2007). Flavonol profiles of Vitis vinifera red grapes and their single-cultivar wines, *J. Agric. Food Chem.*, **55**(3), 992–1002.

Castillo-Muñoz, N., Gómez-Alonso, S., García-Romero, E., Gómez, M.U., Velders, A.H., and Hermosin-Gutiérrez, I. (2009). Flavonol 3-*O*-Glycosides Series of Vitis vinifers Cv, Petit Verdot Red Wine Grapes, *J. Agric. Food Chem.*, **57**, 209–219.

Cheynier, V. and Rigaud, J. (1986). HPLC separation and characterization of flavonols in the skin of Vitis vinifera var. Cinsault, *Am. J. Enol. Vitic.*, **37**(4), 248–252.

Cheynier, V., Doco, T., Fulcrand, H., Guyot, S., Le Roux, E., Souquet, J.M., Rigaud, J., and Moutounet, M. (1997). ESI–MS analysis of polyphenolic oligomers and polymers, *Analusis Magazine*, **25**(8), 32–37.

de Freitas, V.A.P., Glories, Y., Bourgeois, G., and Virty, C. (1998). Characterisation of oligomeric and polymeric procyanidins from grape seeds by liquid secondary ion mass spectrometry, *Phytochemistry*, **49**(5), 1435–1441.

de Hoffmann, E. (1996). Tandem Mass Spectrometry: a Primer, *J. Mass. Spectrom.*, **31**(2), 125–137.

De Rosso, M., Panighel, A., Dalla Vedova, A., Stella, L., and Flamini, R. (2009). Changes in Chemical Composition of a Red Wine Aged in Acacia, Cherry, Chestnut, Mulberry and Oak Wood Barrels. *J. Agric. Food Chem.*, **57**(5), 1915–1920.

Favretto, D. and Flamini, R. (2000). Application of electrospray ionization mass spectrometry to the study of grape anthocyanins, *Am. J. Enol. Vitic.*, **51**(1), 55–64.

Flamini, R. (2003). Mass spectrometry in grape and wine chemistry. Part I: Polyphenols, *Mass. Spectrom. Rev.*, **22**(4), 218–250.

Flamini, R. and De Rosso, M. (2006). Mass Spectrometry in the Analysis of Grape and Wine Proteins, *Exp. Rev. Proteom.*, **3**, 321–331.

Flamini, R., Dalla Vedova, A., Cancian, D., Panighel, A., and De Rosso, M. (2007). GC/MS-Positive Ion Chemical Ionization and MS/MS study of volatile benzene compounds in five different woods used in barrel-making, *J. Mass Spectrom.*, **42**, 641–646.

Fulcrand, H., Benabdeljalil, C., Rigaud, J., Cheynier, V., and Moutounet, M. (1998). A new class of wine pigments generated by reaction between pyruvic acid and grape anthocyanins, *Phytochemistry*, **47**(7), 1401–1407.

Fulcrand, H., Remy, S., Souquet, J.M., Cheynier, V., and Moutounet, M. (1999). Study of wine tannin oligomers by on-line liquid chromatography electrospray ionization mass spectrometry, *J. Agric. Food Chem.*, **47**(3), 1023–1028.

Gabetta, B., Fuzzati, N., Griffini, A., Lolla, E., Pace, R., Ruffilli, T., and Peterlongo, F. (2000). Characterization of proanthocyanidins from grape seeds, *Fitoterapia*, **71**(2), 162–175.

Gao, L., Girard, B., Mazza, G., and Reynolds, A.G. (1997). Changes in anthocyanins and color characteristics of Pinot Noir during different vinification processes, *J. Agric. Food Chem.*, **45**, 2003–2008.

Gómez-Ariza, J.L., García-Barrera, T., and Lorenzo, F. (2006). Anthocyanins profile as fingerprint of wines using atmospheric pressure photoionisation coupled to quadrupole time-of-flight mass spectrometry, *Anal. Chim. Acta*, **570**(1), 101–108.

González-Manzano, S., Santos-Buelga, C., Pérez-Alonso, J.J., Rivas-Gonzalo, J.C., and Escribano-Bailón, M.T. (2006). Characterization of the mean degree of polymerization of proanthocyanidins in red wines using liquid chromatography-mass spectrometry (LC–MS), *J. Agric. Food Chem.*, **54**(12), 4326–4332.

Hayasaka, Y. and Asenstorfer, R.E. (2002). Screening for potential pigments derived from anthocyanins in red wine using nanoelectrospray tandem mass spectrometry, *J. Agric. Food Chem.*, **50**(4), 756–761.

Hayasaka, Y., Waters, E.J., Cheynier, V., Herderich, M.J., and Vidal, S. (2003). Characterization of proanthocyanidins in grape seeds using electrospray mass spectrometry, *Rapid Commun. Mass Spectrom.*, **17**(1), 9–16.

Hong, V. and Wrolstad, R.E. (1990). Characterization of anthocyanin-containing colorants and fruit juices by HPLC/photodiode array detection, *J. Agric. Food Chem.*, **38**(3), 698–708.

Košir, I.J., Lapornik, B., Andrenšek, S., Wondra, A.G., Vrhovšek, U., and Kidrič, J. (2004). Identification of anthocyanins in wines by liquid chromatography, liquid chromatography-mass spectrometry and nuclear magnetic resonance, *Anal. Chim. Acta*, **513**(1), 277–282.

Krueger, C.G., Dopke, N.C., Treichel, P.M., Folts J., and Reed, J.D. (2000). Matrix-assisted laser desorption/ionization time-of-flight mass spectrometry of polygalloyl polyflavan-3-ols in grape seed extract, *J. Agric. Food Chem.*, **48**(5), 1663–1667.

Lazarus, S.A., Adamson, G.E., Hammerstone, J.F., and Schmitz, H.H. (1999). High performance liquid chromatography/mass spectrometry analysis of proanthocyanidins in foods and beverages, *J. Agric. Food Chem.*, **47**(9), 3693–3701.

Lee, J.H., Johnson, J.V., and Talcott, S.T. (2005). Identification of ellagic acid conjugates and other polyphenolics in Muscadine grapes by HPLC–ESI–MS, *J. Agric. Food Chem.*, **53**(15), 6003–6010.

Li, H.-J. and Deinzer, M.L. (2007). Tandem Mass Spectrometry for Sequencing Proanthocyanidins, *Anal. Chem.*, **79**(4), 1739–1748.

Mazzuca, P., Ferranti, P., Picariello, G., Chianese, L., and Addeo, F. (2005). Mass spectrometry in the study of anthocyanins and their derivatives: differentiation of Vitis vinifera and hybrid grapes by liquid chromatography/ electrospray ionization mass spectrometry and tandem mass spectrometry, *J. Mass Spectrom.*, **40**: 83–90.

Monagas, M., Núñez, V., Bartolomé, B., and Gómez-Cordovés, C. (2003). Anthocyanin derived pigments in Graciano, Tempranillo, and Cabernet Souvignon wines produced in Spain, *Am. J. Enol. Vitic.*, **54**(3), 163–169.

Monagas, M., Suárez, R., Gómez-Cordovés, C., and Bartolomé, B. (2005). Simultaneous determination of nonanthocyanin phenolic compounds in red wines by HPLCDAD/ESI-MS, *Am. J. Enol. Vitic.*, **56**(2), 139–147.

Niessen, W.M.A. and Tinke, A.P. (1995). Liquid Chromatography-Mass Spectrometry. General principles and instrumentation, *J. Chromatogr. A*, **703**(1–2), 37–57.

Núñez, V., Gómez-Cordovés, C., Bartolomé, B., Hong, Y.J., and Mitchell, A.E. (2006). Non-galloylated and galloylated proanthocyanidin oligomers in grape seeds from Vitis vinifera, L. cv. Graciano, Tempranillo and Cabernet Sauvignon, *J. Sci. Food Agr.*, **86**(6), 915–921.

Pati, S., Losito, I., Gambacorta, G., La Notte, E., Palmisano, F., and Zambonin, P.G. (2006). Simultaneous separation and identification of oligomeric procyanidins and anthocyanin-derived pigments in raw red wine by HPLC-UV-ESI-MSn, *J. Mass Spectrom.*, **41**(7), 861–871.

Pérez-Magariño, S., Revilla, I., Gonzáles-SanJosé, M.L., and Beltrán, S. (1999). Various applications of liquid chromatography-mass spectrometry to the analysis of phenolic compounds. *J. Chromatogr. A*, **847**(1–2), 75–81.

Revilla, E., Ryan, J.M., and Martín-Ortega, G. (1998). Comparison of several procedures used for the extraction of anthocyanins from red grapes, *J. Agric. Food Chem.*, **46**(11), 4592–4597.

Salas, E., Atanasova, V., Poncet-Legrand, C., Meudec, E., Mazauric, J.P., and Cheynier, V. (2004). Demonstration of the occurrence of flavanol–anthocyanin adducts in wine and in model solutions, *Anal. Chim. Acta*, **513**(1), 325–332.

Somers, T. (1971). The polymeric nature of wine pigments. *Phytochemistry*, **10**(9), 2175–2186.

Stella, L., De Rosso, M., Panighel, A., Dalla Vedova, A., Flamini, R., and Traldi, P. (2008). Collisionally induced fragmentation of [M–H]⁻ species of resveratrol and piceatannol investigated by deuterium labelling and accurate mass measurements, *Rapid Comm. Mass Spectrom.*, **22**, 3867–3872.

Sun, B., Leandro, M.C., de Freitas, V., and Spranger, M.I. (2006). Fractionation of red wine polyphenols by solid phase extraction and liquid chromatography, *J. Chromatogr. A*, **1128**(1–2), 27–38.

Sun, B., Reis Santos, C.P., Leandro, M.C., De Freitas, V., and Spranger, M.I. (2007). High-performance liquid chromatography/electrospray ionization mass spectrometric characterization of new product formed by the reaction between flavanols and malvidin 3-glucoside in the presence of acetaldehyde, *Rapid Comm. Mass Spectrom.*, **21**, 2227–2236.

Timberlake, C.F. and Bridle, P. (1976). Interactions between anthocyanins, phenolic compounds, and acetaldehyde and their significance in red wines. *Am. J. Enol. Vitic.*, **27**(3), 97–105.

Vidal, S., Francis, L., Guyot, S., Marnet, N., Kwiatkowski, M., Gawel, R., Cheynier, V., and Waters, E.J. (2003). The mouth-feel properties of grape and apple proanthocyanidins in a wine-like medium, *J. Sci. Food Agric.*, **83**(6), 564–573.

Vidal, S., Meudec, E., Cheynier, V., Skouroumounis, G., and Hayasaka, Y. (2004). Mass spectrometric evidence for the existence of oligomeric anthocyanins in grape skins, *J. Agric. Food Chem.*, **52**(23), 7144–7151.

Vivas, N., Nonier, M.F., and Vivas De Gaulejac, N. (2004). Structural characterization and analytical differentiation of grape seeds, skins, steams and Quebracho tannins, *Bull. O.I.V.*, **883–884**, 643–659.

Wang, H., Race, E.J., and Shrikhande, A.J. (2003). Characterization of Anthocyanins in Grape Juices by Ion Trap Liquid Chromatography–Mass Spectrometry, *J. Agric. Food Chem.*, **51**, 1839–1844.

Wu, Q., Wang, M., and Simon, J.E. (2005). Determination of proanthocyanidins in fresh grapes and grape products using liquid chromatography with mass spectrometric detection, *Rapid Commun. Mass Spectrom.*, **19**(14), 2062–2068.

Yang, Y. and Chien, M. (2000). Characterization of grape procyanidins using high-performance liquid chromatography/mass spectrometry and matrix-assisted laser desorption time-of-flight mass spectrometry, *J. Agric. Food Chem.*, **48**(9), 3990–3996.

7

COMPOUNDS RELEASED IN WINE FROM WOOD

7.1 INTRODUCTION

Aging wine and spirits in wooden barrels is an industrial process widely used in enology. It is carried out to stabilize the color, improve limpidity, and to enrich the sensorial characteristics of the product. Oxygen permeation through the wood promotes redox processes, formation of new pigments with consequent stabilization of the color, and loss of astringency of wines (Ribéreau-Gayon et al., 1998). Aging in barrels is also used in the production of spirits, such as Armagnac, Whisky, Brandy, and Grappa (Puech, 1981; MacNamara et al., 2001; Delgado et al., 1990; Mattivi et al., 1989a; 1989b). Oak is the main wood used in making barrels for oenology, but to a lesser extent Chestnut and Cherry are also used. More rarely, Acacia and Mulberry, can be used (Salagoity-Auguste et al., 1986). Qualitative and semiquantitative profiles of volatile compounds identified in 50% hydroalcoholic extracts of these types of wood not subjected to any toasting treatments are reported in Table 7.1 (De Rosso et al., 2008; 2009).

Other compounds released from wood into the wine belong to classes of ellagitannins, lactones, coumarins, polysaccharides, hydrocarbons and fatty acids, terpenes and norisoprenoids, steroids, carotenoids,

Mass Spectrometry in Grape and Wine Chemistry, by Riccardo Flamini and Pietro Traldi
Copyright © 2010 John Wiley & Sons, Inc.

and furan compounds (Puech et al., 1999; Pérez-Coello et al., 1999; Guichard et al., 1995; Feuillat et al., 1997; Masson et al., 1996; 2000; Matricardi and Waterhouse, 1999; Hale et al., 1999; Sauvageot and Feuillat, 1999; Ibern-Gómez et al., 2001; Chatonnet et al., 1992). Main compounds characterized by sensorial proprieties are vanillin (45)

TABLE 7.1. The GC/MS Qualitative and Semiquantitative Data of Volatile Compounds Identified in 50% Hydroalcohol Extract of Different Types of Wood Used in Making Barrels for Wine and Spirits Aging[a]

Compound	Acacia	Chestnut	Cherry	Mulberry	Oak
Aldehydes and Ketones					
Furfural	*	*			*
Benzaldehyde			**		
Methylbenzaldehyde			*		
Hydroxybenzaldehyde	*		*		
Anisaldehyde	*				
Cinnamaldehyde			*		
Vanillin	**	***	*	**	**
Syringaldehyde	***	***	**		***
Coniferaldehyde	**	**		*	**
Acetophenone	*	**	**	*	*
Benzophenone				*	*
Acetovanillone	**	**			
3-Methoxyacetovanillone	**				
2-Butanone-4-guaiacol				**	
2,4-Dihydroxybenzaldeide	***			**	
3-Buten-2-one-4-phenyl			**		
Alcohols and Phenols					
α-Terpineol	*	**	*	*	*
3-Oxo-α-ionol		**	**		**
β-Phenylethanol	*	*	*	*	*
Benzenepropanol			*		
α-Methylbenzenepropanol			**		
Coniferyl alcohol					*
Benzotriazole	*	**	*	*	**
4-Methylphenol				*	
4-Ethylphenol			*		
4-Methylguaiacol				*	
Ethylguaiacol				*	
Vinylguaiacol		*		tr	*
Eugenol		***	tr	tr	***
Methoxyeugenol	*	***	*		**
3-Methoxyphenol				*	
Dimethoxyphenol				**	
Trimethoxyphenol	**		**	**	*
1,2,3-Trimethoxybenzene	**				

TABLE 7.1. (*Continued*)

Compound	Acacia	Chestnut	Cherry	Mulberry	Oak
	Acids and Esters				
Ethyl benzoate			*		
2,5-Dihydroxy ethyl benzoate	**				
Methyl salicylate					*
trans-β-Methyl-γ-octalactone					***
cis-β-Methyl-γ-octalactone					***
Homovanillic acid		**	*		**
Capronic acid	**	**	*	*	**
Caprylic acid	*	*	*	*	**
Lauric acid	*	**	*	*	**
Myristic acid	**	**	*	**	**
Pentadecanoic acid	**	**	*	**	*
Palmitic acid	***	***	**	***	***
Margaric acid	**	*	*	**	*
Stearic acid	**	**	**	**	**
Oleic acid	*	**	*	**	**
Linoleic acid	**	**	*	***	**
Linolenic acid	*			**	

[a](Not subjected to any toasting treatment). Data expressed as µg/g of 1-heptanol (internal standard).
*0.1–0.9 µg/g wood;
**1–10 µg/g wood;
***>10 µg/g wood; tr, trace (De Rosso et al., 2008).

(vanilla note; sensory threshold 0.3 ppm) and eugenol (35) (clove, spicy; sensory threshold 0.5 ppm; Boidron et al., 1988). Toasting of wood made for making barrels induces formation of a great number of volatile and odoriferous compounds. In general, furan and pyran derivatives formed with heating wood are characterized from a toasty caramel aroma (Cutzach et al., 1997; Chatonnet, 1999). Among the compounds formed with toasting were (**1**) 3,5-dihydroxy-2-methyl-4*H*-pyran-4-one, (**2**) 3-hydroxy-2-methyl-4*H*-pyran-4-one or maltol, (**3**) 2,3-dihydro-3,5-dihydroxy-6-methyl-4*H*-pyran-4-one (DDMP), (**4**) 4-hydroxy-2,5-dimethylfuran-3(2*H*)-one (furaneol), (**5**) 2,3-dihydro-5-hydroxy-6-methyl-4*H*-pyran-4-one (dihydromaltol), (**6**) 2-hydroxy-3-methyl-2-cyclopenten-1-one (or cyclotene) and 5-(acetoxymethyl)furfural. The structures of compounds **1–6** are shown in Fig. 7.1. Formation of these molecules in the presence of proline infers that Maillard reactions occur. The GC/MS–EI (70 eV) mass spectra of some of them are reported in the Table 7.2. A complete list of compounds identified in toasted oak wood extracts is reported in Table 7.3.

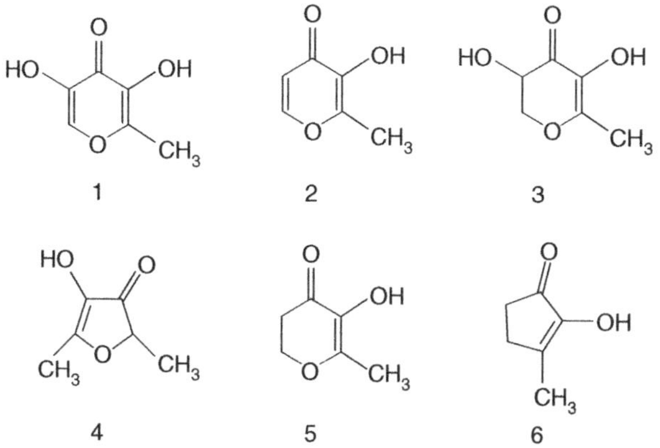

Figure 7.1. Structures of volatile compounds characterized from "toasty caramel" aroma released in wine from toasted woods during aging. (1) 3,5-dihydroxy-2-methyl-4*H*-pyran-4-one; (2) 3-hydroxy-2-methyl-4*H*-pyran-4-one; (3) 2,3-dihydro-3,5-dihydroxy-6-methyl-4*H*-pyran-4-one (DDMP); (4) 4-hydroxy-2,5-dimethylfuran-3 (2*H*)-one (furaneol); (5) 2,3-dihydro-5-hydroxy-6-methyl-4*H*-pyran-4-one (dihydromaltol); (6) 2-hydroxy-3-methyl-2-cyclopenten-1-one (or cycloteme) (Cutzach et al., 1997).

TABLE 7.2. Principal Fragments Observed in the GC/MS–EI (70 eV) Mass Spectra of Compounds in Fig. 7.1 With Their Relative Abundance[a]

Compound	(*m/z*)
3,5-Dihydroxy-2-methyl-4*H*-pyran-4-one (1)	142(100), 55(37), 68(30), 43(29), 85(18), 96(11)
2,3-Dihydro-3,5-dihydroxy-6-methyl-4*H*-pyran-4-one (3)	43(100), 144(36), 101(32), 73(21), 55(18)
4-Hydroxy-2,5-dimethylfuran-3(2*H*)-one (4)	43(100), 128(71), 57(64), 85(21), 55(21)
2,3-Dihydro-5-hydroxy-6-methyl-4*H*-pyran-4-one (5)	43(100), 128(76), 72(26), 57(24), 85(8)

[a]Cutzach et al., 1997.

Some compounds from wood may induce defects to the wine. Carbonyl compounds, such as (*E*)-2-nonenal, (*E*)-2-octenal, 3-octen-1-one, and 1-decanal are responsible for the sawdust smell sometimes found in the wine after ageing in new 225-L oak wood barrels (barriques); (*E*)-2-nonenal is reputed as being mainly responsible for the sawdust smell of wine (Chatonnet and Dubourdieu, 1998).

TABLE 7.3. Compounds Identified in Toasted Oak Wood Extracts[a]

Aliphatic Compounds

Acetic acid
1,2-Propanediol
3-Methoxy-1,2-propanediol
Ethyl 2,3-dihydroxy-butanedioate
Propanoic acid
Butanoic acid
Hexanoic acid
Heptanoic acid
Octanoic acid
Decanoic acid
Dodecanoic acid
Dodecanol
2-Dodecyloxyethanol
Ethyl linoleate
Ethyl oleate
Hexadecanoic acid
Hexadecanol
Ethyl hexadecanoate
Hexyl hexanedioate
Dioctyl hexanedioate
Tetradecanoic acid
Tetradecanol
Tridecane
2-Hydroxy-1-methylcyclopenten-3-one

Heterocycles

1-(2-Furanyl)-ethanone
1,3-Benzothiazole
1H-Pyrrole-2-carboxaldehyde
2,5-Diformylfuran
2,5-Dimethylpyrazine
2,5-Furandicarboxaldehyde
2,6-Dimethylpyrazine
2-Acetylfuran (2-furanyl-1-ethanone)
2-Ethyl-3-hydroxy-4H-pyran-4-one (ethylmaltol)
2-Furancarboxaldehyde (furfural)
2-Furancarboxylic acid
2-Furanmethanol
2H-Pyran-2-one
2-Pentylfurane
3,4-Dimethyl-2(5H)-furanone (2,3-dimethyl-4-hydroxy-2-butenoic lactone)
3H-Furan-2-one
3-Hydroxy-2-methyl-4H-pyran-4-one (maltol)
5-Hydroxy-2-methyl-4H-pyran-4-one (allomaltol)
5-Hydroxymethylfurancarboxyaldehyde (5-hydroxymethylfurfural)
5-Methylfurancarboxyaldehyde (5-methylfurfural)

TABLE 7.3. (*Continued*)

Heterocycles

7-Hydroxy-6-methoxy-2*H*-benzopyranone
cis-β-Methyl-γ-octalactone
Dihydro-2(3*H*)furanone (γ-butyrolactone)
Dihydromaltol (2,3-dihydro-5-hydroxy-6-methyl-4*H*-pyran-4-one)
Furaneol (2,5-dimethyl-4-hydroxy-(2*H*)-furan-3-one)
Furylhydroxymethylketone [1-(2-furanyl)-2-hydroxy ethanone]
Hydroxymaltol (3,5-dihydroxy-2-methyl-4*H*-pyran-4-one)
Methyl-4(*H*)-pyran-4-one (2-methyl-4*H*-pyran-4-one)
Scopoletin (7-hydroxy-6-methoxy-2*H*-1-benzopyran-2-one)
trans-β-Methyl-γ-octalactone
γ-Ethoxy-butyrolactone

Benzene Compounds

1-(4-Hydroxy-3-methoxyphenyl)propanone (propiovanillone)
1-(4-Hydroxy-3-methoxyphenyl)-2-propanone (vanillyl propan-2-one; HMPP)
1-(4-Hydroxy-3-methoxyphenyl)butanone (butyrovanillone)
1-(4-Hydroxy-3-methoxyphenyl)-2-butanone (HMPB)
1-(4-Hydroxy-3,5-dimethoxyphenyl)butanone (butyrosyringone)
1-(4-Hydroxy-3,5-dimethoxyphenyl)-2- or 3-butanone (syringyl 2- or 3-butanone)
1-(4-Hydroxy-3,5-dimethoxyphenyl)propanone (propiosyringone)
1-(4-Hydroxy-3,5-dimethoxyphenyl)-2-propanone (syiringyl propan-2-one)
cis- or *trans*-2,6-Dimethoxy-4-(1-propenyl)phenol (*cis*- or *trans*-propenylsyringol)
2-(4-Hydroxy-3-methoxyphenyl)acetaldehyde (HMPA)
2-(4-Hydroxy-3,5-dimethoxyphenyl)acetaldehyde (HDMPA)
1-(4-Hydroxy-3-methoxyphenyl)ethanone (acetovanillone)
1-(4-Hydroxy-3,5-dimethoxyphenyl)ethanone (acetosyringone)
1-(4-Hydroxy-3-methoxyphenyl)propanal (vanillyl propanal)
1-(4-Hydroxy-3,5-dimethoxyphenyl)propanal (syringyl propanal)
1-(4-Hydroxy-3-methoxyphenyl)ethanal (homovanillin)
1,2,3-Trimethoxyphenyl-5-(1-propenyl)
1,2,3-Trimethoxyphenyl-5-(2-propenyl)
1,2,3-Trimethoxyphenyl-5-propenoic acid methyl ester
1,2-Dihydroxy-3,4-dimethoxybenzene
1,2-Dimethoxyphenyl-4-(1-propenyl)
2,6-Dimethoxy-4-(2-propenyl)phenol (allylsyringol)
2,6-Dimethoxyphenol (syringol)
2-Methoxy-4-(2-propenyl)phenol (eugenol)
2-Methoxyphenol (guaiacol)
4-Ethyl-2-methoxyphenol (ethylguaiacol)
4-Methyl-2-methoxyphenol (methyl guaiacol)
4-Propyl-2-methoxyphenol (4-propylguaiacol)
4-Vinyl-2-methoxyphenol (vinylguaiacol)
2-Phenoxyethanol
2-Phenylethanol
4-Ethylphenol
3,5-Dimethoxy-4-hydroxycinnamaldehyde (sinapic aldehyde)

TABLE 7.3. (*Continued*)

Benzene Compounds

3,5-Dimethoxy-4-hydroxycinnamyl alcohol (sinapic alcohol)
3,5-Dimethoxy-4-hydroxy-dihydro cinnamyl alcohol (dihydrosynapic alcohol)
3-Methoxy-4-hydroxycinnamaldehyde (coniferaldehyde)
3-Methoxy-4-hydroxycinnamyl alcohol (coniferyl alcohol)
4-(Ethoxymethyl)-2-methoxyphenol
4,5-Dimethoxyphenyl-2-(propenyl)
4-Hydroxy-3,5-dimethoxy-benzeneacetic acid, methyl ester (methyl homosyringate)
4-Hydroxy-3,5-dimethoxy-benzoic acid (syringic acid)
4-Hydroxy-3,5-dimethoxy-benzoic acid ethyl ester (ethyl syringate)
4-Hydroxy-3,5-dimethoxy-benzoic acid methyl ester (methyl syringate)
4-Hydroxy-3,5-dimethoxybenzaldehyde (syringaldehyde)
4-Hydroxy-3-methoxy-benzeneacetic acid methyl ester (methyl homovanillate)
4-Hydroxy-3-methoxy-benzoic acid (vanillic acid)
4-Hydroxy-3-methoxy-benzoic acid ethyl ester (ethyl vanillate)
4-Hydroxy-3-methoxy-benzoic acid ethyl ether (vanillyl ethyl ether)
4-Hydroxy-3-methoxy-benzoic acid methyl ester (methyl vanillate)
4-Hydroxy-3-methoxy-benzoic acid methyl ether (vanillyl methyl ether)
4-Methyl-2-(2'-methyl-1-propenyl)-phenol
4-Methyl-2,6-dimethoxyphenol (methylsyringol)
Benzaldehyde
cis- or *trans*-2-Methoxy-4-(1-propenyl)phenol (*cis*- or *trans*-isoeugenol)
Dimethoxy-4-(2-propenyl)-phenol
Dimethyl-aminobenzaldahyde (4-(dimethylamino)benzaldehyde)
Hydroxybenzaldehyde
m-Cresol (3-methylphenol)
o-Cresol (2-methylphenol)
p-Cresol (4-methylphenol)
Phenol
Phenylacetaldehyde
Phenylmethanol
Sinapalcohol
Vanillin

[a]Cutzach et al., 1997; Chatonnet et al., 1999; Pérez-Coello et al., 1999; Cadahía et al., 2003; Vichi et al., 2007.

7.2 THE GC/MS OF WOOD VOLATILE COMPOUNDS

Sample preparation for GC/MS analysis of volatile compounds in wines and extracts was usually performed by liquid–liquid extraction with dichloromethane (Cutzach et al., 1997; Pérez-Coello et al., 1999; Cadahía et al., 2003). Direct extraction of volatiles from the wood by headspace (HS) solid-phase microextraction (SPME) using a polydimethylsiloxane (PDMS) fiber allowed to analyze compounds,

TABLE 7.4. The HS–SPME and GC/MS Conditions Used for Analysis of Wood Volatile Compounds in Wine and Wood Extracts[a]

Sample volume	1-g dry wood ground for 2 min
Vial volume	30 mL
SPME fiber	PDMS 100-µm coating thickness, 1 cm length
Sample heating	80 °C for 30 min
Extraction temperature	room temperature
Extraction time	30 min
Desorption temperature	250 °C
Desorption time	5 min
GC column	Poly(ethylene)glycol (PEG) bound-phase fused-silica capillary (50 m × 0.25 mm i.d.; 0.25-µm film thickness)
Injection	Splitless
Oven program	40 °C Isotherm for 5 min, 3 °C/min to 230 °C, isotherm 25 min

[a]Chatonnet et al., 1999.

such as acetic acid, furfural, 5-methyl-2-furfural, guaiacol, *cis*- and *trans*-β-methyl-γ-octalactone, 4-methyguaiacol, phenol, eugenol, and vanillin (Chatonnet et al., 1999). The HS–SPME and GC/MS conditions are reported in Table 7.4. Figure 7.2 shows the chromatogram relative to analysis of oak wood extracted with a model wine solution (12% ethanol and pH 3.5) performed using the GC/MS conditions reported in Table 7.4.

7.3 THE GC/PICI–MS/MS OF WOOD VOLATILE PHENOLS AND BENZENE ALDEHYDES IN WINE

7.3.1 Sample Preparation

A 30-mL wine sample is adjusted to 90 mL with water and added to 200 µL of 1-heptanol ethanolic solution at concentration 185 mg/L as internal standard. A volume of 45 mL of this solution is passed through a 1-g C_{18} Sep-Pak cartridge previously washed with dichloromethane and activated by successive passages of methanol and water. After the sample was passed through, the stationary phase is washed with 10 mL of water to remove salts and more polar compounds. Analytes are recovered with 6 mL of dichloromethane. The organic phase is dried over Na_2SO_4, filtered, and the volume is reduced to ~300 µL under a nitrogen flow before analysis (Flamini et al., 2007).

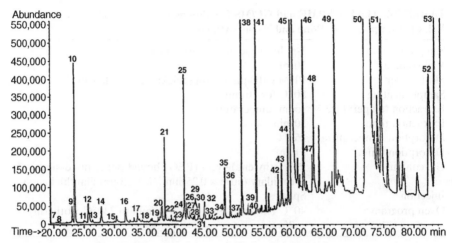

Figure 7.2. The GC/MS SCAN chromatogram of volatile compounds in a toasted oak wood extract. **(7)** 2,5-dimethylpyrazine; **(8)** 2,6-dimethylpyrazine; **(9)** acetic acid; **(10)** furfural; **(11)** furanyl-1-ethanone; **(12)** benzaldehyde; **(13)** propionic acid; **(14)** 5-methylfurfural; **(15)** butyrolactone; **(16)** hydroxybenzaldehyde; **(17)** 3,4-dimethylfuranone-2(5*H*); **(18)** 2(3*H*)-furanone; **(19)** cycloten; **(20)** hexanoic acid; **(21)** guaiacol; **(22)** *trans*-methyloctalactone; **(23)** 2-phenylethanol; **(24)** benzothiazole; **(25)** *cis*-methyloctalactone-4-methylguaiacol; **(26)** maltol; **(27)** 2,5-diformylfuran; **(28)** *o*-cresol; **(29)** phenol; **(30)** 4-ethylguaiacol; **(31)** 1*H*-pyrolcarboxaldehyde; **(32)** octanoic acid; **(33)** *p*-cresol; **(34)** *m*-cresol; **(35)** eugenol; **(36)** isomaltol; **(37)** 4-vinylguaiacol; **(38)** syringol; **(39)** decanoic acid; **(40)** isoeugenol; **(41)** 4-methylsyringol; **(42)** dodecanoic acid; **(43)** 5-hydroxymethylfurfural; **(44)** 4-allylsyringol; **(45)** vanillin; **(46)** acetovanillon; **(47)** tetradecanoic acid; **(48)** propiovanillon; **(49)** butyrovanillon; **(50)** syringaldehyde; **(51)** acetosyringon; **(52)** propiosyringon; **(53)** coniferaldehyde. (Reproduced from *Journal of Agricultural and Food Chemistry*, 1999, 47, p. 4311, Chatonnet et al., with permission of American Chemical Society).

7.3.2   The GC/MS Analysis

In general, PICI using methane as a reagent gas yields a high yield of protonated molecular ions of volatile phenols (Flamini and Dalla Vedova, 2004). Figure 7.3 shows the reconstructed ion chromatogram of [M+H]⁺ ions in the analysis of an oak wood extract. Sixteen benzoic and cinnamic volatile compounds reported in Table 7.5 were identified in the 50% hydroalcoholic extracts of non-toasted Oak, Acacia, Chestnut, Cherry, and Mulberry woods used for making barrels. The CID experiments of the [M+H]⁺ ions using He as a collision gas showed the principal fragments reported in Table 7.6. The fragmentation pattern of cinnamaldehyde **(57)**, coniferaldehyde **(53)** and sinapinaldehyde **(58)** are shown in Fig. 7.4; fragmentations of eugenol **(35)** and methoxyeugenol **(55)** are reported in Fig. 7.5.

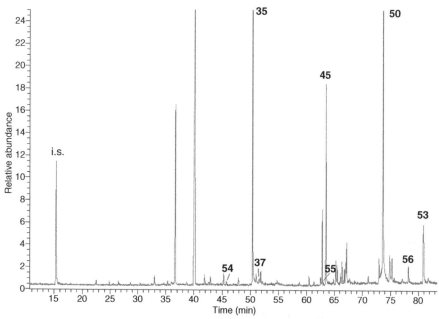

Figure 7.3. Analysis of a 50% ethanol untoasted oak wood extract: GC/PICI reconstructed ion chromatogram of benzene compounds $[M+H]^+$ signals. (**54**) anisaldehyde (*m/z* 137); (**35**) eugenol (*m/z* 165); (**37**) vinylguaiacol (*m/z* 151); (**45**) vanillin (*m/z* 153); (**55**) methoxyeugenol (*m/z* 195); (**50**) syringaldehyde (*m/z* 183); (**56**) trimethoxyphenol (*m/z* 185); (**53**) coniferaldehyde (*m/z* 179). Internal standard-i.s. 1-heptanol (*m/z* 55). Analytical conditions: PEG fused silica capillary column (30 m × 0.25 mm i.d.; df 0.25 µm); injection port 240 °C; volume injected 1 µL (splitless); program oven temperature: 3 min at 70 °C, 2 °C/min to 160 °C, 3 °C/min to 230 °C, 25 min at 230 °C; transfer line temperature 280 °C; carrier gas He at constant flow 1.3 mL/min. (Reprinted from *Journal of Mass Spectrometry* 42, Flamini et al., 2007, GC/MS–Positive Ion Chemical Ionization and MS/MS study of volatile benzene compounds in five different woods used in barrel-making, pp. 641–646, with permission from John Wiley & Sons, Ltd.)

TABLE 7.5. Volatile Benzene Compounds Identified in 50%-Ethanol Solution Wood Extracts of Oak, Acacia, Chestnut, Cherry, and Mulberry Used for Making Barrels

Compound	*m/z* [M + H]⁺	µg/g Wood				
		Acacia	Chestnut	Cherry	Mulberry	Oak
Benzaldehyde[a]	107			2.25		
Methylbenzaldehyde[a]	121			1.05		
Anisaldehyde	137	0.81	0.05			0.01
Hydroxybenzaldehyde	123	1.13		1.12	0.33	0.59
Cinnamaldehyde	133	0.01		0.05		
Coniferaldehyde[b]	179	0.17	0.25	0.03	0.25	0.28
Coniferyl alcohol[b]	181					0.02

TABLE 7.5. (*Continued*)

Compound	m/z [M + H]$^+$	Acacia	Chestnut	Cherry	Mulberry	Oak
		\multicolumn{5}{c}{µg/g Wood}				
Eugenol	165		0.73			2.01
Methoxyeugenol	195		0.20			0.01
Vinylguaiacol^c	151					0.07
Vanillin	153	1.65	5.15	0.13	0.05	1.96
Syringaldehyde	183	10.30	4.23	0.37	0.53	9.25
Trimethoxybenzene^c	169	0.29				
Guaiacol	125				0.04	
2,6-Dimethoxyphenol^c	155			0.11		0.04
Trimethoxyphenol^c	185	0.34		29.94	2.07	0.20

^{a}Quantified on anisaldehyde calibration curve.
^{b}Quantified on cinnamaldehyde calibration curve.
^{c}Quantified on guaiacol calibration curve (Flamini et al., 2007).

TABLE 7.6. Principal Fragments Produced by CID of Volatile Benzene Compound [M + H]$^+$ Ions Using He As a Collisional Gas^a

Compound	Precursor Ion m/z [M + H]$^+$	MS/MS m/z Fragment Ions (Abundance >5%)
Benzaldehyde	107	79 (b.p.)
Methylbenzaldehyde	121	93;43 (b.p.)
Hydroxybenzaldehyde	123	95 (b.p.);91;81
Guaiacol	125	110 (b.p.);96;93;91;65
Cinnamaldehyde	133	115 (b.p.);105;91;79;55
Anisaldehyde	137	122;109 (b.p.);94
Vinylguaiacol	151	136;123;119 (b.p.);115;95;91;81
Vanillin	153	138;125 (b.p.);93
2,6-Dimethoxyphenol	155	140 (b.p.);123;95;91;65
Eugenol	165	150 (b.p.);137;133;105
Trimethoxybenzene	169	154;138 (b.p.);126
Coniferaldehyde	179	164;161;147 (b.p.);133;119;105;55
Coniferyl alcohol	181	166 (b.p.);153;138
Syringaldehyde	183	168;155 (b.p.);140;123;95
Trimethoxyphenol	185	170;153 (b.p.);125
Methoxyeugenol	195	180 (b.p.);167;163;135;107
trans-Sinapinaldehyde	209	194;191;177 (b.p.);149;121;107;93

^{a}Dumping gas flow 0.3 mL/min; excitation voltage 225 mV. b.p.: base peak of fragmentation spectrum. MS conditions: Ion trap operating in positive-ion chemical ionization (PICI) mode, reagent gas methane at flow 0.8 mL/min, ion source temperature 200 °C, scan range m/z 40–550 (Flamini et al., 2007).

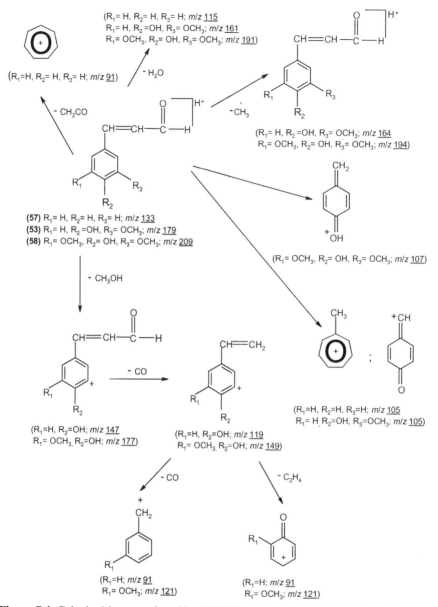

Figure 7.4. Principal ions produced by MS/MS of cinnamaldehyde (**57**), coniferaldehyde (**53**), and sinapinaldehyde (**58**) [M+H]$^+$ ions. (Reprinted from *Journal of Mass Spectrometry* 42, Flamini et al., 2007, GC/MS–Positive Ion Chemical Ionization and MS/MS study of volatile benzene compounds in five different woods used in barrel-making, p. 644, with permission from John Wiley & Sons, Ltd.)

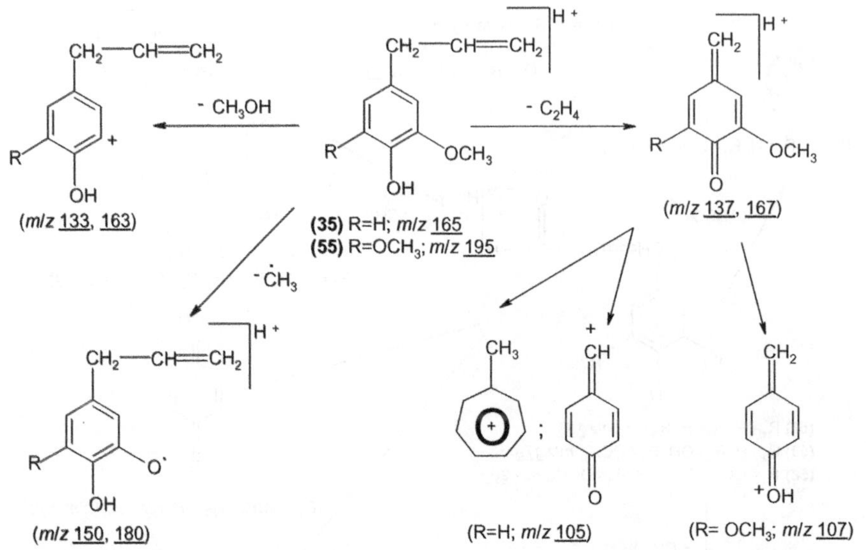

Figure 7.5. Principal ions produced by MS/MS of eugenol (**35**) and methoxyeugenol (**55**) [M+H]⁺ ions. (Reprinted from *Journal of Mass Spectrometry* 42, Flamini et al., 2007, GC/MS–Positive Ion Chemical Ionization and MS/MS study of volatile benzene compounds in five different woods used in barrel-making, p. 645, with permission from John Wiley & Sons, Ltd.)

REFERENCES

Boidron, J.N., Chatonnet, P., and Pons, M. (1988). The influence of wood on certain odorous substances in wines, *Connaiss. Vigne Vin*, **22**, 275–294.

Cadahía, E., de Simón, F.B., and Jalocha, J. (2003). Volatile compounds in Spanish, French, and American oak woods after natural seasoning and toasting, *J. Agric. Food. Chem.*, **51**, 5923–5932.

Chatonnet, P., Cutzach, I., Pons, M., and Dubourdieu, D. (1999). Monitoring toasting intensity of barrels by chromatographic analysis of volatile compounds from toasted oak wood, *J. Agric. Food. Chem.*, **47**, 4310–4318.

Chatonnet, P. and Dubourdieu, D. (1998). Identification of substances responsible for the "sawdust" aroma in oak wood, *J. Sci. Food Agric.*, **76**, 179–188.

Chatonnet, P. (1999). Discrimination and control of toasting intensity and quality of oak wood barrels, *Am. J. Enol. Vitic.*, **50**(4), 479–494.

Chatonnet, P., Dubourdieu, D., and Boidron, J.N. (1992). Influence of fermentation and ageing conditions of dry white wines in barrels on their composition in substances yielded by oak wood, *Sci. Alim.*, **12**, 665–685.

Cutzach, I., Chatonnet, P., Henry, R., and Dubourdieu, D. (1997). Identification of volatile compounds with a "toasty" aroma in heated oak used in barrelmaking, *J. Agric. Food. Chem.*, **45**, 2217–2224.

De Rosso, M., Cancian, D., Panighel, A., Dalla Vedova, A., and Flamini, R. (2008). Chemical compounds released from five different woods used to make barrels for aging wines and spirits: volatile compounds and polyphenols, *Wood Sci. Technol.* (DOI: 10.1007/s00226-008-0211-8).

De Rosso, M., Panighel, A., Dalla Vedova, A., Stella, L., and Flamini, R. (2009). Changes in Chemical Composition of a Red Wine Aged in Acacia, Cherry, Chestnut, Mulberry and Oak Wood Barrels, *J. Agric. Food Chem.*, **57**(5), 1915–1920.

Delgado, T., Gómez-Cordovés, C., and Villarroya, B. (1990). Relationship between phenolic compounds of low molecular weight as indicators of the aging conditions and quality of brandies, *Am. J. Enol. Vitic.*, **41**(4), 342–345.

Feuillat, F., Moio, L., Guichard, E., Marinov, M., Fournier, N., and Puech, J.L. (1997). Variation in the concentration of ellagitannins and *cis*- and *trans*-β-methyl-γ-octalactone extracted from oak wood (Quercus robur L., Quercus petracea Liebl.) under model wine cask conditions, *Am. J. Enol. Vitic.*, **48**(4), 509–515.

Flamini, R. and Dalla Vedova, A. (2004). Fast determination of the total free resveratrol content in wine by direct-exposure-probe, positive-ion chemical ionization and collision-induced-dissociation mass spectrometry, *Rapid Commun. Mass Spectrom.*, **18**, 1925–1931.

Flamini, R., Dalla Vedova, A., Cancian, D., Panighel, A., and De Rosso, M. (2007). GC/MS-Positive Ion Chemical Ionization and MS/MS study of volatile benzene compounds in five different woods used in barrel-making, *J. Mass Spectrom.*, **42**, 641–646.

Guichard, E., Fournier, N., Masson, G., and Puech, J.L. (1995). Stereoisomeres of β-methyl-γ-octalactone quantification in brandies as a function of wood origin and treatment of the barrels, *Am. J. Enol. Vitic.*, **46**(4), 419–423.

Hale, M.D., McCafferty, K., Larmie, E., Newton, J., and Swan, J.S. (1999). The influence of oak seasoning and toasting parameters on the composition and quality of wine, *Am. J. Enol. Vitic.*, **50**(4), 495–502.

Ibern-Gómez, M., Andrés-Lacueva, C., Lamuela-Raventós, R.M., and Lao-Luque, C., Buxaderas, S. and de la Torre-Boronat, M.C. (2001). Differences in phenolic profile between oak wood and stainless steel fermentation in white wines, *Am. J. Enol. Vitic.*, **52**(2), 159–164.

MacNamara, K., van Wyk, C.J., Brunerie, P., Augustyn, O.P.H., and Rapp, A. (2001). Flavour components of Whiskey. III. Ageing changes in the low-volatility fraction, *S. Afr. J. Enol. Vitic.*, **22**(2), 82–91.

Masson, E., Baumes, R., Moutounet, M., and Puech, J.L. (2000). The effect of kiln-drying on the levels of ellagitannins and volatile compounds of european oak (Quercus petracea Liebl.) stave wood, *Am. J. Enol. Vitic.*, **51**(3), 201–213.

Masson, G., Puech, J.L., and Moutounet, M. (1996). The chemical composition of barrel oak wood, *Bull. O.I.V.*, **785–786**, 634–655.

Matricardi, L. and Waterhouse, A.L. (1999). Influence of toasting technique on color and ellagitannins of oak wood in barrel making, *Am. J. Enol. Vitic.*, **50**(4), 519–526.

Mattivi, F., Versini, G., and Sarti, S. (1989a). Study on the presence of scopoletin in commercial wood-aged brandies, *Riv. Vitic. Enol.*, **3**, 23–30.

Mattivi, F., Versini, G., and Sarti, S. (1989b). The presence of Scopoletin as an indicator of ageing parameters in distillates, *Riv. Vitic. Enol.*, **2**, 3–10.

Pérez-Coello, M.S., Sanz, J., and Cabezudo, M.D. (1999). Determination of volatile compounds in hydroalcoholic extracts of french and american oak wood, *Am. J. Enol. Vitic.*, **50**(2), 162–165.

Puech, J.L., Feuillat, F., and Mosedale, J.R. (1999). The Tannins of oak heartwood: structure, properties, and their influence on wine flavor, *Am. J. Enol. Vitic.*, **50**(4), 469–478.

Puech, J.L. (1981). Extraction and evolution of lignin products in Armagnac matured in oak, *Am. J. Enol. Vitic.*, **32**(2), 111–114.

Ribéreau-Gayon, P., Glories, Y., Maujean, A., and Dubourdieu, D. (1998). Trattato di enologia II—*Chimica del vino stabilizzazione e trattamenti*, *Dunod*, Paris, 389–394.

Salagoity-Auguste, M.H., Tricard, Chr, Marsal, F., and Sudraud, P. (1986). Preliminary investigation for the differentiation of enological tannins according to botanical origin: Determination of gallic acid and its derivatives, *Am. J. Enol. Vitic.*, **37**, 301–303.

Sauvageot, F. and Feuillat, F. (1999). The influence of oak wood (Quercus robur L., Q. Petracea Liebl.) on the flavor of burgundy pinot noir an examination of variation among individual trees, *Am. J. Enol. Vitic.*, **50**(4), 447–455.

Vichi, S., Santini, C., Natali, N., Riponi, C., López-Tamames, E., and Buxaderas, S. (2007). Volatile and semi-volatile components of oak wood chips analysed by Accelerated Solvent Extraction (ASE) coupled to gas chromatography–mass spectrometry (GC—MS), *Food Chem.*, **102**, 1260–1269.

8

COMPOUNDS RESPONSIBLE FOR WINE DEFECTS: OTA, TCA AND TBA, GEOSMIN, 1-OCTEN-3-ONE, 2-METHOXY-3,5-DIMETHYLPYRAZINE, BIOGENIC AMINES, ETHYL CARBAMATE, "GERANIUM", AND "MOUSY" TAINTS

Food contaminants can be defined as compounds dangerous to consumer health or that affect the organoleptic characteristics of the product. This chapter presents the principal contaminants of grapes and wines, and the MS methods for their detection.

8.1   OCHRATOXIN A IN GRAPE AND WINE

Ochratoxin A (OTA) is dangerous to human health, and its legal limits in grape and wine are fixed. The molecule consists of an isocoumarin derivative linked to phenylalanine through a carboxyl group (structure reported in Fig. 8.1).

This toxin is a secondary metabolite of *Penicillium verrucosum*, *Aspergillus ochraceus*, *A. carbonarius*, and *A. niger* fungi, whose development is promoted by favorable environmental conditions,

Mass Spectrometry in Grape and Wine Chemistry, by Riccardo Flamini and Pietro Traldi
Copyright © 2010 John Wiley & Sons, Inc.

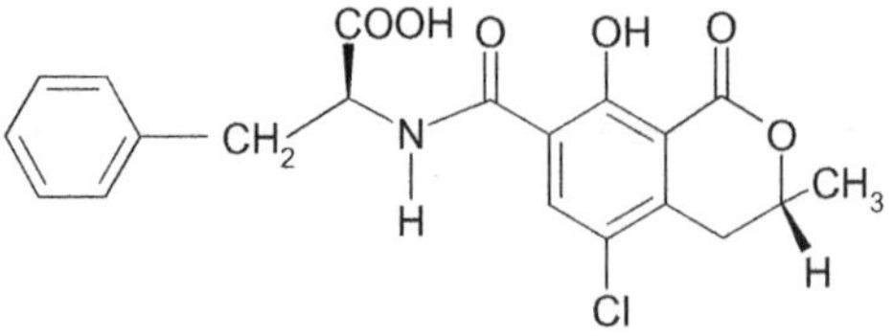

(OTA)

Figure 8.1. Structure of (R)-N-[(5-chloro-3,4-dihydro-8-hydroxy-3-methyl-1-oxo-1H-2-benzopyran-7-yl)carbonyl]-L-phenylalanine) (OTA).

particularly during the products storage. It has nephrotoxic effects, induces renal damages (Schwerdt et al., 1999), and may promote renal tumors (Castegnaro et al., 1998; Pfohl-Leszkowicz et al., 1998). Moreover, it was reported to interfere with mitochondrial respiratory function and pH homeostasis (Sauvant et al., 1998), to inhibit tRNA-synthetase accompanied by the reduced protein synthesis and enhanced lipid peroxidation via free radical generation (Hohler, 1998). The fungal growth occurs on the surface of grape and is promoted by favorable environmental conditions, often occurring during the harvest, such as high humidity and temperature. Since the fungi is present on the cluster surface, there is a higher risk of contamination for red wines, produced by skin maceration in the juice during fermentation, and for sweet wines, compared to to white wines.

The World Health Organization (WHO) set a provisionally tolerable weekly intake level for OTA at 100 ng/kg of body weight, considering its potentially carcinogenic effect (JEFCA, 2001). The tolerated maximum concentrations are based on a tolerable daily intake of 5 ng/kg of body weight suggested by the Scientific Committee on Food (Commission of the European Communities, 1998). The CE Regulation n° 123/2005 fixed the OTA legal maximum limit in grape, wine, and grape juice at 0.002 ppm (Flamini and Panighel, 2006 and references cited therein).

8.1.1 The LC/MS of OTA

In general, a satisfying method for quantitative analysis has to provide high sensitivity, low limits of detection (LOD), low limits of quantification (LOQ), a linearity range of at least three to four orders of magnitude, high precision (repeatability and reproducibility of data), and accuracy (experimental data as close as possible to the "true value").

In particular, methods for analysis of OTA in wine, including both sample preparation and analysis, have to provide LOQ of at least 0.6 µg/L (the legal limit is 2 ppb), LOD <0.2 µg/L, a linearity range of 0.1–100 ppb, and an extraction yield from red and white wines at least 84.6 and 88.4%, respectively (International Organization of Vine and Wine, 2006). Solid-phase extraction (SPE) sample preparation coupled with LC/MS analysis provides the performances required. Recently, LC/MS methods by direct injection of the sample were developed as well.

Spot-contamination of the fungi on the surface of grape cluster and its heterogeneity (it is constituted of berries and stalks) make it difficult to dispose of a representative sample for analysis. The problem arises when grapes are collected in the vineyard, and even when berries are selected from clusters for the sample preparation. Because the fungi is present on the entire surface of the cluster, a representative sample is prepared by using both the juice exited from the berries (which washes the cluster surfaces) and by performing solvent extraction of all solid parts of clusters: a 100-g sample constituted of berries, stalks, and juice is extracted with 100 mL of chloroform for 24 h with stirring. No sample homogenization is made in order to avoid extraction of interfering substances (Garcia-Moruno et al., 2004; Tonus et al., 2005). A volume of 90 mL of the organic phase is recovered and filtered, brought to dryness under vacuum at 30 °C, and the residue is dissolved in the LC mobile phase.

Alternatively, extraction can be performed with a sodium hydrogen carbonate and polyethylene glycol (PEG) solution (1% PEG 8000 and 5% $NaHCO_3$ in H_2O) followed by purification of the extract using an immunoaffinity column (IAC) specific for OTA (Serra et al., 2004). Berries are slightly homogenized, a 50-g sample is brought to 150 mL with a $NaHCO_3$/PEG solution, and the mixture is stirred for 30 min, then centrifuged at 4 °C. The supernatant is filtered through a glass microfiber (1.5 µm) and a volume of 20 mL of solution is purified by IAC. The use of PEG seems to reduce interfering substances in the analysis. Also, acidified methanol was used for extraction of OTA from grape: 50 g of berries are homogenized and extracted for 2 min with 50-mL methanol added with 5-mL orthophosphoric acid. The mixture is filtered thorough a 1.5-µm glass microfiber and diluted to 100 mL with a 1% PEG–5% $NaHCO_3$ solution prior to IAC purification (MacDonald et al., 1999).

For wine analysis, the sample is preliminary centrifuged and degassed by ultrasound. The use of IAC is the more common method of sample preparation (Burdaspal and Legarda, 1999; Visconti et al., 1999;

Castellari et al., 2000). A 10-mL sample is added to an equal volume of a $PEG/NaHCO_3$ solution and the pH is adjusted to 8.5 by addition of concentrated NaOH. The solution is filtered through a glass microfiber to remove solids. A 10 mL volume of sample is purified by passage through IAC and the stationary phase is washed with 5 mL of a $NaCl/NaHCO_3$ aqueous solution (2.5% NaCl and 0.5% $NaHCO_3$) and 5 mL of H_2O after sample loading. Ochratoxin A is recovered from the cartridge with 2 mL of methanol containing acetic acid 2% (v/v). The solution is brought to dryness at 50 °C under a nitrogen stream and the residue is dissolved in the LC mobile phase before analysis (Sáez et al., 2004).

Sample preparation can be performed also by using a C_{18} cartridge: 10 mL of wine is passed through a 100-mg cartridge previously activated by passage of 5-mL methanol and 5-mL water, the stationary phase is washed with 2-mL water, then dried. The OTA is eluted by 1.25 mL of methanol (Zöllner et al., 2000). Alternatively, OTA can be recovered by methanol/acetic acid 99.5:0.5 (Sáez et al., 2004) or 3 mL of ethyl acetate/methanol/acetic acid 95:5:0.5 (Chiodini et al., 2006) solutions.

Sample preparation by liquid–liquid extraction can be performed by mixing 5 mL of wine with 10 mL of 3.4% orthophosphoric acid (85%) aqueous solution and 1.18 g NaCl, and performing two successive extractions with 5 mL of chloroform (Zimmerli and Dick, 1996). Organic phases are combined and the solvent is removed under vacuum. The residue is dissolved into 5 mL of a phosphate buffer solution (NaCl 120 mM, KCl 2.7 mM, phosphate buffer 10 mM, pH 7.4) and the solution is purified using IAC.

Commonly, OTA in grape juice and wine is determined by LC and fluorescence detection, but this approach requires a prior sample purification by IAC (Burdaspal and Legarda, 1999; Visconti et al., 1999; Castellari et al., 2000). Moreover, fluorescence does not allow unambiguous identification of the analyte and OTA presence is confirmed by synthesis of methyl ester, or by OTA enzymatic cleavage yielding ochratoxin α (7-carboxy-5-chloro-3,4-dihydro-8-hydroxy-3-methylisocoumarin) (Tonus et al., 2005; Filali et al., 2001). By performing LC/ESI–MS analysis a LOD of 0.5 ppb can be achieved and the compound can be identified on the basis of the distinct Cl isotopic cluster distribution centered on the ^{35}Cl-containing $[M+H]^+$ ion at m/z 404 and the ^{37}Cl-containing ion at m/z 406 (MacDonald et al., 1999; Zöllner et al., 2000). Additional confirmation can be achieved by comparing the ion intensities of the m/z 404/406 peaks of the sample with those recorded in standard analysis (MacDonald et al., 1999). Positive-ion LC/ESI–MS/MS using a C_{18}

column and ion trap (IT) system provides unambiguous confirmation of OTA presence in the sample up to 0.2 ng/mL (Medina et al., 2006). Confirmation can be based on the $[M+H]^+$ ion at m/z 404 and the most abundant product ion $[(M+H)–HCOOH]^+$ at m/z 358 using the following ESI parameters: nebulizer gas N_2 60 psi, dry gas 10 L/min, dry temperature 220 °C, capillary voltage 3 kV. Commercially available Zearalanone (ZAN) can be used as an internal standard.

Analysis using a triple quadrupole (3Q) system was performed by multiple reaction monitoring (MRM) selecting for OTA the fragmentation m/z 404 $\rightarrow$ 239 ($[M+H-Phe]^+$ species), m/z 404 $\rightarrow$ 257 and m/z 406 $\rightarrow$ 241, and m/z 321 $\rightarrow$ 123/189 for ZAN (Zöllner et al., 2000). The CID experiments carried out with an IT system are performed on the OTA protonated molecular ion $[M+H]^+$ at m/z 404, and by selected reaction monitoring (SRM) of the resultant product ions $[M+H-H_2O-CO]^+$ at m/z 358 and $[M+H-H_2O]^+$ at m/z 386 (Shephard et al., 2003). A scheme of OTA fragmentation is reported in Fig. 8.2 (Lau et al., 2000).

Quantitative analysis also can be performed by a stable isotope dilution assay using d_5-OTA as the standard (Lindenmeier et al., 2004; MacDonald et al., 1999). The ^{35}Cl-containing $[M+H]^+$ ion at m/z 404 and the ^{37}Cl-containing analog at m/z 406 are monitored: comparison between the m/z 404/406 peak area ratio in the sample with standard solutions provides an additional confirmation.

Recently, direct injection nano-LC/ESI–MS analysis of grape extracts (1 μL of sample introduced onto a reverse-phase capillary with a solvent flow rate 200 nL/min) was proposed. The LOD and LOQ of 1 and 2 pg/g, respectively, were reported (Timperio et al., 2006).

In general, coupling SPE by C_{18} with LC/ESI–MS/MS analysis provides LOD and LOQ comparable with IAC clean-up coupled to LC/fluorescence. Coupling of IAC with LC/MS showed no advantages in terms of sensitivity and accuracy (Leitner et al., 2002). The main advantages of MS are to provide structural information and the possibility of using the cheaper C_{18} cartridges. The latter is not possible in performing LC/fluorescence analysis due to the interfering substances in the chromatogram that have to be removed by IAC.

8.1.2 The LC/SACI–MS Analysis of OTA

Recently, a LC/MS method for analysis of OTA in grape extracts and wine by surface-activated chemical ionization and multistage fragmentation mass spectrometry (LC/SACI–MS^n) was developed (Flamini et al., 2007). The SACI method (see section 1.4) is an

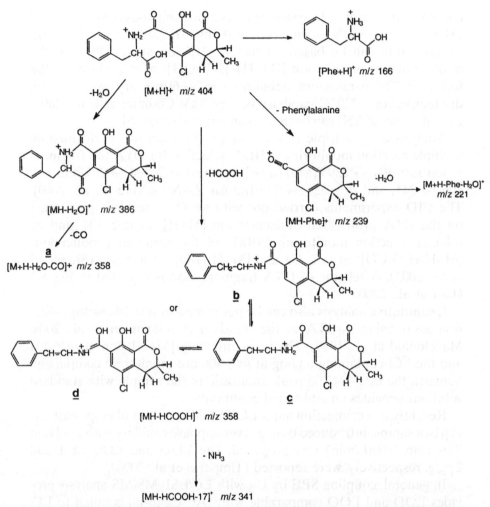

Figure 8.2. The ESI–MS/MS fragmentation pathways of OTA. (Reprinted from *Journal of Mass Spectrometry* 35, Lau et al., Quantitative determination of ochratoxin A by liquid chromatography/electrospray tandem mass spectrometry, p. 28, Copyright © 2000, with permission from John Wiley & Sons, Ltd.)

ionization source suitable for analysis of medium-high polar compounds in a wide range of m/z ratios (100–4000 Da). This device was used for analysis of compounds, such as aminoacids, drugs, and steroids, in conjunction with high-flow gradient chromatography. The technique is based on the presence of an ionization chamber of metallic catalyst set at a low potential. Solvent from the chromatographic column is vaporized by a nitrogen flow and directed to the metallic surface. At the positively charged surface, the solvent molecules transfer a proton to the analyte according to their proton affinity, forming the $[M+H]^+$ or $[M+nH]^{n+}$ ions. An analogous mechanism occurs when the surface is charged negatively: consequently there is an increased proton affinity of the solvent and the occurrance of a proton transfer from the analyte molecule to the solvent with formation of $[M-H]^-$ or $[M-nH]^{n-}$ species. Ionization using traditional ESI and atmospheric-pressure chemical ionization (APCI) techniques induces extra charge of the solvent due to the high potential used (2–5 kV). SACI shows lower chemical noise related to the solvent-charged clusters, and the high ionization yields obtained increases the signal/noise (S/N) ratio for many compounds (Cristoni et al., 2003).

OTA analysis is performed by direct injection of the wine sample into the column, without sample preconcentration or purification steps. The most abundant OTA daughter ion produced by MS^2 experiments using as precursor the ion $[M+H]^+$ (^{35}Cl-containing molecule) is at m/z 358 (Scheme 1 in Figure 8.3); the most abundant daughter ion from the precursor ion $[M+H]^+$ of the internal standard ZAN is at m/z 303 (Scheme 2). Performing MS^3 experiments improves specificity of the method. Figure 8.4 shows the MS^3 spectra from the $[M+H]^+$ ion of OTA at m/z 404 (above) and of ZAN at m/z 321 (below) recorded in the analysis of a wine spiked with two compounds at 10 ng/mL. Additive confirmation can be provided from MS^3 experiments performed on the isotopic species ^{37}Cl. MS/MS of the $[M+H]^+$ species at m/z 406, corresponding to OTA molecule ^{37}Cl-containing, produces the ion m/z 343, the successive MS^3 experiment gives the ion at m/z 241.

The use of high-flow chromatographic conditions avoids a matrix effect, and an LOQ at least 20-fold lower than the maximum legal limit (2 ppm) is achieved. Figure 8.5 shows the LC–MS^3 extracted ion chromatograms recorded in the analysis of a 0.1 ng/mL OTA-spiked wine sample and of an extract from a natural OTA-contaminated grape sample. The figure shows the peaks resulting from the sum of the m/z signals used for quantitative analysis: m/z 239+341 for OTA (above) and m/z 207+189+163 for ZAN (below). Under the chromatographic conditions reported below the two compounds elute from the column with the same retention time. This approach is highly sensitive and has a good precision with an error percentage ranging between 1 and 4% (calculated for 15 wine samples spiked with OTA at 10 ng/mL).

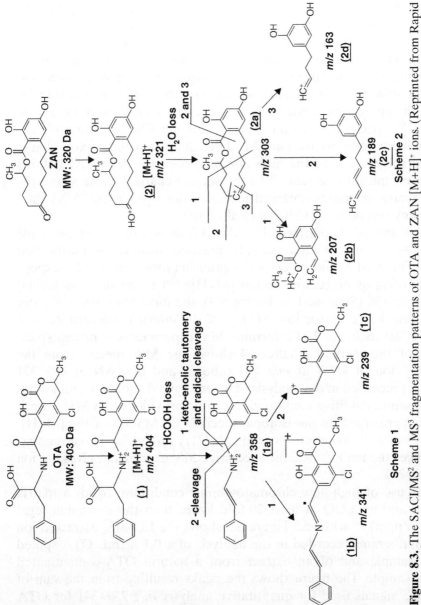

Figure 8.3. The SACI/MS² and MS³ fragmentation patterns of OTA and ZAN [M+H]⁺ ions. (Reprinted from Rapid Communications in Mass Spectrometry 21, Flamini et al., A new sensitive and selective method for analysis of ochratoxin A in grape and wine by direct liquid chromatography/surface activated chemical ionization-tandem mass spectrometry, p. 3738, Copyright © 2007, with permission from John Wiley & Sons, Ltd.)

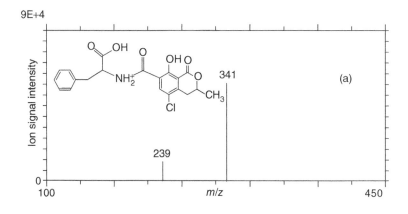

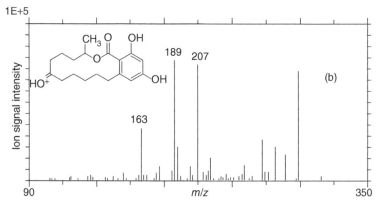

Figure 8.4. The SACI/MS[3] spectra of the OTA daughter ion at m/z 358 (above) and of the ZAN daughter ion at m/z 303 (below). The OTA m/z 239+341 and ZAN m/z 163+189+207 signals are used for quantitative analysis. (Spectra acquired in positive-ion mode; collision energy applied to the parent ion 80% of maximum value; MS[3] of daughter ions at 100% of maximum collision energy). (Reprinted from Rapid Communications in Mass Spectrometry 21, Flamini et al., A new sensitive and selective method for analysis of ochratoxin A in grape and wine by direct liquid chromatography/surface activated chemical ionization-tandem mass spectrometry, p. 3740, Copyright © 2007, with permission from John Wiley & Sons, Ltd.)

8.2 THE SPME–GC/MS/MS ANALYSIS OF TCA AND TBA IN WINE

Due to the physical properties of being an excellent seal for liquids, cork stoppers are the principal means of closure used for bottled wines. Unfortunately, some compounds released from cork stoppers may

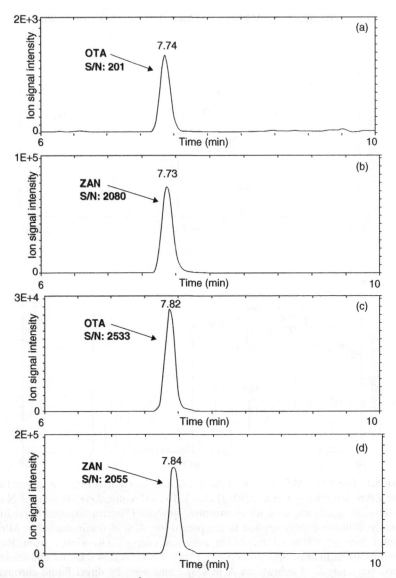

Figure 8.5. The LC/SACI–MS³ extracted ion chromatograms of a wine spiked with OTA 0.1 ng/mL (**a**) and ZAN 10 ng/mL as internal standard (**b**), and of a naturally contaminated grape extract containing OTA 1.3 ng/mL (**c**), and spiked with ZAN 10 ng/mL (**d**) (signals m/z 239+341 for OTA and m/z 207+189+163 for ZAN are recorded). Analytical conditions: LC column C_{18} (250 × 3 mm; 5 μm), binary solvent composed of (A) $H_2O/0.1\%$ formic acid/sodium acetate 0.6 mM and (B) methanol/0.1% formic acid. Gradient program: 50% A for 1 min, from 50 to 20% of A in 7 min, isocratic for 4 min, from 20 to 50% of A in 3 min, isocratic for 3 min (flow rate 0.5 mL/min). SACI vaporizer temperature 400 °C; entrance capillary temperature 150 °C; surface voltage 50 V; surface temperature 110 °C; nebulizing sheath gas N_2 at a flow rate of 9 L/min; curtain gas 2 L/min; spray needle voltage set to 0 V. (Reprinted from Rapid Communications in Mass Spectrometry 21, Flamini et al., A new sensitive and selective method for analysis of ochratoxin A in grape and wine by direct liquid chromatography/surface activated chemical ionization-tandem mass spectrometry, p. 3741, Copyright © 2007, with permission from John Wiley & Sons, Ltd.)

confer defects to the wine causing off-flavor or "corkiness" in wine. Due to its very low odor threshold (1.4–10 ppt), 2,4,6-trichloroanisole (TCA) has been identified as the major cause of corkiness in wine. Corks may be contaminated by chloroanisoles during transport from packaging and shipping containers by microbiological methylation of chlorophenols during the corks bleaching with hypochlorite, and microbial contamination of raisins (Aung et al., 1996). Chloroanisoles may also be present as residues of pesticides and insecticides used in the cork forest (Burttshel et al., 1951; Lee and Simpson, 1993). Irradiation of TCA causes the formation of haloanisoles, such as 2-chloroanisole, 4-chloroanisole, 2,4-dichloroanisole, and 2,6-dichloroanisole (Careri et al., 2001; Flamini and Panighel, 2006 and references cited therein). The relationship between TCA and 2,4,6-trichlorophenol (TCP) is shown in Fig. 8.6.

Although TCA is considered to be the primary cause of cork taint, other compounds found in corks can contribute to wines taint, such

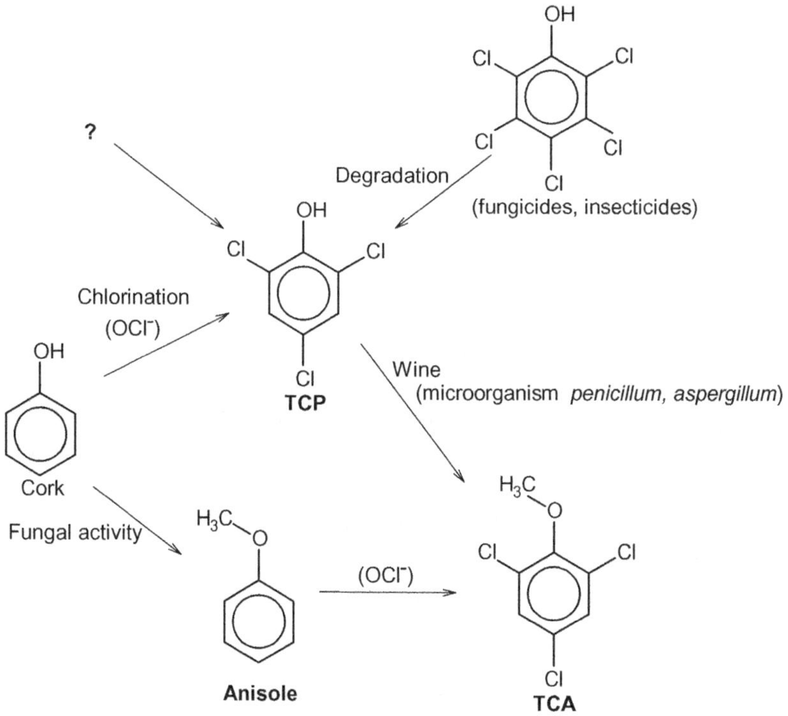

Figure 8.6. Origin of 2,4,6-trichloroanisole in cork used for making stoppers for bottled wines.

as 2,3,4,6-tetrachloroanisole, pentachloroanisole, 2,4,6-tribromoanisole (TBA), guaiacol, geosmin, 2-methylisoborneol, octen-3-ol, 1-octen-3-one, and 2-methoxy-3,5-dimethylpyrazine (Amon et al., 1989; Simpson et al., 2004). The TBA causes earthy-musty off-flavors in water and also may be a cause of cork taint in wine where it has a sensory threshold of 7.9 ng/L (Chatonnet et al., 2004). Wine may be tainted from contaminated containers, oak barrels, or wooden structures in cellars after disinfection by chlorinated compounds. Tribromophenol (TBP), widely used as a fungicide, is a replacement for pentachloropenol (PCP), whose use has been restricted by the European Union (Council Directive 91/173/EEC, 1991), since it might degrade into TBA by the same fungus that methylates chlorophenols (Whitfield et al., 1997).

8.2.1 Sample Preparation

The SPME–GC/MS is a fast and sensitive method for determination of TCA in wine. Extraction is performed using a 100-μm polydimethylsiloxane (PDMS) 1-cm length fiber in the headspace (HS) of a 10-mL sample, transferred in a 20-mL vial and addition of 2-g NaCl, for 20 min at 30–35 °C (Evans et al., 1997). The fiber is then desorbed for 3 min in the GC/MS injector port at 250 °C.

Alternatively, analysis of TCA and TCP in wine can be performed by performing an SPE sample preparation using a 500-mg C_{18} cartridge previously conditioned by consecutive washings with ethyl acetate, ethanol, and aqueous ethanol at 10%. A 50-mL sample is passed through the cartridge, the stationary phase is dried and TCA and TCP are recovered using 0.5 mL of dichloromethane. The first 200 μL of eluate containing TCA and TCP are mixed with 200 μL of acetonitrile in order to have a 125-fold concentrated sample. Recoveries of TCA range between 86 and 102%, and of TCP between 82 and 103%. By this method, the LOD and LOQ reported for TCA are 0.1 and 2 ng/L, respectively and for TCP they are 0.7 and 4 ng/L, respectively (Soleas et al., 2002).

Simultaneous analysis of TCA and TBA was performed by extraction of 50 mL of wine with a 50-mg ethylvinylbenzene–divinylbenzene copolymer cartridge and recovering analytes with 0.6 mL of dichloromethane (Insa et al., 2005).

Also, a stir bar sorptive extraction (SBSE) method for GC/MS analysis of TCA in wine was proposed using a stir bar with 0.5-mm thickness and 10-mm length coated with PDMS (Hayasaka et al., 2003).

8.2.2 The GC/MS Analysis

Chloroanisoles (2,4-dichloroanisole, TCA, 2,3,4,6-tetrachloroanisole, pentachloroanisole, TCP, 2,3,4,6-tetrachlorophenol, pentachlorophenol) in wines or cork stopper extracts are usually analyzed using a 5% diphenyl–95% dimethyl polysiloxane GC column (e.g., 30 m × 0.25 mm i.d., 0.25-µm film thickness). By using a singular quadrupole mass spectrometer recording signals in SIM mode, TCA is quantified on the sum of signals m/z 195+197+199+210+212+214, with the last two coming from molecules containing one or two ^{37}Cl atoms, respectively.

The signals recorded for analysis of TCP are at m/z 196, 198 and 200, for tetrachlorophenol at m/z 229, 231, 244 and 246. By performing SPME-GC/MS-SIM single quadrupole analysis the LOD and LOQ achieved for TCA are 0.2 and 0.4 ng/L, respectively (Lizarraga et al., 2004). The GC/MS–electron impact (EI 70 eV) fragmentation spectra of TCA is reported in Fig. 8.7.

A GC/MS–EI chromatogram recorded in the analysis of TCA and TCP in wine is shown in Fig. 8.8; below the chromatographic conditions used are reported.

The TCA can be determined using an ion trap system performing collision-induced dissociation (CID). Quantification is based on the daughter ion signals of the $M^{+\bullet}$ species at m/z 210 and 212 used as precursor ions. Depending on the system used, CID can be performed in either resonant or non-resonant mode. In the former condition, the most intense daughter ions are at m/z 195 and 197, in non-resonant mode the principal signals are at m/z 167 and 169. The CID of a wine spiked with

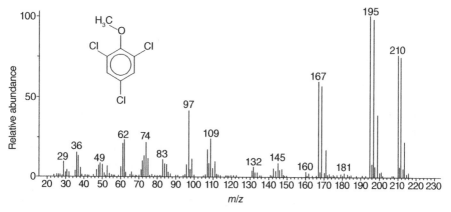

Figure 8.7. The GC/MS–EI fragmentation spectrum (70 eV) of 2,4,6-trichloroanisole.

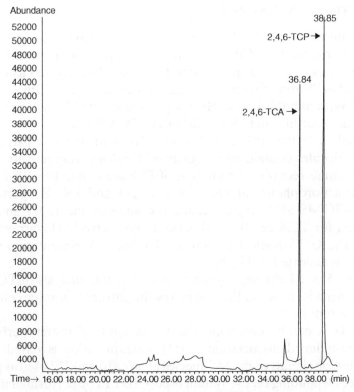

Figure 8.8. The GC/MS–SIM analysis of TCA and TCP in a wine extract. Analytical conditions: Injector and detector temperatures 200 and 240 °C, respectively; oven temperature program: 50 °C for 5 min, 1.5 °C/min to 100 °C, isotherm for 3 min, 30 °C/ min to 250 °C, isotherm for 5 min. Carrier gas He; column head pressure 8 psi. (Reproduced from *Journal of Agricultural and Food Chemistry*, 2002, 50, p. 1034, Soleas et al., with permission of American Chemical Society.)

1 ppt of TCA provides the sum of signals at 195+197 (resonant) or m/z 167+169 (non-resonant) with an S/N ratio of 20. Figure 8.9 shows the chromatograms recorded in the analysis of a red wine spiked with 1 ppt TCA performed in resonant (above) and non-resonant (below) mode (Flamini and Larcher, 2008).

Two different sensitive and selective HS–SPME–GC/MS approaches for simultaneous analysis of TCA and TBA in wine using negative chemical ionization MS (GC/NCI–MS) and high-resolution mass spectrometry (GC–HRMS), were developed (Jönsson et al., 2006). Experimental conditions and performance of the methods are summarized in the Table 8.1.

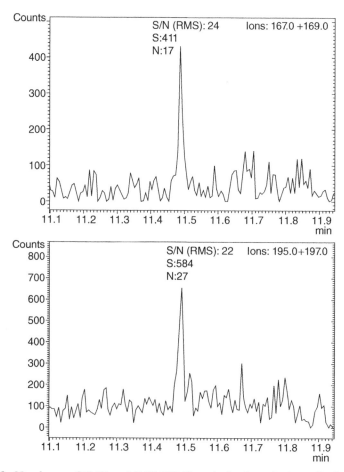

Figure 8.9. Headspace–SPME and GC/MS/MS analysis of a red wine spiked with TCA 1 ppt performed in non-resonant (signal *m/z* 167+169 above) and resonant (signal *m/z* 195+197 below) mode. Precursor ion *m/z* 211.9, isolation window 5 uma; excitation amplitude 80 V (Flamini and Larcher, 2008).

8.3   GEOSMIN

Geosmin [octahydro-4,8*a*-dimethyl-4*a*(2*H*)-naphthalenol] is characterized by a distinctive earthy, musty odor, and a very low sensory threshold (1–10 ng/L). In wine, usually it is present as a metabolite from *Streptomyces*, and *Botritis cinerea* and *Penicillium expansum* moulds growth on grapes.

TABLE 8.1. Experimental Conditions and Performances of Headspace Sampling SPME/GC Negative Chemical Ionization-MS and SPME/GC High-Resolution-MS Methods Used for Analysis of TCA and TBA in Wine[a]

SPME fiber	PDMS 100 μm
Sample volume	3 mL
Vial volume	4 mL
Addition to the sample	NaCl (30% w/w)
Internal standard	2,3,6-TCA
Extraction temperature	Room temperature
Extraction time	30 min with stirring
Desorption temperature	250 °C
Desorption time	5 min
GC column	5% Diphenyl–95% dimethyl polysiloxane-fused silica capillary (30 m × 0.25 mm i.d.; 0.25-μm film thickness)
Carrier gas	He 1.1 mL/min
Oven program	70 °C isotherm for 3 min, 5 °C/min to 180 °C, 20 °C/min to 300 °C
GC–NCI–MS conditions	Reagent gas methane pressure 4.5×10^{-2} Pa; SIM mode detection of ions m/z 174, 176, 210, 212 for TCA and the internal standard and m/z 79, 81, 344, 346 for TBA; quantification ion m/z 174 for TCA and IS and m/z 79 for TBA
GC–HRMS conditions	EI at 35 eV; instrument tuned to resolution 10.000; SIM for quantification of ion 209.9406 for TCA and IS and 343.7870 for TBA

	LOQ
HS–SPME–GC–NCI–MS (SIM)	
TCA (column RT 13.7 min)	0.3 ng/L
TBA (column RT 20.7 min)	0.2 ng/L
HS-SPME–GC–HRMS (SIM)	
TCA	0.03 ng/L
TBA	0.03 ng/L

[a]Jönsson et al., 2006.

8.3.1 Extraction from Wine and Grape Juice

Liquid–liquid extraction can be performed using as an internal standard 2-undecanone (150 μL of a 100-mg/L ethanolic solution added to 1.5 L of sample). The sample is extracted three times with 60, 40, and 40 mL of pentane for 10 min with stirring. The extracts are collected together and the resulting solution is concentrated to 10 mL at 4 °C under vacuum, then concentrated to 500 μL, and purified on silica gel (70–230 mesh, 60 Å), activated at 120 °C. The wine extract is then passed through the silica column (100 × 10 mm) and four 40-mL fractions are recovered using pentane (I), pentane/dichloromethane

80:20 (II), pentane/dichloromethane 60:40 (III), and pentane/dichloromethane 50:50 (IV). Geosmin is recovered in fraction II. This solution is concentrated to 500 μL under a nitrogen flow before GC/MS analysis (Darriet et al., 2000).

Also, HS–SPME has been used for analysis of geosmin in wine: 5 mL of sample is transferred in a 20-mL vial and saturated with 3 g of NaCl. The solution is diluted 1:1 by volume with water acidified to pH 3. Sampling is carried out by a PDMS 100-μm fiber at 40 °C while keeping the solution stirred for 30 min. Desorption of fiber is performed in the GC injector port at 260 °C (Dumoulin and Riboulet, 2004).

8.3.2 The GC/MS Analysis

Usually a PEG fused silica capillary column (e.g., 50 m × 0.25 mm i.d., 0.25-μm film thickness) is used with the following oven temperature program: 45 °C isotherm for 1 min, 3 °C/min to 230 °C, isotherm for 10 min. An example of a GC chromatogram for a *Cabernet Sauvignon* wine analysis is shown in Fig. 8.10. The GC/MS–EI (70 eV) mass spectrum of geosmin is reported in Fig. 8.11 (Darriet et al., 2000).

Mass spectrum ions used for identification of geosmin are at m/z 111, 168, and 182. Quantification is performed on the ion at m/z 112 recorded in SIM mode. For internal standard 2-undecanone, the signal recorded is

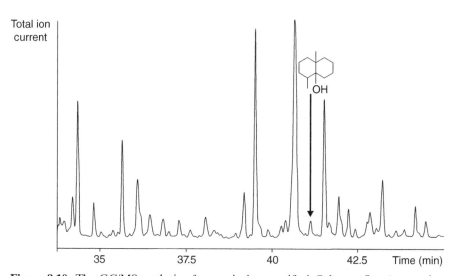

Figure 8.10. The GC/MS analysis of geosmin in a purified *Cabernet Sauvignon* wine extract performed using a PEG capillary column. (Reproduced from *Journal of Agricultural and Food Chemistry*, 2000, 48, p. 4836, Darriet et al., with permission of American Chemical Society.)

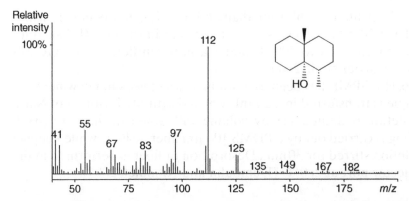

Figure 8.11. The GC/MS–EI (70 eV) mass spectrum of geosmin [octahydro-4,8a-dimethyl-4a(2H)-naphthalenol, $C_{12}H_{22}O$, MW 82.30248]. (Reproduced from *Journal of Agricultural and Food Chemistry*, 2000, 48 p. 4837, Darriet et al., with permission of American Chemical Society.)

at m/z 58 (Darriet et al., 2000). Performing MS/MS analysis of the signal at m/z 112 quantification of geosmin is done on the daughter ion at m/z 97. The SPME–GC/MS analysis using geosmine-d^5 as internal standard gave a LOD of 5 ng/L (Dumoulin et al., 2004).

8.4   ANALYSIS OF 1-OCTEN-3-ONE

1-Octen-3-one is a compound that can be present in relevant levels in the must obtained from mildew-infected grapes. This compound is characterized by very low sensory thresholds that can be responsible for the fungus odor in the must (Darriet et al., 2002). Normally, during fermentation the molecule is completely reduced to the lesser powerful 3-octanone by the yeasts, but it was found to be present in dry wines (Cullere et al., 2006).

A method proposed for determination of 1-octen-3-one in wines is by performing an on cartridge derivatization of the sample followed by GC/MS/MS analysis (Culleré et al., 2006). A 90-mL volume of wine is passed through a 90-mg ethylvinylbenzene–divinylbenzene copolymer SPE cartridge and, after removing the major volatiles by washing with 9 mL of a 40% methanol/water solution containing 1% $NaHCO_3$, the analyte adsorbed on the stationary phase is derivatized by passing through the cartridge 2 mL of an O-(2,3,4,5,6-pentafluorobenzyl) hydroxylamine (PFBOA) 5 mg/mL aqueous solution. To allow the reaction to occur, the cartridge is kept imbibed with the reagent for 15 min at room

TABLE 8.2. Analytical Conditions for GC/MS–IT Analysis of 1-Octen-3-one in Wine[a]

GC column	100% Dimethyl polysiloxane fused silica capillary (60 m × 0.25 mm i.d.; 1.00-μm film thickness)
Carrier gas	He 1.0 mL/min
Oven program	40 °C isotherm for 5 min, 10 °C/min to 140 °C, 3 °C/min to 235 °C, 235 °C isotherm for 20 min
Injection	2 μL splitless
GC/MS–IT conditions	mass range m/z 45–350 recorded in full-scan mode non-resonant fragmentation of parent ion m/z 140; applied excitation amplitude 50 V; daughter ions m/z 77, 79, 94 for *trans*-oxime, m/z 77, 79 for *cis*-oxime
MS/MS acquisition conditions	Isomer (E) Isomer (Z)
Acquisition interval (min)	41.60–42.15 42.15–43.15
Retention time (min)	41.8 42.8
Recorded mass range	75–95 75–85
Quantification masses	77 + 79 + 94 77 + 79

[a]Culleré et al., 2006.

temperature. Excess of PFBOA is removed by washing the cartridge with 20 mL of a 0.05 M sulfuric acid solution, and the two syn and anti 1-octen-3-one oximes are recovered with 2 mL of pentane. As the internal standard 2-octanol is added (30 μL of a 40-mg/L solution), the solution is concentrated to 100 μL prior analysis. Recovery is quantitative and the method provides an LOD of 0.75 ng/L, below the sensory threshold of the compound (15 ng/L). The analytical conditions used are reported in Table 8.2.

8.5 ANALYSIS OF 2-METHOXY-3,5-DIMETHYLPYRAZINE IN WINE

This compound is responsible for a "fungal must" taint reported in the wine cork industry. Characterized by unpleasant, musty, moldy odor, and a sensory threshold in white wine of 2.1 ng/L, this compound has been assessed as the second cause of cork taint (Simpson et al., 2004). Bacteria capable of producing 2-methoxy-3,5-dimethylpyrazine could be present in areas where the cork is processed or stored.

For wine analysis, 3-g sodium tetraborate is added to 100 mL of sample and liquid–liquid extraction is performed with pentane (2 × 10 mL). To purify the extract, the organic solution is washed with water (5 mL), then extracted with cold sulfuric acid 1 M (2 × 10 mL). The acid

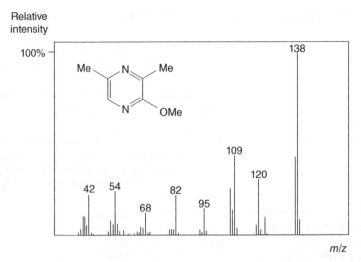

Figure 8.12. The EI mass spectrum (70 eV) of 2-methoxy-3,5-dimethylpyrazine. GC/ MS conditions: PEG fused silica capillary column (30 m × 0.25 mm; 0.25-μm film thickness); carrier gas He at flow rate 1.2 mL/min; oven temperature program: 50 °C for 1 min, increased to 220 °C at 10 °C/min, isotherm for 10 min; injector temperature 200 °C; transfer line temperature 250 °C. (Reproduced from *Journal of Agricultural and Food Chemistry*, 2004, 52 p. 5426, Simpson et al., with permission of American Chemical Society.)

extract is washed with pentane (5 mL), basified by addition of saturated sodium hydrogen carbonate (40 mL), and further extracted with pentane (2 × 8 mL). This solution is finally washed with water (5 mL), dried over anhydrous MgSO₄, and concentrated to 100 μL. The GC/ MS–SIM analysis is performed by monitoring the ions at *m/z* 109, 120, 137, and 138 (Simpson et al., 2004). The mass spectrum of 2-methoxy-3,5-dimethylpyrazine is reported in Fig. 8.12.

8.6 BIOGENIC AMINES IN GRAPE AND WINE

Biogenic amines are dangerous to human health and legal limits are fixed in grape and wine (structures are reported in Fig. 8.13). These compounds were found in fermented foods and beverages, such as cheeses, beer and fish, and meat products (Stratton et al., 1991; Shalaby, 1996). In wine, the most abundant are histamine, tyramine, putrescine, and phenylethylamine (Radler and Fath, 1991; Lehtonen, 1996).

A relationship between putrescine, cadaverine, and histamine was suggested as responsible for numerous cases of food intoxication

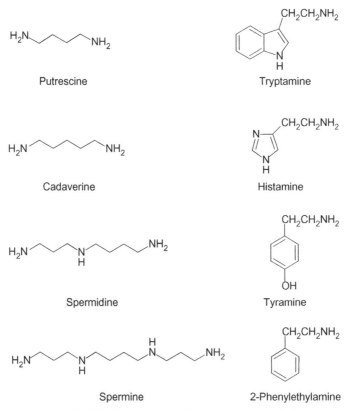

Figure 8.13. Structures of principal biogenic amines.

(Lovenberg, 1974). Besides, when putrescine and cadaverine are cooked they may be converted into pyrrolidine and piperidine, respectively (Yamamoto et al., 1982). These secondary amines, as well as spermidine and spermine, may undergo nitrosation forming the extremely carcinogenic compound nitrosamine. The aromatic amines β-phenylethylamine, tyramine, isopentylamine, and 3-(2-aminoethyl) indole (tryptamine), are responsible for dietary disturbances, including migraines and hypertension (Stratton et al., 1991; Anderson et al., 1993). In wine, these compounds are present as odorless salts, but at the pH in the mouth they may have repulsive smells.

Histamine is probably the most important amine with physiological effects for human health. Ingested daily in small amounts, it normally degrades, but it still can induce a drastic response in a number of sensitive people inducing symptoms, such as skin redness, headache, nausea, stomach disorder, and respiratory troubles, this is a pathology known as Histamine Intolerance. Usually people affected by this disorder cannot convert biogenic amines into harmless products, and histamine levels >500–1000 mg/kg must be considered dangerous to human health (Taylor, 1985). The simultaneous consumption of foods containing high biogenic amines and alcoholic beverages increases risks because ethanol reduces the ability of the human detoxification system to degrade histamine by the diamine oxidase enzyme. Hypertensive crises have been observed in psychiatric patients treated with drugs inhibiting monoamine oxidase following the consumption of alcoholic beverages (Kalač and Křížek, 2003).

Spermine, spermidine, and cadaverine can be present in grape berries, seeds, and vine leaves. Histamine, tyramine, and 1-methylhistamine can be present in vine leaves in trace amounts (Adams et al., 1990; Radler and Fath, 1991; Geny et al., 1997; Nicolini et al., 2003). More than 30 biogenic amines, produced by enzymatic degradation or decarboxylation of aminoacids during fermentation, were identified in wine (Ngim et al., 2000). Microorganisms, such as *lactobacilli* and *pediococci*, are mainly responsible for biogenic amines in wines. In additive, *oenococci* are able to produce amines (Delfini, 1989; Farias et al., 1993; Leitao et al., 2000). The ability of lactic acid bacteria in wine to decarboxylate histidine to histamine and tyrosine to tyramine, has been demonstrated (Lonvaud-Funel, 2001). Aliphatic primary amines in wines produced from botrytized grapes are higher than from intact grape berries, in particular 2-methyl-butyl amine, 3-methyl-butyl amine, and phenylethylamine (Kiss et al., 2006; Eder et al., 2002).

The maximal level of tolerance of histamine in wine has been established at 10 mg/L in Switzerland, 8 mg/L in France, 5–6 mg/L in Belgium, and 2 mg/L in Germany, however, the level for histamine-free wines should be <0.5 mg/L (Bauza et al., 1995; Lehtonen, 1996).

8.6.1 Preparation of Samples

For extraction of biogenic amines from grapes, the sample is added to an adequate aliquot of 10% $HClO_4$, homogenized, centrifuged, and the supernatant is filtered (Kiss et al., 2006). Alternatively, extraction can be performed by directly crushing and pressing grape (Bertoldi et al., 2004; Nicolini et al., 2003). Wines are usually just degassed and filtred.

Polyphenolic compounds can interfere in the analysis of red wines, and amino acids in the analysis of grape juices. Consequently several methods for isolating biogenic amines from wines and juices have been proposed: liquid-liquid extraction with butanol of the sample preliminarily concentrated and adjusted to pH 1.5 (Almy et al., 1983); in general, for SPE is preferred strong cation exchange (SCX) under acid conditions with respect to the use of strong anion exchange (SAX) or C_{18} cartridges (Yamamoto et al., 1982; Busto et al., 1995). Removal of phenolics before SPE can be achieved by sample treatment with polyvinylpyrrolidone (PVP) (Busto et al., 1994).

8.6.2 Analysis of Biogenic Amines

Biogenic amines are usually detected by LC with a pre- or postcolumn derivatization with o-phthalaldehyde in the presence of mercaptoethanol, and fluorimetric detection of derivatives. A sample derivatization also has to be done to perform GC/MS analysis of grape juice or wine. Amines are distilled from the alkalized sample and trapped in an acidified solution. After concentration under vacuum, salts of ethylamine, dimethylamine, ethylamine, diethylamine, n-propylamine, isobutylamine, α-amylamine, isoamylamine, pyrrolidine, and 2-phenethylamine are derivatized with trifluoroacetic (TFA) anhydride. Their derivatives are extracted with ethyl ether. GC/MS is performed using a capillary fused silica PEG column with an oven temperature programmed for 8 min at 70 °C, 1 °C/min to 160 °C, isotherm for 90 min (Daudt and Ough, 1980).

Alternatively, after a sample clean-up by anionic exchange, SPE derivatization can be performed using heptafluorobutyric anhydride (HFBA) at 80 °C for 60 min and derivatives are extracted with dichloromethane. By this approach, simultaneous determination of principal diamines, polyamines, and aromatic amines in wine and grape juices can be achieved (Fernandes and Ferreira, 2000), and 1,3-diaminopropane, putrescine, cadaverine, spermidine, spermine, β-phenylethylamine, and tyramine are also determined. The GC/MS–SIM analysis is performed recording one target ion signal and at least two qualifying ions of HFBA derivatives. Quantification is based on the signals at m/z 104 for β-phenylethylamine, m/z 480 for putrescine, m/z 494 for cadaverine, m/z 316 for tyramine, and m/z 254 for spermine and spermidine. Amphetamine, d_8-putrescine, 1,7-diaminoheptane, norspermidine, and norspermine are used as internal standards. The method has high reproducibility and LOD <10 µg/L. The GC of derivatives can be performed using a 5% phenyl–95% dimethlypolysiloxane capillary column (30 m × 0.25 mm, 0.25 µm) with an oven temperature program starting at 80 °C for 1 min, 15 °C/min to 210 °C, 20 °C/min to 290 °C, and 290 °C for 5 min.

The GC/MS analysis of biogenic primary alkylamines in wine was performed also by derivatization with pentafluorobenzaldeide (PFB) (Ngim et al., 2000). Derivatization of the sample at pH 12 is carried out for 30 min at 24 °C with a reagent concentration of 10 mg/mL. Derivatives are extracted with hexane, and analysis can be performed with a similar column used for HFBA derivatives, recording the signals at m/z 208 and 211. These signals correspond to α-cleavage products of undeuterated PFB-imines and methyl-d_3-PFB-imine, respectively. The signal at m/z 213 corresponds to the molecular ion of pentafluoronitrobenzene. A chromatogram of analysis of PFB-amines in a *Cabernet Sauvignon* wine is reported in Fig. 8.14. Compared with conventional LC, this method has higher selectivity, sensibility, and resolution.

Recently, a LC/ESI–MS method for analysis of tyramine, tryptamine, 2-phenylethylamine, histamine, cadaverine, putrescine, spermidine, and spermine in wine without any sample pretreatment, was

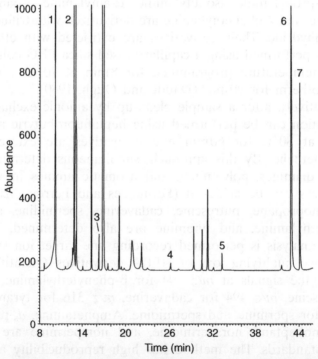

Figure 8.14. GC/MS analysis of amine-pentafluorobenzyl derivatives in a *Cabernet Sauvignon* wine. Analytical conditions: 5% phenyl–95% dimethlypolysiloxane capillary column (30 m × 0.25 mm, 0.25 μm); oven temperature program: 45 °C for 4 min, 15 °C/min to 280 °C, isotherm for 15 min. (**1**) methyl-d_3-amine and methylamine (coeluted), (**2**) ethylamine, (**3**) *n*-propylamine, (**4**) *n*-hexylamine, (**5**) 2-phenylethylamine, (**6**) 1,4-diaminobutane, and (**7**) 1,5-diaminopentane. (Reproduced from *Journal of Agricultural and Food Chemistry*, 2000, 48, p. 3314, Ngim et al., with permission of American Chemical Society.)

TABLE 8.3. Elution Program for LC/ESI–MS Analysis of Tyramine, Tryptamine, 2-Phenylethylamine, Histamine, Cadaverine, Putrescine, Spermidine, and Spermine in Wine[a]

Time (min)	% Solvent B	Flow (mL/min)
0.0–7.5	50	0.6
7.5–15.0	70	0.6
15.0–20.0	90	0.5
20.0–23.0	100	0.5
23.0–26.0	50	0.6

[a]Solvent A: water + ammonium acetate 5 mM + perfluoroheptanoic acid (PFHA) 5 mM; solvent B: methanol + ammonium acetate 5 mM + PFHA 5 mM (Millán et al., 2007).

proposed (Millán et al., 2007). After dilution and filtration, wines are directly injected into the column using heptylamine as an internal standard. Analysis is performed using a reverse-phase LC C_8 column (150 × 4.6 mm, 5 μm) at 30 °C with flow-rate and solvent gradient reported in Table 8.3. Under this condition the analysis takes ~20 min. Detection is performed operating in the positive-ion and MS/MS mode. Data acquisition and MS/MS parameters are reported in Table 8.4. Due to its high selectivity and unequivocal identification of compounds, quantitative analysis was performed in MS/MS mode.

For most amines, the most abundant ions are formed by loss of an ammonia group and for spermine and heptylamine the principal product ion comes from the loss of 1,3-propyldiamine and formation of an adduct with water, respectively. The LODs calculated in a synthetic wine were in the range between 0.5 and 40 μg/L. The higher values resulted for phenylethylamine, cadaverine, putrescine, and spermine (10–40 μg/L).

8.7 ETHYL CARBAMATE IN WINE

Ethyl carbamate (EC) is a potential human mutagen and carcinogen. The presence of EC in alcoholic beverages, especially dessert wines and spirits, can be up to several hundred micrograms per liter (Conacher et al., 1987). During fermentation, arginine is metabolized from yeasts forming urea, and this compound is, by reaction with ethanol, the EC precursor (Monteiro et al., 1989; Ough et al., 1988; 1990). Moreover, the arginine metabolism of wine lactic bacteria induces formation of citrulline, another EC precursor (Tegmo-Larsson et al., 1989; Liu and Pilone, 1998). The U.S. wine industry established a voluntary target for

TABLE 8.4. Data Acquisition Parameters of MRM Transitions Used for LC/ESI-MS Analysis of Biogenic Amines in Wine[a]

Compound	Retention Time (min)	Molecular Mass (MW)	Precursor Ion [M+H]⁺	MRM Transition (m/z)	Fragmentation Amplitude V	Fragmentation Width (m/z)
Tyramine	8.4	137.2	138	$138 \rightarrow 121$	0.50	4.0
Tryptamine	10.9	160.2	161	$161 \rightarrow 144$	0.45	4.0
2-Phenylethylamine	11.4	121.2	122	$122 \rightarrow 105$	0.60	4.0
Histamine	12.9	111.1	112	$112 \rightarrow 95$	1.25	10.0
Cadaverine	13.2	102.2	103	$103 \rightarrow 86$	1.30	10.0
Putrescine	13.2	88.2	89	$89 \rightarrow 72$	1.00	10.0
Heptylamine (I.S.)	14.5	115.1	116	$116 \rightarrow 58$	0.55	4.0
Spermidine	16.6	145.2	146	$146 \rightarrow 129$	0.65	4.0
Spermine	18.3	202.3	203	$203 \rightarrow 129$	0.65	4.0

[a]ESI conditions: isolation width 2.0, drying gas N₂ at 300°C and flow 9.0 L/min, nebulizer gas N₂ at pressure 40 psi, capillary voltage 3200 V (Millán et al., 2007).

266

EC (15 μg/L or less in table wines; 60 μg/L or less in fortified wines), the U.S. Food and Drug Administration published recommendations to minimize EC in wine (Butzke and Bisson, 1997).

8.7.1 The EC Analysis

Liquid–liquid extraction of EC from alcoholic beverages and wines can be performed using dichloromethane after saturation of the sample with NaCl (Conacher et al., 1987; Daudt et al., 1992), or extraction with diethyl ether after adjusting the sample to pH 9 using *n*-butyl carbamate or cyclopentyl carbamate as an internal standard (Fauhl and Wittkowski, 1992; Ferreira and Fernandes, 1992).

The International Association of Official Analytical Chemists adopted an SPE method with the use of prepacked diatomaceous cartridges (50 mL) performing elution of the analyte with dichloromethane for determination of EC in alcoholic beverages (AOAC, 1995).

Another SPE method was developed with the use of cross-linked copolymer styrene–divinylbenzene cartridges (ENV+), to perform the recovery of the analyte with ethyl acetate after removing ethanol from wine under vacuum and using $^{13}C^{15}N$-labeled EC as an internal standard (Jagerdeo et al., 2002).

A SPME/GC/MS method for analysis of EC in wine was developed using a 65-μm PEG/DVB fiber (Whiton and Zoecklein, 2002). Propyl carbamate was added to 7 mL of wine as an internal standard, the fiber was exposed for 30 min to headspace of the sample at 22 °C, and the analyte was desorbed from the fiber into a GC injection port at 250 °C. An LOD of 9.6 μg/L was achieved by this method.

The GC/MS–SIM analysis performed by monitoring the EC ion at *m/z* 62 (and at *m/z* 64 for labeled EC), provides LOD and LOQ of 0.1 and 1 μg/L, respectively (AOAC, 1995; Jagerdeo et al., 2002). Figure 8.15 shows the EI-70-eV mass spectrum of EC. The GC/MS conditions are reported below.

8.8 WINE GERANIUM TAINT

Potassium sorbate is used as a yeast inhibitor for the stabilization of table wines containing residual sugar. When conditions permit the growth of lactic acid bacteria, wines treated with sorbic acid can develop an odor resembling crushed geranium leaves (Burkhardt, 1973; Radler, 1976; Wurdig et al., 1975). This result due to bacterial reduction

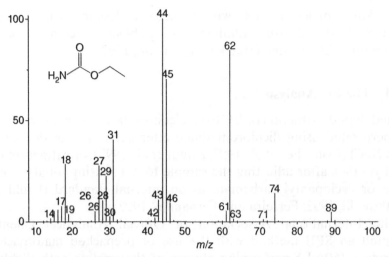

Figure 8.15. The EI (70 eV) mass spectrum of ethyl carbamate. Conditions for GC/MS analysis: PEG fused silica capillary column (30 m × 0.25 mm, 0.25 μm); oven temperature program: 40 °C for 0.75 min, increased to 60 °C at 10 °C/min, then to 150 °C at 3 °C/min.

of sorbic acid to (*E*,*E*)-2,4-hexadien-1-ol (sorbyl alcohol) (Edinger and Splittstoesser, 1986; Wurdig et al., 1975), the precursor of 2-ethoxyhexa-3,5-diene, which is the compound responsible for a geranium odor (Crowell and Guymon, 1975; von Rymon-Lipinski et al., 1975).

8.8.1   2-Ethoxyhexa-3,5-diene Analysis

The wine sample is extracted by ethyl acetate and the organic solution is dried with magnesium sulfate. The GC/MS analysis of the extract can be performed using a fused silica methyl silicone column and the following oven temperature program: 60 °C isotherm for 2 min, then raised to 250 °C at 6 °C/min. Identification of analyte is based on the library mass spectrum shown in Fig. 8.16 (Chisholm and Samuels, 1992).

8.9   MOUSY OFF-FLAVOR OF WINES

Mousy taint is a microbiological defect of wine due to *Brettanomyces/Dekkera* yeasts, as well as certain lactic acid bacteria, such as *Leuconostoc oenos* (*Oenococcus oeni*) and *Lactobacillus*. *Brettanomyces* yeasts are frequently found in wooden casks. Mousy taint can occur particularly

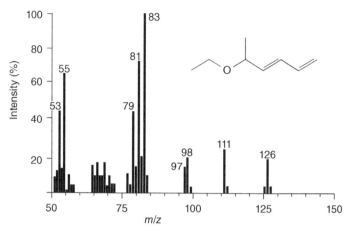

Figure 8.16. The EI (70 eV) mass spectrum of 2-ethoxyhexa-3,5-diene. (Reproduced from *Journal of Agricultural and Food Chemistry*, 1992, 40, p. 632, Chisholm et al., with permission of American Chemical Society.)

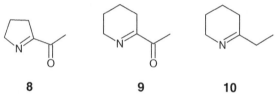

8   **9**   **10**

Figure 8.17. Chemical structures of *N*-heterocycles responsible for mousy off-flavor of wines: (**8**) 2-acetyl-1-pyrroline (APY), (**9**) 2-acetyltetrahydropyridine (ATHP), (**10**) 2-ethyltetrahydropyridine (ETHP).

in wines that are low in acid, that are oxidative, and that have a residual sugar content. 2-Ethyltetrahydropyridine (ETHP), 2-acetyl-1-pyrroline (APY), 2-acetyl-3,4,5,6-tetrahydropyridine (ATHP), and 2-acetyl-1,4,5,6-tetrahydropyridine (the latter are two tautomeric forms probably pH dependent), have been identified in mousy wines (structures in Fig. 8.17) (Lay, 2003; duToit and Pretorius, 2000; Costello et al., 2001; Strauss and Heresztyn, 1984; Herderich, et al., 1995).

2-Acetyl-1-pyrroline was reported to be the major contributor to mousy off-flavor (Herderich, et al., 1995), with an aroma impact of one order of magnitude greater than ATHP (Buttery et al., 1982), but it is a relatively unstable compound and was found in wine in trace quantities up to 7.8 μg/L (Grbin et al., 1996). At the pH of wine these compounds are not volatile and as a consequence they have a low sensory impact. However, when mixed with the neutral pH of saliva they can become very apparent on the palate as mouse cage or mouse urine (Snowdon et al., 2006).

8.9.1   Extraction and Analysis of Mousy *N*-Heterocycles

After addition of 4-acetylpyridine as an internal standard, 250 mL of wine is adjusted to pH 2.5 and extracted with Freon 11 (3 × 100 mL) in order to remove acidic and neutral compounds to reduce the GC/MS background interference. Continuous liquid–liquid extraction of basic mousy *N*-heterocycles is then performed overnight with Freon 11 at pH 8.0 (Herderich et al., 1995). After the extract is dried over Na_2SO_4, 3-acetylpyridine can be added as a second internal standard, and the organic phase is concentrated by distillation at 37 °C, replacing Freon 11 with 0.5 mL of dichloromethane. Prior to GC/MS, the extract is further concentrated to 10 µL of isooctane.

Analysis is performed with a PEG capillary column and an oven temperature program from 60 to 220 °C at 3 °C/min. Analytes are detected by recording the signals at *m/z* 111 and 110 for ETHP, *m/z* 111 and 83 for APY, *m/z* 125 and 82 for ATHP. Quantification of ATHP, ETHP, and APY is performed recording the signals at *m/z* 111, 111 and 125, respectively, and the signal at *m/z* 121 for 4-acetylpyridine.

For the chromatographic conditions used, in addition to the two tautomeric forms of ATHP, two isomers of both ETHP and APY can be yielded as well (Costello and Henschke, 2002).

A SPME–GC/MS method applied to analysis of APY in rice was developed performing sampling at 80 °C with a CAR/DVB/PDMS fiber exposed to the sample headspace for 15 min (Grimm et al., 2001).

REFERENCES

Adams, D.O., Franke, K.E., and Christensen, L.P. (1990). Elevated putrescine levels in grapevine leaves that display symptoms of potassium deficiency, *Am. J. Enol. Vitic.*, **41**, 121–125.

Almy, J., Ough, C.S., and Crowell, E.A. (1983). Identification of two new volatile amines in wine, *J. Agric. Food Chem.*, **31**, 911–913.

Amon, J.M., Vandepeer, J.M., and Simpson, R.F. (1989). Compounds responsible for cork taint in wine, *Aust. New Zeal. Wine Ind. J.*, **4**, 62–69.

Anderson, M., Hasan, F., McCroden, J., and Tipton, K. (1993). Monoamine oxidase inhibitors and the cheese effect, *Neurochem. Res.*, **18**(11), 1145–1149.

AOAC (1995). Official Methods of Analysis of AOAC International, 16th ed, Vol. II. Chapt. 28. Method 994.07. *AOAC*, Gaithersburg, MD.

Aung, L.H., Smilanick, J.L., Vail, P.V., Hartsell, P.L., and Gomez, E. (1996). Investigations into the origin of chloroanisoles causing musty off-flavor of raisins, *J. Agric. Food Chem.*, **44**(10), 3294–3296.

Bauza, T., Blaise, A., Mestres, P., Teissedre, P.L., Daumas, F., and Cabanis, J.C. (1995). Teneurs en amines biogènes et facteurs de leurs variations dans les vins des côtes du Rhône, de la vallèe du Rhône et de Provence, *Sci. Aliments*, **15**, 367–380.

Bertoldi, D., Larcher, R., and Nicolini, G. (2004). Content of some free amines in grapes from Trentino. *Ind. delle Bevande*, **33**, 437–441.

Burdaspal, P.A. and Legarda, T.M. (1999). Ochratoxin A in wine and grape musts and juices produced in Spain and other European countries, *Alimentaria*, **299**, 107–113.

Burkhardt, R. (1973). Occasional occurrence of unpleasant odor and flavor of musts (sweet reserve) and wines after treatment with potassium sorbate, *Ges. Dtsch. Chem. Fachgruppe Lebensm. Gericht. Chem. Mitteilungsbl.*, **27**, 259–261.

Burttshel, R.H., Rosen, A.A., Middleton, F.M., and Ettinger, M.B. (1951). Chlorine derivatives of phenol causing taste and odor, *J. Am. Water Works Ass.*, **51**, 205–214.

Busto, O., Guasch, J., and Borrull, F. (1995). Improvement of a solid-phase extraction method for determining biogenic amines in wines, *J. Chromatogr. A*, **718**, 309–317.

Busto, O., Valero, Y., Gulasch, J., and Borrull, F. (1994). Solid phase extraction applied to the determination of biogenic amines in wines by HPLC, *Chomatographia*, **38**, 571–578.

Buttery, R.G., Ling, L.C., and Juilano, B.O. (1982). 2-Acetyl-1-pyrroline: an important aroma component of cooked rice, *Chem. Ind. (London)*, 958–959.

Butzke, C.E. and Bisson L.F. (1997). Ethyl carbamate preventative action manual. U.S. Food and Drug Administration, Center for Food Safety and Applied Nutrition, http://vm.cfsan.fda.gov/~frf/ecaction.html

Careri, M., Mazzoleni, V., Musci, M., and Molteni, R. (2001). Study of electron beam irradiation effects on 2,4,6-trichloroanisole as a contaminant of cork by gas chromatography-mass spectrometry, *Chromatographia*, **53**(9–10), 553–557.

Castegnaro, M., Mohr, U., Pfohl-Leszkowicz, A., Esteve, J., Steinmann, J., Tillmann, T., Michelon, J., and Bartsch, H. (1998). Sex-and strain-specific induction of renal tumors by ochratoxin A in rats correlates with DNA adduction, *Int. J. Cancer*, **77**, 70–75.

Castellari, M., Fabbri, S., Fabiani, A., Amati, A., and Galassi, S. (2000). Comparison of different immunoaffinity clean-up procedures for high-performance liquid chromatographic analysis of ochratoxin A in wines, *J. Chromatogr. A*, **888**, 129–136.

Chatonnet, P., Bonnet, S., Boutou, S., and Labadie, M.D. (2004). Identification and responsibility of 2,4,6-tribromoanisole in musty, corked odors in wine, *J. Agric. Food. Chem.*, **52**, 1255–1262.

Chiodini, A.M., Scherpenisse, P., and Bergwerff, A.A. (2006). Ochratoxin A contents in wine: comparison of organically and conventionally produced products, *J. Agric. Food Chem.*, **54**(19), 7399–7404.

Chisholm, M.G. and Samuels, J.M. (1992). Determination of the Impact of the Metabolites of Sorbic Acid on the Odor of a Spoiled Red Wine, *J. Agric. Food Chem.*, **40**, 630–633.

Commission of the European Communities. (1998). Directorate General XXIV, Scientific Committee on Food, Outcome of Discussions 14, expressed on 17 September.

Conacher, H.B.S., Page, B.D., Lau, B.P.Y., Lawrence, J.F., Bailey, R, Calway P., Hanchay, J.P., and Mori, B. (1987). Capillary column gas chromatographic determination of ethyl carbamate in alcoholic beverages with confirmation by gas chromatography/mass spectrometry, *J. Ass. Official. Anal. Chem.*, **70**, 749–751.

Costello, P.J. and Henschke, P.A. (2002). Mousy off-flavor of wine: Precursors and biosynthesis of the causative N-heterocycles 2-ethyltetrahydropyridine, 2-acetyltetrahydropyridine, and 2-acetyl-1-pyrroline by *Lactobacillus hilgardii* DSM 20176, *J. Agric. Food Chem.*, **50**(24), 7079–7087.

Costello, P.J., Lee, T.H., and Henschke, P.A. (2001). Ability of lactic acid bacteria to produce N-heterocycles causing mousy off-flavour in wine, *Aust. J. Grape Wine R.*, **7**, 160–167.

Council Directive 91/173/EEC of 21 March 1991, amending for the ninth time Directive 76/769/EEC on the approximation of the laws, regulations and administrative provisions of the Member States relating to restrictions on the marketing and use of certain dangerous substances and preparations, Official Journal L 085. 05/04/1991 0034–0036.

Cristoni, S., Bernardi, L.R., Biunno, I., Tubaro, M., and Guidugli, F. (2003). Surface-activated no-discharge atmospheric pressure chemical ionization, *Rapid. Commun. Mass Spectrom.*, **17**(17), 1973–1981.

Crowell, E.A. and Guymon, J.F. (1975). Wine constituents arising from sorbic acid addition, and identification of 2-ethoxyhexa-3,5-diene as source of geranium-like off-odor, *Am. J. Enol. Vitic.*, **26**, 97–102.

Culleré, L., Cacho, J., and Ferreira, V. (2006). Validation of an analytical method for the solid phase extraction, in cartridge derivatization and subsequent gas chromatographic-ion trap tandem mass spectrometric determination of 1-octen-3-one in wines at ngL^{-1} level, *Anal. Chim. Acta The Netherlands*, **563**(1–2), 51–57.

Darriet, P., Pons, M., Lamy, S., and Dubourdieu, D. (2000). Identification and quantification of geosmin, an earthy odorant contaminating wines, *J. Agric. Food Chem.*, **48**, 4835–4838.

Darriet, P., Pons, M., Henry, R., Dumont, O., Findeling, V., Cartolaro, P., Calonnec, A., and Dubourdieu, D. (2002). Impact odorants contributing to the fungus type aroma from grape berries contaminated by powdery mildew

(Uncinula necator); incidence of enzymatic activities of the yeast saccharomyces cerevisiae, *J. Agric. Food Chem.*, **50**(11), 3277–3282.

Daudt, C.E., Ough, C.S., Stevens, D., and Herraiz, T. (1992). Investigations into ethyl carbamate, n-propyl carbamate, and urea in fortified wines, *Am. J. Enol. Vitic.*, **43**(4), 318–322.

Daudt, C.E. and Ough, C.S. (1980). A method for detecting volatile amines in grapes and wines, *Am. J. Enol. Vitic.*, **31**, 356–360.

Delfini, C. (1989). Ability of wine lactic acid bacteria to produce histamine, *Sci. Aliments*, **9**, 413–416.

Dumoulin, M. and Riboulet, J.-M. (2004). Analyse de géosmine dans les vins: apport de la microextraction en phase solide (SPME) couplée à l'analyse par GC–MS, *Rev. Franç. Œnol.*, **208** septembre/octobre, 28–30.

duToit, M. and Pretorius, I.S. (2000). "Microbial spoilage and preservation of wine: Using weapons from nature's own arsenal—A review", *South Afr. J. Enol. Vitic.*, **21**, 74–96.

Eder, R., Brandes, W., and Paar, E. (2002). Influence of grape rot and fining agents on the contents of biogenic amines in musts and wines. *Mitt. Klosterneuburg, Rebe Wein, Obstbau FrüchteVerwertung*, **52**, 204–217.

Edinger, W.D. and Splittstoesser, D.F. (1986). Production by lactic acid bacteria of sorbic alcohol, the precursor of the geranium odor compound, *Am. J. Enol. Vitic.*, **37**, 34–38.

Evans, T.J., Butzke, C.E., and Ebeler S.E. (1997). Analysis of 2,4,6-trichloroanisole in wines using solid-phase microextraction coupled to gas chromatography-mass spectrometry, *J. Chromatogr. A*, **786**, 293–298.

Farias, M.E., Manca de Nadra, M.C., Rollan, G.C., and Strasser de Saad, A.M. (1993). Histidine decarboxylase activity in lactic acid bacteria from wine, *J. Int. Sci. Vigne Vin*, **27**, 191–199.

Fauhl, C. and Wittkowski, R. (1992). Determination of ethyl carbamate in wine by GC–SIM–MS after continous extraction with diethyl ether, *J. High. Res. Chromatogr.*, **15**, 203–205.

Fernandes, J.O. and Ferreira, M.A. (2000). Combined ion-pair extraction and gas chromatography-mass spectrometry for the simultaneous determination of diamines, polyamines and aromatic amines in Port wine and grape juice. *J. Chromatogr. A*, **886**, 183–195.

Ferreira, M.A. and Fernandes, J.O. (1992). The application of an improved GC-MS procedure to investigate ethyl carbamate behaviour during the production of Madeira wines, *Am. J. Enol. Vitic.*, **43**(4), 339–343.

Filali, A., Ouammi, L., Betbeder, A.M., Baudrimont, I., Soulaymani, R., Benayada, A., and Creppy, E.E. (2001). Ochratoxin A in beverages from Morocco: a preliminary survey, *Food Addit. Contam.*, **18**(6), 565–568.

Flamini, R. and Larcher, R. (2008). Grape and Wine Contaminants: Ochratoxin A, Biogenic Amines, Trichloroanisole and Ethylphenols, In Hyphenated

Techniques in Grape & Wine Chemistry, Riccardo Flamini (Ed.). *John Wiley & Sons, Ltd.*, pp. 129–172.

Flamini, R. and Panighel, A. (2006). Mass spectrometry in grape and wine chemistry. Part II: The Consumer Protection, *Mass Spectrom. Rev.*, **25**(5), 741–774.

Flamini, R., Dalla Vedova, A., De Rosso, M., and Panighel, A. (2007). A new sensitive and selective method for analysis of ochratoxin A in grape and wine by direct liquid chromatography/surface activated chemical ionization-tandem mass spectrometry, *Rapid Commun. Mass Spectrom.*, **21**, 3737–3742.

Garcia-Moruno, E., Sanlorenzo, C., Di Tommaso, D., and Di Stefano, R. (2004). Un metodo per la determinazione dell'ocratossina A nelle uve: applicazione allo studio dell'influenza del sistema di appassimento delle uve sul contenuto in ocratossina A, *Riv. Vitic. Enol.*, **1/2**, 3–11.

Geny, L., Broquedis, M., Martin-Tanguy, J., and Bouard, J. (1997). Free, conjugated, and wall-bound polyamines in various organs of fruiting cuttings of Vitis vinifera L. CV. Cabernet Sauvignon, *Am. J. Enol. Vitic.*, **48**, 80–84.

Grbin, P.R., Costello, P.J., Herderich, M., Markides, A.J., Henschke, P.A., and Lee, T.H. (1996). Developments in the sensory, chemical and microbiological basis of mousy taint in wine, In *Proceedings of the Ninth Australian Wine Industry Technical Conference*, 16–19, July, Stockely, C. S., Sas, A. N., Johnstone, R. S., and Lee, T. H. (Eds.), Winetitles: Adelaide, Australia, pp. 57–61.

Grimm, C.C., Bergman, C., Delgado, J.T., and Bryant, R. (2001). Screening for 2-Acetyl-1-pyrroline in the headspace of rice using SPME/GC-MS, *Agric. Food Chem.*, **49**, 245–249.

Hayasaka, Y., MacNamara, K., Baldock, G.A., Taylor, R.L., and Pollnitz, A.P. (2003). Application of stir bar sorptive extraction for wine analysis, *Anal. Bioanal. Chem.*, **375**, 948–955.

Herderich, M., Costello, P.J., Grbin, P.R., and Henschke, P.A. (1995). The occurrence of 2-acetyl-1-pyrroline in mousy wines, *Nat. Prod. Lett.*, **7**, 129–132.

Hohler, D. (1998). Ochratoxin A in food and feed: occurence, legislation and mode of action, *Z. Ernaehrungswiss*, **37**, 2–12.

Insa, S., Anticó, E., and Ferreira, V. (2005). Highly selective solid-phase extraction and large volume injection for the robust gas chromatography–mass spectrometric analysis of TCA and TBA in wines, *J. Chromatogr. A*, **1089**, 235–242.

International Organization of Vine and Wine (2006). Measuring ochratoxin A in wine after going through an immunoaffinity column and HPLC with fluorescence detection, In *Compendium of international methods of wine and must analysis. International Organization of Vine and Wine*, Vol. 2.

Jagerdeo, E., Dugar, S., Foster, G.D., and Schenck, H. (2002). Analysis of ethyl carbamate in wines using solid-phase extraction and multidimensional gas chromatography/mass spectrometry, *J. Agric. Food. Chem.*, **50**(21), 5797–5802.

JEFCA (Joint FAO/WHO Expert Committee on Food Additives) (2001). Ochratoxin A. In: *Safety Evaluation of Certain Mycotoxins in Food*: WHO Food Additives Series 47; FAO Food and Nutrition Paper 74; *WHO*: Geneva, Switzerland; p. 366.

Jönsson, S., Uusitalo, T., van Bavel, B., Gustafsson, I.-B., and Lindström, G. (2006). Determination of 2,4,6-trichloroanisole and 2,4,6-tribromoanisole on $ng L^{-1}$ to $pg L^{-1}$ levels in wine by solid-phase microextraction and gas chromatography-high-resolution mass spectrometry, *J. Chromatogr. A*, **1111**, 71–75.

Kalač P. and Krížek, M. (2003). A review of biogenic amines and polyamines in beer, *J. Inst. Brew.* **109**(2), 123–128.

Kiss, J., Korbász, M., and Sass-Kiss, A. (2006). Study of amine composition of botrytized grape berries, *J. Agric. Food Chem.*, **54**, 8909–8918.

Lau, B.P.Y., Scott, P.M., Lewis, D.A., and Kanhere, S.R. (2000). Quantitative determination of ochratoxin A by liquid chromatography/electrospray tandem mass spectrometry, *J. Mass. Spectrom.*, **35**, 23–32.

Lay, H. (2003). Research on the development of "Mousy taint" in wine and model solutions, *Mitteilungen Klosterneuburg, Rebe und Wein, Obstbau und Früchteverwertung*, Austria, **54**(7–8), 243–250.

Lee, T.H. and Simpson, R.F. (1993). *Wine microbiology and biotechnology*, GN Fleet (ed.) Harwood Academic Publishers, Philadelphia, pp. 353–372.

Lehtonen, P. (1996). Determination of amines and amino acids in wine—A review, *Am. J. Enol. Vitic.*, **47**, 127–133.

Leitao, M.C., Teixeira, H.C., Barreto Crespo, M.T., and San Romao, M.V. (2000). Biogenic amines occurrence in wine. Amino acid decarboxylase and proteolytic activities expression by *Oenococcus oeni*, *J. Agric. Food Chem.*, **48**, 2780–2784.

Leitner, A., Zöllner, P., Paolillo, A., Stroka, J., Papadopoulou-Bouraoui, A., Jaborek, S., Anklam, E., and Lindner, W. (2002). Comparison of methods for the determination of ochratoxin A in wine, *Anal. Chim. Acta*, **453**, 33–41.

Lindenmeier, M., Schieberle, P., and Rychlik, M. (2004). Quantification of ochratoxin A in foods by a stable isotope dilution assay using high-performance liquid chromatography-tandem mass spectrometry, *J. Chromatogr. A*, **1023**, 57–66.

Liu, S.Q. and Pilone, G.J. (1998). A review: Arginine metabolism in wine lactic acid bacteria and its pratical significance, *J. Appl. Microb.*, **84**, 315–327.

Lizarraga, E., Irigoyen, A., Belsue, V., and González-Peñas, E. (2004). Determination of chloroanisole compounds in red wine by headspace solid-phase microextraction and gas chromatography-mass spectrometry, *J. Chromatogr. A*, **1052**(1–2), 145–149.

Lonvaud-Funel, A. (2001). Biogenic amines in wines: Role of lactic acid bacteria, *FEMS Microbiol. Lett.*, **199**, 9–13.

Lovenberg, W. (1974). Psycho and vasoactive compounds in food substances, *J. Agric. Food Chem.*, **22**(1), 23–26.

MacDonald, S., Wilson, P., Barnes, K., Damant, A., Massey, R., Mortby, E., and Shepherd, M.J. (1999). Ochratoxin A in dried vine fruit: method development and survey, *Food Addit. Contam.*, **16**(6), 253–260.

Medina, A., Valle-Algarra, F.M., Gimeno-Adelantado, J.V., Mateo, R., Mateo, F., and Jiménez, M. (2006). New method for determination of ochratoxin A in beer using zinc acetate and solid-phase extraction silica cartridges, *J. Chromatogr. A*, **1121**, 178–183.

Millán, S.M., Sampedro, C., Unceta, N., Goicolea, M.A., and Barrio, R.J. (2007). Simple and rapid determination of biogenic amines in wine by liquid chromatography–electrospray ionization ion trap mass spectrometry, *Anal. Chim. Acta*, **584**, 145–152.

Monteiro, F.F., Trousdale, E.K., and Bisson, L.F. (1989). Ethyl carbamate formation in wine: Use of radioactively labeled precursors to demonstrate the involvement of urea, *Am. J. Enol. Vitic.*, **40**, 1–8.

Ngim, K.K., Ebeler, S.E., Lew, M.E., Crosby, D.G., and Wong, J.W. (2000). Optimized procedures for analyzing primary alkylamines in wines by pentafluorobenzaldehyde derivatization and GC-MS, *J. Agric. Food Chem.*, **48**, 3311–3316.

Nicolini, G., Larcher, R., and Bertoldi, D. (2003). Free amines in grape juices of Vitis vinifera L. wine varieties", *J. Comm. Sci.*, **42**(2), 67–78.

Ough, C.S., Crowell, E.A., and Mooney, L.A. (1988). Formation of ethyl carbamate precursors during grape juice (Chardonnay) fermentation. I. Addition of amino acids, urea, and ammonia: effects of fortification on intracellular and extracellular precursors, *Am. J. Enol. Vitic.*, **39**(3), 243–249.

Ough, C.S., Stevens, D., Sendovski, T., Huang, Z., and An, D. (1990). Factors contributing to urea formation in commercially fermented wines, *Am. J. Enol. Vitic.*, **41**, 68–73.

Pfohl-Leszkowicz, A., Pinelli, E., Bartsch, H., Mohr, U., and Castegnaro, M. (1998). Sex- and strain-specific expression of cytochrome P450s in ochratoxin A-induced genotoxicity and carcinogenicity in rats, *Mol. Carginogen.*, **23**, 76–85.

Radler, F. (1976). Degradation of sorbic acid by bacteria, *Bull. O.I.V.*, **49**, 629–635.

Radler, F. and Fath, K.-P. (1991). Histamine and other biogenic amines in wines, In J. Rantz (Ed.). *Proceedings of International Symposium on*

Nitrogen in Grapes and Wine, American Society for Enology and Viticulture, Davis, CA.

Sáez, J.M., Medina, A., Gimeno-Adelantado, J.V., Mateo, R., and Jiménez, M. (2004). Comparison of different sample treatments for the analysis of ochratoxin A in must, wine and beer by liquid chromatography, *J. Chromatogr. A*, **1029**(1–2), 125–133.

Sauvant, C., Silbernagel, S., and Gekle, M. (1998). Exposure to ochratoxin A impairs organic anion transport in proximal-tubule-derived opossum kidney cells, *J. Pharmacol. Exp. Ther.*, **287**, 13–20.

Schwerdt, G., Freudinger, R., Mildenberger, S., Silbernagl, S., and Gekle, M. (1999). The nephrotoxin ochratoxin A induces apoptosis in cultured human proximal tubule cells, *Cell. Biol. Toxicol.*, **15**, 405–415.

Serra, R., Mendonça, C., Abrunhosa, L., Pietri, A., and Venâncio, A. (2004). Determination of ochratoxin A in wine grapes: comparison of extraction procedures and method validation, *Anal. Chim. Acta*, **513**(1), 41–47.

Shalaby, A.R. (1996). Significance of biogenic amines to food safety and human health, *Food Res. Int.*, **29**, 675–690.

Shephard, G.S., Fabiani, A., Stockenström, S., Mshicileli, N., and Sewram, V. (2003). Quantitation of Ochratoxin A in South African Wines, *J. Agric. Food Chem.*, **51**(4), 1102–1106.

Simpson, R.F., Capone, D.L., and Sefton, M.A. (2004). Isolation and identification of 2-Methoxy-3,5-dimethylpyrazine, a potent musty compound from wine corks, *J. Agric. Food Chem.*, **52**, 5425–5430.

Snowdon, E.M., Bowyer, M.C., Grbin, P.R., and Bowyer, P.K. (2006). Mousy off-flavour: A review, *J. Agric. Food Chem.*, **54**(18), 6465–6474.

Soleas, G.J., Yan, J., Seaver, T., and Goldberg, D.M. (2002). Method for the gas chromatographic assay with mass selective detection of trichloro compounds in corks and wines applied to elucidate the potential cause of cork taint, *J. Agric. Food Chem.*, **50**(5), 1032–1039.

Stratton, J., Hutkins, R., and Taylor, S. (1991). *Biogenic amines* in cheese and other fermented foods, *J. Food Protect.*, **54**, 460–470.

Strauss, C.R. and Heresztyn, T. (1984). 2-Acetyltetrahydropyridines a cause of the "mousy" taint in wine, *Chem. Ind. (London)*, 109–110.

Taylor, S.L. (1985). Histamine Poisoning Associated with Fish, Cheese and Other Foods, FAO/WHO Monograph VPH/F05/85.1, 1–47.

Tegmo-Larsson, I.M., Spittler, T.D., and Rodriguez, S.B. (1989). Effect of malolactic fermentation on ethyl carbamate formation in Chardonnay wine, *Am. J. Enol. Vitic.*, **40**(2), 106–108.

Timperio, A.M., Magro, P., Chilosi, G., and Zolla, L. (2006). Assay of ochratoxin A in grape by high-pressure liquid chromatography coupled on line with an ESI-mass spectrometry, *J. Chromatogr. B*, **832**(1), 127–133.

Tonus, T., Dalla Vedova, A., Spadetto, S., and Flamini, R. (2005). Study of chemical methods for monitoring of ochratoxin A (OTA) in grape, *Riv. Vitic. Enol.*, **1**, 3–13.

Visconti, A., Pascale, M., and Centone, G. (1999). Determination of ochratoxin A in wine by means of immunoaffinity column clean-up and high-performance liquid chromatography, *J. Chromatogr. A*, **864**, 89–101.

von Rymon-Lipinski, G.W., Luck, E., Oeser, H., and Lomker, F. (1975). Formation and causes of the "geranium off-odor", *Mitt. Klosterneuburg*, **25**, 387–394.

Whitfield, F.B., Hill, J.L., and Shaw, K.J. (1997). 2,4,6-Tribromoanisole: a Potential Cause of Mustiness in Packaged Food, *J. Agric. Food Chem.*, **45**, 889–893.

Whiton, R.S. and Zoecklein, B.W. (2002). Determination of ethyl carbamate in wine by solid-phase microextraction and gas chromatography/mass spectrometry, *Am. J. Enol. Vitic.*, **53**(1), 60–63.

Wurdig, G., Schlotter, H.A., and Klein, E. (1975). Origin of geranium aroma in wines treated with sorbic acid, *Connaiss. Vigne Vin*, **9**, 43–55.

Yamamoto, S., Itano, H., Kataoka, H., and Makita, M. (1982). Gas-liquid chromatographic method for analysis of di- and polyamines in foods, *J. Agric. Food Chem.*, **30**(3), 435–439.

Zimmerli, B. and Dick, R. (1996). Ochratoxin A in table wine and grape juice: occurence and risk assessment, *Foods Add. Contam.*, **13**(6), 655–688.

Zöllner, P., Leitner, A., Luboki, D., Cabrera, K., and Lindner, W. (2000). Application of a Chromolith SpeedROD RP-18e HPLC column: determination of ochratoxin A in different wines by high-performance liquid chromatography-tandem mass spectrometry, *Chromatographia*, **52**(11/12), 818–820.

9

PESTICIDES IN GRAPE AND WINE

9.1 INTRODUCTION

The high concern about health risks connected with the use of fungicides, insecticides, and herbicides, as well as the possible presence of
their residues in processed foods and drinks, led to the development
of several European Community (EC) Directives stating maximum
residue limits (MRLs) tolerated for each food commodity. The main
parasites affecting grape in vineyards are powdery mildew (*Uncinula
necator*), downy mildew (*Plasmopora viticola*), gray mold (*Botryitis
cinerea*), and European grapevine moth (*Lobesia botrana*). A large
number of pesticides are commonly used to control these diseases.
Some maximum limits of contaminants in grape fixed by regulations,
are reported in Table 9.1.

In Italy, the Ministry of Health controls the registration process of
each pesticide before entering the market: the culture, rate, preharvest
interval (PHI), and MRL defined as "residues of active ingredients
and relative impurities that are present in products destined to human
and animal feeding, and resulting from the pesticide use inclusive of
those substances of toxicological significance deriving from metabolism
and degradation of the active ingredients," are issued. It is commonly

Mass Spectrometry in Grape and Wine Chemistry, by Riccardo Flamini
and Pietro Traldi
Copyright © 2010 John Wiley & Sons, Inc.

TABLE 9.1. Maximum Residue Limits in Grape[a]

Pesticide	Grape (mg/kg)	Wine (mg/L)
Azoxystrobin	2	0.5
Bromuconazole	0.5	0.2
Buprofezin	1	0.5
Cyazofamid	1	0.05
Cyproconazole	0.2	0.02
Cyprodinil	5	0.5
Dazomet		0.02
Diethofencarb	1	0.3
Etofenprox	1	0.1
Etoxazole	0.02	0.01
Fenamidone	0.5	0.5
Fenazaquin	0.2	0.01
Fenhexamid	5	1.5
Fenpropidin	2	0.5
Flazasulfuron	0.01	0.01
Fluazinam	1	0.02
Fludioxonil	2	0.5
Hesaconazole	0.1	0.01
Indoxacarb	0.5	0.02
Iprodione	10	2
Iprovalicarb	2	1
Mepanipyrim	3	1
Metalaxyl-M	1	0.2
Metam-Sodium	2	0.2
Methoxyfenozide	1	0.05
Myclobutanil	1	0.1
Procymidone	5	0.5
Pyrimethanil	3	2
Quinoxyfen	0.5	0.01
Spinosad	0.2	0.01
Spiroxamine	1	0.5
Tebuconazole	1	0.5
Tebufenozide	0.5	0.1
Tebufenpyrad	0.3	0.1
Teflubenzuron	1	0.01
Thiamethoxam	0.5	0.5
Trifloxystrobin	3	0.3
Ziram	2	0.2
Zoxamide	5	0.5

[a]From Cabras and Caboni, 2008.

considered that a pesticide is "not present" when the residue is <0.01 mg/kg. The MRLs of pesticides on grapes currently registered in Italy are reported in Table 9.2. Structures of the principal pesticides used in viticulture are reported in Fig. 9.1.

TABLE 9.2. Maximum Residue Limits of Pesticides on Grapes Registered in Italy[a]

Pesticide (mg/kg)	MRL	Pesticide (mg/kg)	MRL	Pesticide (mg/kg)	MRL
Abamectin	0.01	Esfenvalerate	0.1	Methoxyfenozide	1
Acrinathrin	0.1	Ethephon	0.1	Metiram	2
Alcalines solphites	10	Etofenprox	0.05	Myclobutanil	1
Alphamethrin	0.3	Etoxazole	1	Oxadiazon	0.05
Azadirachtin	0.5	Famoxadone	0.02	Oxyfluorfen	0.05
Azinphos-methyl	1	Fenamidone	2	Paraquat	0.05
Azociclotin	0.3	Fenamiphos	0.5	Penconazole	0.2
Azoxystrobin	2	Fenarimol	0.02	Phosalone	1
Benalaxyl	0.2	Fenazaquin	0.3	Phosetyl–Al	2
Benfuracarb	0.05	Fenbuconazole	0.2	Piperonyl butoxide	3
Bifenthrin	0.2	Fenbutatin oxide	0.2	Pirimicarb	0.2
Bifentrin	0.2	Fenhexamid	2	Pirimiphos-methyl	2
Bromopropylate	2	Enitrothion	5	Procymidone	5
Bromuconazole	0.5	Fenoxycarb	0.5	Propargite	2
Buprofezin	1	Fenpropidin	0.2	Propiconazole	0.5
Calcium polysulfide	50	Fenpropimorph	2	Propineb	2
Captan	10	Fenpyroximate	0.05	Propyzamide	0.02
Carbaryl	3	Flazasulfuron	0.3	Pyraclostrobin	2
Carbendazim	2	Fluazifop-P-butyl	0.01	Pyrethrins	1
Chloropicrin	0.05	Fluazinam	0.1	Pyridaben	0.1
Chlorothalonil	3	Fludioxonil	2	Pyrimethanil	3
Chlorpropham	0.05	Flufenoxuron	2	Quinoxyfen	0.5
Chlorpyrifos	0.5	Flusilazole	0.1	Rotenone	0.05
Chlorpyrifos-methyl	0.2	Fluvalinate	0.01	Spinosad	0.2
Clofentezine	1	Folpet	0.5	Spiroxamine	1
Cyanamide	0.05	Glufosinate ammon.	10	Sulphur	50
Cyazofamid	1	Glyphosate	0.1	Tebuconazole	1
Cycloxidim	0.1	Glyphosate trimesium	0.1	Tebufenozide	0.5
Cyfluthrin	0.3	Hesaconazole	0.1	Tebufenpyrad	0.3
Cyhexatin	0.3	Hexythiazox	0.5	Teflubenzuron	1
Cymoxanil	0.1	Indoxacarb	0.5	Tetraconazole	0.5
Cypermethrin	0.5	Iprodione	10	Thiamethoxam	0.5
Cyproconazole	0.2	Iprovalicarb	2	Thiodicarb	1
Cyprodinil	5	Kresoxim-methyl	1	Thiram	3.8
Deltamethrin	0.1	Lambda cyalothrin	0.2	Tiophanate-methyl	2
Diazinon	0.02	Lufenuron	0.5	Tolyfluanid	5
Dichlobenil	0.1	Mancozeb	2	Triadimenol	2
Dichlorvos	0.1	Maneb	2	Trichlorfon	0.5
Dicofol	2	Mcpa	0.1	Trifloxystrobin	3
Diethofencarb	1	Mecoprop	0.1	Trifluralin	0.05
Dimethomorph	0.5	Mepanipyrim	3	Vinclozolin	5
Diquat	0.05	Metalaxil-M	1	White mineral oil	0
Dithianon	0.6	Metam-sodium	2	Zeta cypermethrin	0.5
Diuron	0.05	Methidathion	0.5	Ziram	2
Dodine	0.2	Methiocarb	0.05	Zoxamide	5
Endosulfan	0.5	Methomyl	1		

[a]From Cabras and Caboni, 2008.

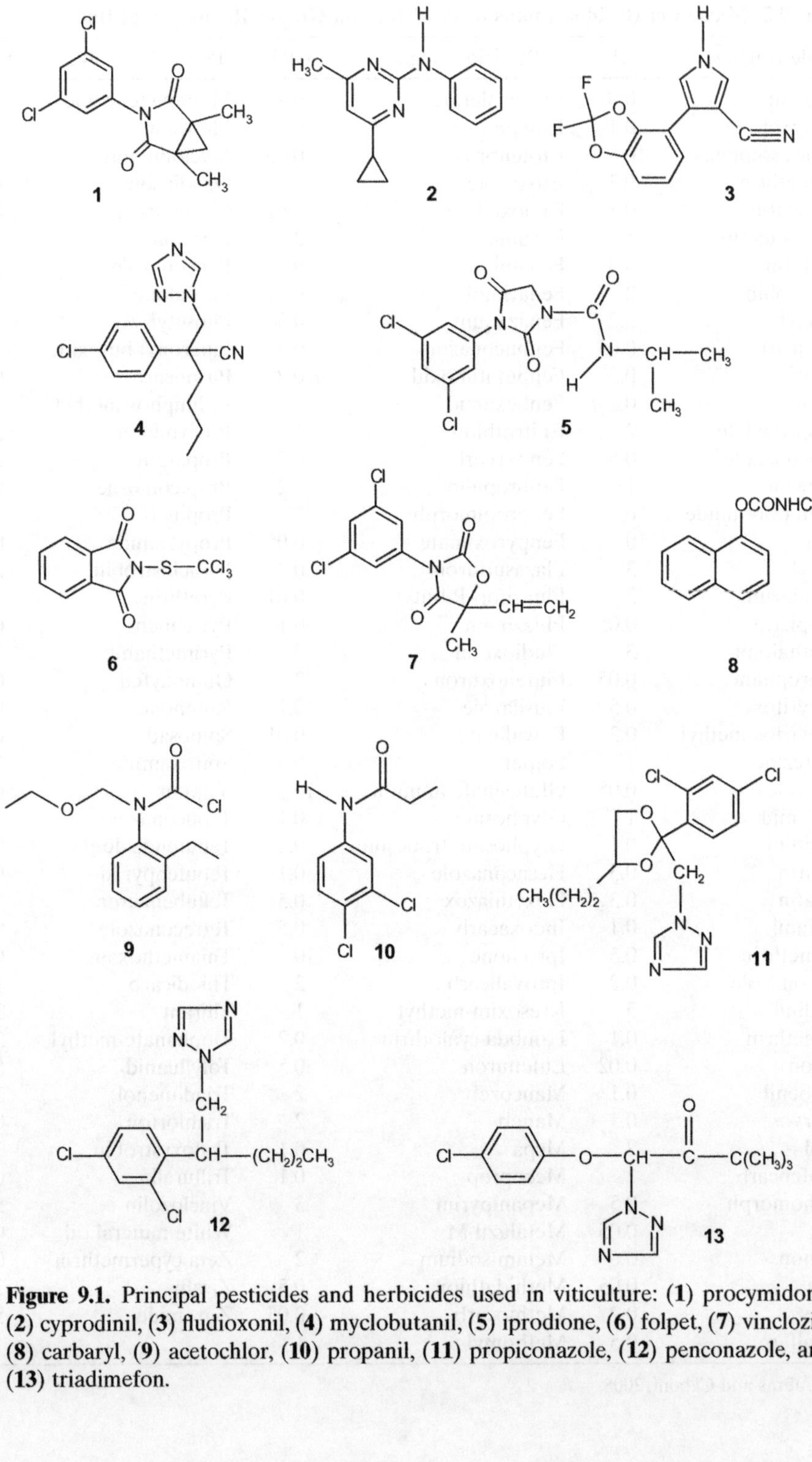

Figure 9.1. Principal pesticides and herbicides used in viticulture: (1) procymidone, (2) cyprodinil, (3) fludioxonil, (4) myclobutanil, (5) iprodione, (6) folpet, (7) vinclozin, (8) carbaryl, (9) acetochlor, (10) propanil, (11) propiconazole, (12) penconazole, and (13) triadimefon.

The pesticide residues are affected by environmental conditions, such us temperature, wind, rain, and solar irradiance. Consequently, MRLs can vary between countries because of the different climatic conditions. Currently, the EU is working for the harmonization of MRLs for raw food, but no limits are fixed for transformed foods. In Italy, when there is no legal limit for transformed food, the amount of raw food for a transformed food unit (e.g., ~1.5 kg of grape to produce 1 L of wine) and the incidence of technological process should be taken into account. Unfortunately, in the absence of specific data on changes of the residue occurring with the transformation, the only reference data is the MRL of raw food. Some countries, such as the United States, adopt the same MRL of grape for wine (Cabras and Caboni, 2008).

Dicarboxyimide fungicides have been used widely in viticulture against *Botrytis cinerea*. Vineyards are treated at the final stage of vegetation to prevent attack on grape before the harvest. Among them, vinclozin and iprodione are currently employed in Italy (Matisová et al., 1996; Cabras et al., 1983). These fungicides have a reduced toxicity, but 3,5-dichloroaniline, the probable common final product of degradation, is reputed to be as hazardous as the other aromatic amines. Because the vinification process lowers the level of pesticides, in wines they are significantly lower than in grapes, and methods of analysis must be very sensitive.

Folpet [N-(trichloromethylthio)phthalimide] is a fungicide used in vineyards, particularly against downy mildew (*Plasmopara viticola*), powdery mildew (*Uncinula necator*), and gray mold (*Botrytis cinerea*) (Tomlin, 1994). In general, the presence of fungicide residues in grape must may inhibit alcoholic fermentation. Studies conducted to assess the natural hydrolysis of folpet residues in must showed that folpet residues are fully decomposed by sunlight, and with winemaking the compound is degraded completely. At the end of fermentation, phthalimide is the main degradation product in wine (Cabras et al., 1997a; Hatzidimitriou et al., 1997). Moreover, the fungicide also may be added to the wine as an illegal preservative. Consequently, there is a relevant interest in the development of methods to determine the folpet residues in wine.

Captan is another fungicide used in viticulture, particularly in the past. Tetrahydrophthalimide (THPI) found in grape samples is mainly due to captan degradation promoted by the must acidity during either sampling or analysis, and only a minor part is present on the grape. Mechanisms of photodegradation and codistillation with water evaporation are mainly responsible for disappearance of captan from grapes. During winemaking, captan is degraded quantitatively to THPI, and at the end of fermentation only THPI, a very stable degradation product, is present in wine (Angioni et al., 2003).

Also, triazoles (triadimefon, penconazole, propiconazole, and myclobutanyl structures are reported in Fig. 9.1) are fungicides widely employed in viticulture to control powdery mildew, molds, and other fungal pathogens. These compounds are classified as acutely toxic. They may affect liver functionality, decrease kidney weights, alter urinary bladder structure, and have acute effects on the central nervous system (Briggs, 1992). Due to their persistence, they may be present in fruit juices and wines. The Italian law fixed their LODs in wine between 100 and 500 µg/kg.

Both the LC/MS and GC/MS methods allow a better understanding of the degradative behavior of individual active ingredients and their metabolites during field experiments and winemaking. In general, they reduce the need for a purification step of grapes and wine extracts in multiresidue methods. A recent description of MS applications in the grape and wine pesticides analysis was reported (Flamini and Panighel, 2006).

9.2   ANALYTICAL METHODS

9.2.1   Sampling and Sample Preparation

Sampling is crucial for the correct quantification of pesticide residues. For official controls, sampling procedures are reported in the EU Directive 2002/63/CE and in the Italian G.U. n.221 of 09/23/2003. The minimum amount of grape should be 2 kg (almost five bunches, each one weighing at least 250 g), and 0.5 L for wine. Analysis of active ingredients and their degradation products can be performed by monoresidue or, in particular for legal controls, multiresidue analytical methods. In general, the former are characterized by a few interference substances, have lower LODs, and high reliability in the compound identification. Multiresidue approaches have the advantage of the simultaneous determination of many active ingredients, but with higher LODs.

In the treatment of the vineyard, the active ingredients are deposited on the grape surface. In multiresidue methods, after grinding the sample is homogenized in representative aliquots and 50–100 g are analyzed. With the dissolution of pesticides in the acidic must (pH 3.0–3.7), some active ingredients, such as captan and folpet, may rapidly degrade. Consequently, the timeframe from grinding to extraction should be as short as possible (Cabras et al., 1997a; Angioni et al., 2003).

9.2.1.1 Extraction Using Solvent.

In the QuEChERS (Quick, Easy, Cheap, Effective, Rugged, Safe) method, liquid-liquid extraction of pesticides from homogenized grape is performed using acetonitrile and by addition of $MgSO_4$ and NaCl to the sample. After extraction, the sample is purified by a dispersive aminopropyl SPE to remove unwanted substances, such us sugars, organic acids, and other compounds interfering in the quantitative analysis (Anastassiades et al., 2003). The method allows extraction of polar pesticides (methamidophos, acephate, omethoate, imazalil, thiabendazole) and also of compounds with lower polarity such as pyrethrins. Homogenized grapes (10 g) are extracted with 10 mL of acetonitrile by shaking for 1 min, then 4 g of anhydrous $MgSO_4$ and 1 g of NaCl are added to the solution and mixed. The organic phase (1 mL) is transferred to a 1.5-mL vial containing 150 mg of anhydrous $MgSO_4$ and centrifuged. For the sample cleanup, 1 mL of the upper acetonitrile layer is mixed with 25 mg of primary secondary amine (PSA) adsorbent and 150 mg of $MgSO_4$. After shaking, the organic layer is centrifuged and used for GC/MS analysis. Recoveries of pesticides range between 85 and 101% (mostly >95%), with repeatability typically <5%.

A fast extraction of pesticides from 75 g of grape sample was performed using 200 mL of ethyl acetate. After filtration, 10 mL of extract was evaporated to dryness, taken up with 1.5 mL of methanol, and analyzed by LC/MS without further clean up. A total of 57 different pesticides were reported with recoveries ranging between 70 and 100% (Jansson et al., 2004).

A method for the multiresidue analysis of 82 pesticides in grapes by liquid–liquid extraction using ethyl acetate was recently studied (Banerjee et al., 2007). Previously homogenized berries (10 g) are mixed with 10 g of sodium sulfate and extracted with 10 mL of ethyl acetate. A volume of 5 mL of extract is cleaned by dispersive solid-phase extraction with 25 g of PSA (40 μm) to remove interfering substances like fatty acids. The cleaned extract (4 mL) is transferred in a 10-mL test tube and to it 200 μL of 10% diethylene glycol (in methanol) is added to prevent losses of analytes during evaporation. The sample is then evaporated to near dryness by a nitrogen stream and the residue is dissolved in 2 mL of a 1:1 (v/v) methanol/0.1% acetic acid solution to prevent pH-dependent degradation, particularly of organophosphates. After centrifugation, the sample is filtered through a 0.2 μm polyvinylidene fluoride membrane and LC/MS/MS analysis was performed. The LOQ were <10 µg/kg and recoveries ranged between 70 and 120% for most of the pesticides studied. Authors reported that accuracy and precision of method are similar to QuEChERS, and that the advantages are that it is cheaper and safer than the typical multiresidue analysis methods used for grape.

Captan and its metabolites (structures in Fig. 9.2) in grape, must, and wine are extracted by an organic solution (Angioni et al., 2003). An amount of 5 g of homogenized sample is transferred to a 30-mL screw-capped flask and added to 4 g of NaCl and 10 mL of a acetone/ petroleum ether 50:50 (v/v) solution. The tube is agitated in a rotatory shaker for 15 min and, after separation of two phases, an aliquot of the organic layer is analyzed by GC/MS. Recoveries of analytes range between 70 and 100%.

For extraction of zoxamide from grape, must, wine, and spirits, 5 g of sample is homogenized and transferred to a 40-mL screw-capped flask. After addition of 2 g of NaCl and 10 mL of hexane, extraction is performed by shaking for 15 min. After separation of phases, an aliquot of the organic layer is poured into a screw-capped tube containing 1 g of anhydrous sodium sulfate, filtered and analyzed by GC/MS. Recoveries range between 80 and 114% (Angioni et al., 2005).

Several methods of extraction using organic solutions were proposed for analysis of the fungicides cyprodinil, fludioxonil, pyrimethanil, tebuconazole, azoxystrobin, fluazinam, kresoxim-methyl, mepanipyrim, and tetraconazole in grapes, musts, and wines (Cabras et al., 1997b; 1998). An amount of 5–10 g of sample (homogenized for grape) is transferred into a 30–40-mL screw-capped tube, added to 2–4 g NaCl

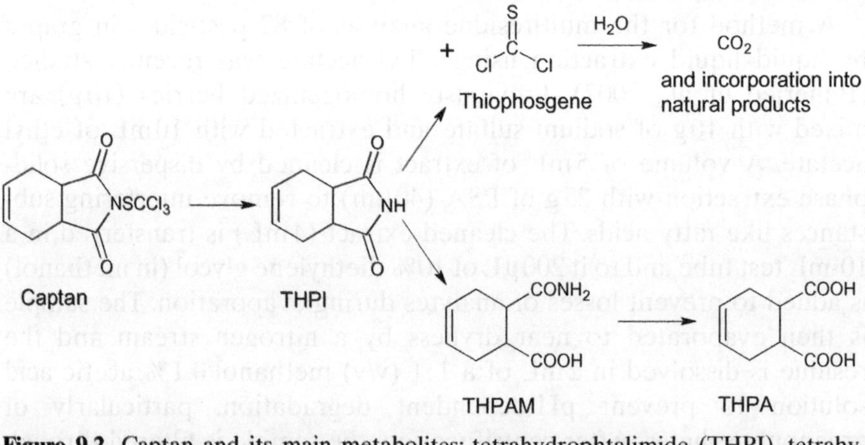

Figure 9.2. Captan and its main metabolites: tetrahydrophtalimide (THPI), tetrahydrophthalamic acid (THPAM), and tetrahydrophtalic acid (THPA).

and methyl parathion or triphenylphosphate as an internal standard. Extraction of cyprodinil, fludioxonil, pyrimethanil, and tebuconazole is performed by 10 mL of an acetone/petroleum ether 1:1 (v/v) solution, and azoxystrobin, fluazinam, kresoxim-methyl, mepanipyrim, and tetraconazole are extracted with 10 mL of acetone/hexane 1:1 (v/v) solution, keeping the solutions stirred for 20–30 min. After separation of two layers, the organic phase is treated with anhydrous sodium sulfate, filtered, 10-fold concentrated, and analyzed.

9.2.1.2 *Solid-Phase Extraction (SPE).*

These methods are valid for analysis of wines allowing isolation of a wide number of pesticides. The less polar compounds are adsorbed on a C_{18} cartridge with the percolation of undesirable polar substances. For a higher pesticide recovery, the wine is usually diluted with water to reduce the eluting effect of ethanol on the analytes through the stationary phase. Adsorbed compounds are then eluted with methylene chloride (Cabras et al., 1992; Holland et al., 1994) or ethyl acetate (Wong and Halverson, 1999). Also, porous carbon was used as a stationary phase performing elution of the analytes with toluene (Matisová et al., 1996). In general, these extraction procedures do not require further purification steps when they are coupled with LC–MS/MS or GC–MS/MS analysis.

An accurate SPE method for the multiresidue analysis of pesticides in wine was developed by using a polymeric cartridge and removing compounds coeluted with analytes by passage through an aminopropyl–$MgSO_4$ cartridge (Wong et al., 2003). The scheme of sample preparation is reported in Fig. 9.3. A 200-mg C_{18} cartridge is attached to the top of the aminopropyl (500 mg) cartridge and, after a sample passes through the cartridges, the analtytes are eluted by three successive passages of ethyl acetate–hexane. The solutions are polled together (~15 mL). The resulting solution is evaporated under a nitrogen stream, the residue is added to 1 mL of 0.1% corn oil–ethyl acetate, transferred to a GC sample vial and added to acenaphthalene-d_{10} as an internal standard. This multiresidue method coupled with GC/MS–SIM analysis is suitable for detection of organohalogen, organonitrogen, organophosphate, and organosulfur pesticides. Recoveries from 10-μg/L spiked red and white wines are >70% for 116 and 124 analytes (out of 153 total pesticides), respectively. By this method, recoveries of propargite are >80% in both red and white wines. Organohalogenated pesticides, such as the N-trihalomethylhalo compounds (captafol and folpet), the dicarboximide pesticides (iprodione and chlozolinate), and the organochlorine compound (endrin aldehyde), showed recoveries <70% for both high- and low-spiked

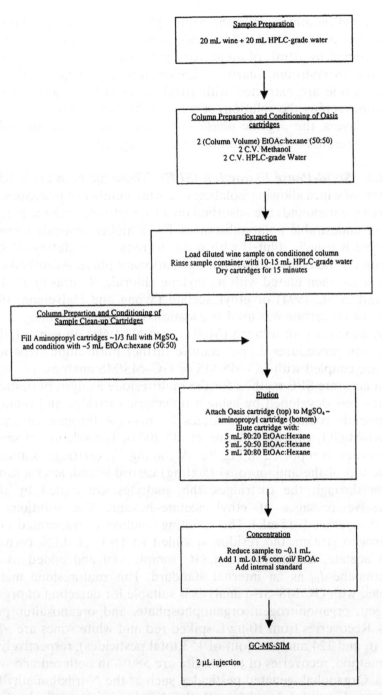

Figure 9.3. Sample preparation method proposed by Wong et al. for multiresidue analysis of pesticides in wine. (Reprinted from *Journal of Agricultural and Food Chemistry* 51, Wong et al., Multiresidue pesticide analysis in wines by solid-phase extraction and capillary gas chromatography-mass spectrometric detection with selective ion monitoring. p. 1150, Copyright © 2003, with permission from American Chemical Society.)

wines. Recoveries >70% have been observed for most organonitrogen pesticides, such as the 1,3,5-triazines and amides (phenylamides, napropamide, and propyzamide), and most of the azole pesticides, such as myclobutanil, triadimefon, and its degradation product triadimenol. In addition to some of the polar organonitrogen pesticides previously mentioned, chlorothalonil, desmetryn, fenpropimorph, hexazinone, and prochloraz showed poor recoveries (<60%) in both red and white wines (data in Table 9.3).

9.2.1.3 Solid-Phase Microextraction (SPME). Coupled with GC/ MS, SPME was proposed for analysis of the insecticides lindane, parathion, carbaryl, malathion, endosulfan, methoxychlor, and methidathion, procymidone, vinclozoline, folpet, and captan (fungicides), and the herbicides terbuthylazine, trifluralin, and phosalone in wine (Vitali

TABLE 9.3. Percentage Recoveries of Pesticides Extracted by Solid-Phase Extraction With a C$_{18}$ Cartridge from Red and White Wines Spiked at 0.10 (high) and 0.01 mg/L (low) Levels[a]

	High Spike (0.10 mg/L)		Low Spike (0.01 mg/L)	
	Wine			
Pesticide	Red	White	Red	White
Organohalogen				
benzilate				
Bromopropylate	89 ± 1	90 ± 1	90 ± 3	90 ± 2
dicarboximide				
Iprodione	47 ± 3	37 ± 2	73 ± 6	77 ± 5
Procymidone	88 ± 2	91 ± 2	79 ± 5	82 ± 6
Vinclozolin	83 ± 2	80 ± 3	83 ± 2	91 ± 4
***N*-trihalomethylhalo**				
Captan	64 ± 3	80 ± 8	89 ± 4	109 ± 3
Folpet	n.d.	9 ± 4	n.d.	57 ± 2
Tolylfluanid	112 ± 5	91 ± 6	107 ± 4	92 ± 16
organochlorine				
Endosulfan-α	90 ± 1	91 ± 1	83 ± 2	86 ± 5
Endosulfan-β	95 ± 1	91 ± 1	86 ± 4	89 ± 5
Pyrethroid				
Cypermethrin I	73 ± 2	82 ± 2	66 ± 4	80 ± 4
Cypermethrin II	74 ± 1	81 ± 2	71 ± 4	85 ± 4
Cypermethrin III	72 ± 2	82 ± 2	69 ± 4	83 ± 4
Cypermethrin IV	71 ± 1	81 ± 2	65 ± 4	78 ± 4
Deltamethrin	78 ± 4	84 ± 2	75 ± 4	89 ± 2
Fluvalinate tau-I	85 ± 4	82 ± 2	89 ± 11	86 ± 2
Fluvalinate tau-II	77 ± 3	83 ± 2	71 ± 4	84 ± 2

TABLE 9.3. (*Continued*)

	High Spike (0.10 mg/L)		Low Spike (0.01 mg/L)	
	Wine			
Pesticide	Red	White	Red	White
Organonitrogen				
2,6-dinitroaniline				
Trifluralin	69 ± 6	70 ± 3	68 ± 4	68 ± 3
amide				
Propyzamide	93 ± 2	95 ± 1	93 ± 2	96 ± 2
anilinopyrimidine				
Cyprodinil	72 ± 3	62 ± 12	64 ± 7	62 ± 12
Pyrimethanil	87 ± 3	94 ± 4	79 ± 6	109 ± 2
Myclobutanil	86 ± 4	96 ± 4	108 ± 8	110 ± 9
Penconazole	91 ± 2	88 ± 5	91 ± 4	87 ± 3
Tebuconazole	83 ± 2	70 ± 14	80 ± 4	83 ± 7
Triadimenol	98 ± 1	89 ± 6	108 ± 10	98 ± 5
benzonitrile				
Chlorothalonil	53 ± 5	30 ± 4	81 ± 5	69 ± 3
Dichlobenil	67 ± 10	77 ± 6	62 ± 3	69 ± 4
carbamate/thiocarbamate				
Carbaryl	86 ± 5	71 ± 9	124 ± 12	100 ± 11
diphenyl ether				
Oxyfluoren	73 ± 3	74 ± 3	72 ± 6	79 ± 2
morpholine				
Fenpropimorph	6 ± 0.1	8 ± 6	21 ± 0.6	27 ± 5

[a]Wong et al., 2003.

et al., 1998). Extraction is performed by immerging a polydimethylsiloxane (PDMS) 100-µm silica fiber into 30 mL of a wine sample saturated with $MgSO_4$ in a 40-mL vial, with stirring for 30 min. Analytes are desorbed from the fiber into the GC injection port at 250 °C. The LODs ranging between 0.1 and 6.0 µg/L are achieved.

Organochlorine fungicides nuarimol, triadimenol, triadimefon, folpet, voinclozolin, and penconazole in wine can be extracted by using a polydimethylsiloxane–divinylbenzene (PDMS/DVB) 60-µm fiber with immersion of the fiber into a 3 mL sample at room temperature for 30 min. Coupled with LC/DAD analysis, LODs ranging between 4 and 27 µg/L and were reported (Millán et al., 2003).

A poly(ethylene)glycol–divinylbenzene (PEG/DVB) 65-µm fiber was used (Natangelo et al., 2002) for sampling of propanil (anilide postemergent herbicide), acetochlor (chloroacetanilide preemergent herbicide), myclobutanil (azole fungicide), and fenoxycarb (carbamate insecticide) in grape juice and wine.

In addition, SPME was applied to control triazole residues in wine (Zambonin et al., 2002). Analysis of triadimefon, propiconazole, myclobutanil, and penconazole was performed by a polyacrylate (PA) 85-μm silica fiber. A 2.5-mL wine sample is diluted 1:1 with H_2O and trasferred in a 7-mL vial. After addition with 0.5 g NaCl extraction is performed by immersing the fiber into the sample with stirring for 45 min. Thermal desorption of analytes into the GC injection port is performed at 250 °C for 5 min.

A method for analysis of polar pesticides in wine by the use of automated in-tube SPME coupled with LC/ESI–MS was proposed (Wu et al., 2002). In-tube SPME is a microextraction and preconcentration technique that can be coupled on-line with high-performance liquid chromatography (HPLC), suitable for the analysis of less volatile and/ or thermally labile compounds. This technique uses a coated open tubular capillary as an SPME device and automated extraction. Using a polypyrrole coating, six phenylurea pesticides (diuron, fluometuron, linuron, monuron, neburon, siduron) and six carbamates (barban, carbaryl, chlorpropham, methiocarb, promecarb, propham) were analyzed in wine. Structures of compounds are reported in Fig. 9.4. Due to the high extraction efficiency of the fiber toward polar compounds, benzene compounds, and anionic species, LODs ranging between 0.01 and 1.2 μg/L were achieved, even if the sample ethanol content affects the recoveries of analytes.

Some carbamates (carbosulfan, benfuracarb, carbofuran, pirimicarb, diethofencarb, and diuron) and phenylurea pesticides (monuron and monolinuron) were sampled from different fruit juices by using 50-μm carbowax–templated resin (CW–TPR) and a 60-μm PDMS/DVB SPME fiber (Sagratini et al., 2007). The fiber desorption into the SPME–LC/MS interface chamber previously filled with 70% methanol and 30% water, was performed in static mode.

9.2.1.4 Stir Bar Sorptive Extraction (SBSE).

The SBSE and thermal desorption (TD) was proposed for multiresidue GC/MS analysis of dicarboximide fungicides vinclozolin, iprodione (as its degradation product 3,5-dichlorophenyl hydantoin), and procymidone in wines (Sandra et al., 2001; Hayasaka et al., 2003). The SBSE uses a stir bar (typically 10 mm in length) incorporated into a glass tube and coated with a high amount (25–125 μL) of PDMS. With stirring, analytes are partitioned between the liquid matrix of sample and the PDMS phase on the stir bar. Recoveries increase according to the volume of PDMS to the sample volume matrix ratio. The stir bar is

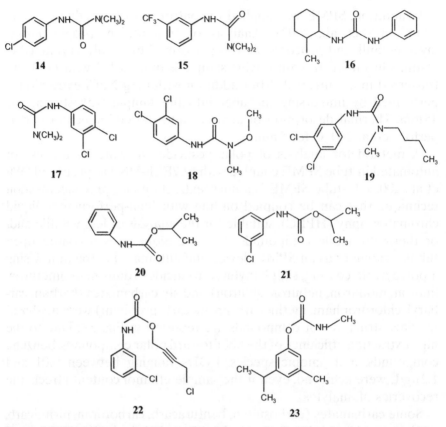

Figure 9.4. Phenylurea pesticides and carbamates detected in wine by automated in-tube SPME and LC/ESI–MS analysis (Wu et al., 2002). (**14**) monuron, (**15**) fluometuron, (**16**) siduron, (**17**) diuron, (**18**) linuron, (**19**) neburon, (**20**) propham, (**21**) chlorpropham, (**22**) barban, (**23**) promecarb (structures of carbaryl and methiocarb are reported in Figs. 9.1 and 9.11, respectively).

then transferred to a compact thermal desorption unit mounted on a programmable temperature vaporization (PTV) GC injector and analytes are thermally desorbed into the GC column. In general, stir bar is reported to increase the sample enrichment and consequently the sensitivity of the method. For sample extraction, 10 mL of wine are transferred in a 20-mL vial and a stir bar containing 25-μL PDMS was stirred into the sample for 40 min. After sampling, the stir bar is rinsed in distilled water, GC thermal desorption is performed in a glass tube (187 mm in length and 4 mm in internal diameter), LC by performing extraction of stir bar using 1-mL acetonitrile with ultrasounds and analysis of extract.

9.2.2   The GC/MS–SIM Analysis of Pesticides

The active ingredients are commonly divided into chemical classes, such as pyrethroids, chloroorganic, and organophosporus. Coextracted substances may affect the analyte signal. To avoid this matrix effect, standard solutions should be prepared by using an extract from a non-contaminated sample, or a calibration curve calculated by the standard addition method.

The multiresidue detection of organohalogen, organonitrogen, organophosphate, and organosulfur pesticides and residues (in total 153 compounds) can be performed with different GC/MS analyses by using three different SIM programs and the analytical conditions reported in Table 9.4 (Wong et al., 2003). Compounds are reported in Table 9.5 with their LOD, target ion, and three qualifier/target ion ratios. Quantitative analysis is performed on the peak area ratio of the target ion divided by the peak area of the internal standard versus concentration of the calibration standards.

For routine monitoring of 21 pesticides in wine, a fully automated SPE–GC/MS method using C_{18} 300-mg cartridges was proposed (Kaufmann, 1997). By recording the m/z signals reported in Table 9.5 in the SIM mode, 4,4′-dichloro-benzophenone, azinphos-methyl, bromo-propylate, captafol, captan, chlorpyrifos, dichlofluanid, dicofol, dimethoate, endosulfan, etrimfos, fenamiphos, fenamirol, folpet, iprodione, malathion, methidathion, parathion-methyl, procymidone, triadimefon, and vinclozolin were analyzed, using ethyl hydrocinnamate (signals recorded at m/z 104 and 178) as an internal standard. Analysis was performed with a 5% diphenyl-95% dimethyl polisiloxane (30 m × 0.25 mm i.d.; 0.25-µm film thickness) capillary column or similar type, with oven temperature starting at 80 °C for 1.5 min, temperature increasing at

TABLE 9.4. The GC/MS Conditions Used in the Multiresidue Analysis of Pesticides in Wines[a]

Column	5% Diphenyl–95% dimethylpolysiloxane
	(30 m × 0.25 mm i.d.; 0.25-µm film thickness)
Injection mode	Splitless
Injected volume	2.0 µL
Carrier gas	He Constant pressure
Injector temperature	250 °C
Oven temperature	70 °C (isotherm for 2 min) → 150 °C (25 °C/min) → 200 °C
	(3 °C/min) → 280 °C (8 °C/min, isotherm for 10 min)
Detector	MS/EI (70 eV); transfer line 280 °C; source 230 °C
MS conditions	SIM mode

[a]Wong et al., 2003.

TABLE 9.5. Wine Pesticides (Total 153) and the Corresponding Target and Qualifier Ions Detected by Solid-Phase Extraction (Figure 9.3) and GC/MS–SIM Analysis[a]

Pesticide	MW	Target (T)	Qualifiers Q₁, Q₂, Q₃	LOD (ppm)	Pesticide	MW	Target (T)	Qualifiers Q₁, Q₂, Q₃	LOD (ppm)
Acephate	183.2	136	94 95 125	25.0	Fenpropimorph	305.5	128	129 303 117	<0.5
Acenaphthalene-d₁₀ (IS)	164.3	164	162 160 80	10.0	Fenson	268.7	77	141 268 51	10.0
Alachlor	269.8	160	188 146 237	1.0	Fenthion	278.3	278	125 109 169	<1.5
Aldrin	364.9	263	265 261 66	1.5	Fenvalerate I	419.9	167	125 181 152	3.0
Allethrin	302.4	123	79 136 107	3.0	Fenvalerate II	419.9	167	125 181 169	3.0
Atrazine	215.7	200	215 202 58	1.0	Flucythrinate I	451.4	199	157 181 107	2.5
Azinphos-ethyl	345.4	132	160 77 105	1.0	Flucythrinate II	451.4	199	158 181 107	2.5
Azinphos-methyl	317.3	160	132 77 105	3.0	Fludioxinil	248.2	248	127 154 182	1.0
Benalaxyl	325.4	148	91 206 204	1.0	Fluvalinate tau-I	502.9	250	252 181 208	0.5
Benfluralin	335.3	292	264 276 293	<1.0	Fluvalinate tau-II	502.9	250	253 181 208	0.5
BHC-α	290.8	181	183 219 217	1.0	Folpet	296.6	147	104 76 260	15.0
BHC-δ	290.8	181	219 183 217	2.0	Fonofos	246.3	109	246 137 110	<1.0
BHC-γ (Lindane)	290.8	181	183 219 111	1.5	Furalaxyl	301.3	95	242 152 146	1.0
Bitertanol I	337.4	170	168 171 57	0.5	Heptachlor	373.3	272	274 100 270	0.5
Bitertanol II	337.4	170	168 171 57	0.5	Heptachlor epoxide	389.3	353	355 351 357	0.5
Bromophos-ethyl	394.1	359	303 357 301	<1.0	Hexachlorobenzene	284.8	284	286 282 288	<0.5
Bromophos-methyl	366.0	331	329 333 125	<1.0	Hexaconazole	352.9	83	214 216 82	1.0
Bromopropylate	428.1	341	183 339 343	0.5	Hexazinone	252.3	171	83 128 71	1.0
Bromoxynil	276.9	277	275 279 88	10.0	Imazalil	297.2	41	215 173 217	6.0
Captafol	349.1	79	80 77 151	25.0	Iprodione	330.2	314	187 189 244	5.0
Captan	300.6	79	80 151 77	10.0	Isofenphos	345.4	213	58 121 255	1.0
Carbaryl	210.2	144	115 116 145	10.0	Malaoxon	314.3	127	99 109 125	3.0
Carbofuran	221.3	164	149 131 123	2.0	Malathion	330.4	173	127 125 93	<1.5
Carbophenothion	342.9	157	342 121 99	<1.5	Metalaxyl	279.3	206	45 160 249	1.0
Chlorbenside	269.2	125	127 268 270	1.0	Methidathion	302.3	145	85 93 125	1.0
cis-Chlordane	409.8	373	375 377 371	<1.0	Methoxychlor	345.7	227	228 152 113	<1.0
trans-Chlordane	409.8	373	376 377 371	<1.0	Metolachlor	283.8	162	238 240 146	<1.0
Chlorfenvinphos	359.6	267	323 269 325	1.0	Mevinphos	224.2	127	192 109 67	<1.5

Compound					Compound				
Chlorothalonil	265.9	266	264 268 270	1.0	Mirex	545.6	272	274 270 237	<1.0
Chlorpyrifos	350.6	197	199 314 97	1.0	Monocrotophos	223.2	127	67 192 97	3.0
Chlorpyrifos-methyl	322.5	286	288 125 290	<1.0	Myclobutanil	280.8	179	150 82 181	1.0
Chlozolinate	332.1	188	259 186 187	1.5	Naled	380.8	109	185 79 145	6.5
Chrysene-d12 (IS)	240.4	240	236 241 238		Napropamide	271.4	72	128 100 271	<1.0
Coumaphos	362.8	362	226 109 210	1.0	Nitralin	345.4	316	274 300 317	0.5
Cyanazine	240.7	212	213 214 68	3.0	Nitrofen	284.1	283	253 283 202	3.0
Cyfluthrin I	434.3	163	206 165 227	1.5	Nitrothal-isopropyl	295.3	236	194 212 254	1.0
Cyfluthrin II	434.3	163	207 165 227	1.5	Norflurazon	303.7	303	145 102 305	1.0
Cyfluthrin III	434.3	163	208 165 227	2.5	Omethoate	213.2	156	110 79 109	6.0
Cyfluthrin IV	434.3	163	206 199 227	2.5	Oryzalin	346.4	317	275 258 58	100.0
Cyhalothrin	449.9	181	197 208 209	1.5	Oxadiazon	345.2	175	177 258 260	0.6
Cypermethrin I	416.3	181	163 165 209	2.0	Oxadixyl	278.3	105	163 45 132	1.5
Cypermethrin II	416.3	181	164 165 209	2.0	Oxyfluorfen	361.7	252	361 302 331	1.0
Cypermethrin III	416.3	163	181 165 209	2.0	Paraoxon	275.2	109	149 275 139	6.0
Cypermethrin IV	416.3	163	182 165 209	2.0	Parathion	291.3	291	109 97 139	1.0
Cyprodinil	225.3	224	225 210 77	<1.5	Parathion-methyl	263.2	263	109 125 79	1.0
o,p'-DDT	354.5	235	237 165 236	<0.5	Penconazole	284.2	248	159 161 250	1.0
p,p'-DDT	354.5	235	238 165 236	<1.0	cis-Permethrin	391.3	183	163 165 184	<0.5
Deltamethrin	505.2	181	253 251 255	8.0	trans-Permethrin	391.3	183	164 165 184	<0.5
Demeton-O	230.3	88	60 89 171	2.5	Phenanthrene-d10 (IS)	188.3	188	189 184 187	
Demeton-S	230.3	88	60 170 89	2.5	Phorate	260.4	75	121 260 97	<1.0
Desmetryn	213.3	213	198 171 58	<1.5	Phosalone	367.8	182	367 121 184	<1.0
Dialifos	393.9	208	173 210 76	1.0	Phosmet	317.3	160	161 77 93	<1.5
Diallate I	270.2	86	234 236 128	<0.5	Prochloraz	376.7	180	70 307 310	6.0
Diallate II	270.2	86	235 236 128	<0.5	Procymidone	284.1	96	283 285 67	1.0
Diazinon	304.3	179	137 199 152	<1.0	Profenophos	373.6	208	339 139 206	3.0
Dichlobenil	172.0	171	173 136 100	<1.5	Prometryn	241.4	241	184 226 105	<1.5
Dichlofluanid	333.2	123	224 167 226	<1.5	Propargite	350.5	135	150 231 34	0.5
4,4'-Dichlorobenzophenone	251.1	139	111 141 250	0.5	Propazine	229.7	214	229 172 58	<1.0
Dichlorvos	221.0	109	185 79 187	<1.0	Propetamphos	281.3	138	194 236 222	<1.0

TABLE 9.5. (*Continued*)

Pesticide	MW	Target (T)	Qualifiers Q₁, Q₂, Q₃	LOD (ppm)	Pesticide	MW	Target (T)	Qualifiers Q₁, Q₂, Q₃	LOD (ppm)
Dicloran	207.0	206	176 178 208	4.0	Propyzamide	256.1	173	175 145 255	1.5
Dicrotophos	237.2	127	67 193 72	3.0	Pyrimethanil	199.3	198	199 77 200	<1.0
Dieldrin	380.9	79	263 277 279	2.0	Quinalphos	298.3	146	157 118 156	50.0
Dimethoate	229.3	87	93 125 143	2.5	Quintozene	295.3	237	249 295 214	<2.0
Dinoseb	240.2	211	163 147 240	150.0	Simazine	201.7	201	186 173 68	3.0
Dioxathion	456.0	97	125 271 153	5.0	Tebuconazole	307.8	125	250 70 83	1.5
Disulfoton	274.4	88	89 97 142	1.0	Tecnazene	260.9	203	261 215 201	1.0
Endosulfan-α	406.9	241	195 239 237	1.5	Terbufos	288.4	231	57 103 153	<1.0
Endosulfan-β	406.9	195	237 241 207	3.0	Terbuthylazine	229.7	214	173 216 229	<1.5
Endrin	380.9	317	263 315 319	3.5	Terbutryn	241.4	226	185 241 170	<1.0
Endrin aldehyde	380.9	67	345 250 347	2.0	Tetrachlorovinphos	366.0	329	331 109 333	<1.0
Endrin ketone	380.9	317	67 315 319	<1.0	Tetradifon	356.1	159	111 229 227	1.0
EPN	323.3	157	169 141 185	<1.0	Thiometon	246.3	88	125 89 93	1.5
Eptam	189.3	128	43 86 132	1.0	Tolylfluanid	347.3	137	238 106 83	2.6
Ethalfluralin	333.3	276	316 292 333	1.0	Triadimefon	293.8	57	208 85 210	1.0
Ethion	384.5	231	153 97 125	1.0	Triadimenol	295.8	112	168 128 70	4.0
Fenamiphos	303.4	303	154 288 217	<1.0	Tri-allate	304.7	86	268 270 128	<1.0
Fenarimol	331.2	139	219 251 107	0.6	Trifluralin	335.3	306	264 290 307	<1.0
Fenitrothion	277.2	277	125 109 260	1.0	Vinclozolin	286.1	212	198 187 285	1.0
Fenpropathrin	349.4	97	181 125 265	0.6					

*a*Molecular weight = MW; LOD = limit of detection (Wong et al., 2003).

20 °C/min to 180 °C, then at 3.5 °C/min to 230 °C, and finally at 7 °C/min to 280 °C. The method provides LODs ranging between 5 and 10 μg/L, linearity regression coefficients >0.99 (except for 4,4'-dichloro-benzophenone and dicofol), and recoveries from spiked wines ranging from 80 to 115% for 17 of the 21 pesticides analyzed.

GC/MS–SIM analysis coupled with QuEChERS method (section 9.2.1.1) is performed using the analytical conditions reported in Table 9.6. Pesticides analyzed are reported in Table 9.7 with their GC retention time and the *m/z* ion used for quantification.

By SBSE-thermal desorption and GC/MS analysis, recording signals in SCAN mode LODs 0.2 μg/L for vinclozolin and procymidone, 2 μg/L for iprodione, LOQs 0.5 μg/L for vinclozolin and procymidone, and 5 μg/L for iprodione, were reported (Sandra et al., 2001). Vinclozolin and procymidone are easily identified, whereas 90% of iprodione degrades in the GC column (or during thermal desorption) at temperatures >200 °C, forming (3,5-dichlorophenyl)hydantoin. Therefore, quantification has to be performed on this metabolite recording the signals in SIM mode at *m/z* 187 (relative abundance 100%) and 244 (66%) with the isotopic ion clusters, and at *m/z* 124 (27%). The LODs reported are 2 ng/L for vinclozolin and procymidone, and 50 ng/L for iprodione. Fragmentation spectra of the three compounds are reported in Fig. 9.5.

The SPME–GC/MS (EI 70 eV) can be used for analysis of triazole residues in wine (Zambonin et al., 2002). Mass spectra of four triazoles are shown in Fig. 9.6. Quantitative analysis is performed on the fragment ions at *m/z* 128, 210, 293 for triadimefon, *m/z* 145, 173, 259 for propiconazole, *m/z* 179, 206, 288 for myclobutanil, and *m/z* 159, 161, 248 for penconazole recording the signals in SIM mode.

TABLE 9.6. The GC/MS Conditions Used for Analysis of Pesticides Coupled with the QuEChERS Method[a]

Column	35% Diphenyl–65% dimethylpolysiloxane
	(30 m × 0.25 mm i.d.; 0.25-μm film thickness)
Injection mode	Splitless
Injected volume	1.5 μL
Carrier gas	He Flow rate 1 mL/min
Injector temperature	250 °C
Oven temperature	95 °C (isotherm for 15 min) → 190 °C (20 °C/min) → 230 °C
	(5 °C/min) → 290 °C (25 °C/min, isotherm for 20 min)
Detector	MS, Transfer line 290 °C, full-scan *m/z* 50–450

[a]Anastassiades et al., 2003.

TABLE 9.7. Retention Times (RT) and Quantification Ions in the GC/MS–SIM Analysis of Pesticides in Grape Extract and Wine Coupled with the QuEChERS Method[a]

Pesticide	RT (min)	Quantification Ion (m/z)
Acephate	7.3	136
Azinphos-methyl	18.8	160
Captan	14.9	79
Carbaryl	12.4	144
Chlorothalonil	11.5	266
Chlorpyrifos	12.1	197
cis-Permethrin	18.5	183
Coumaphos	19.6	362
Cyprodinil	13.4	224
Deltamethrin	23.1	181
Diazinon	9.2	179
Dichlofluanid	12.3	224
Dichlorvos	5.3	185
Dicofol	17.4	251
Dimethoate	10.1	93
Endosulfan sulfate	16.8	272
Fenthion	12.7	268
Folpet	15.1	260
Imazalil	15.7	201
Lindane	9.8	181
Metalaxyl	11.5	206
Methamidophos	5.8	94
Methiocarb	12.5	168
Mevinphos	6.6	127
Omethoate	9.1	156
o-Phenylphenol	7.3	170
Phosalone	18.1	182
trans-Permethrin	18.7	183
Vinclozolin	10.7	285

[a]Analytical conditions are reported in Table 9.6 (Anastassiades et al., 2003).

The method provides LODs between 30 ng/kg for propiconazole and 100 ng/kg for triadimefon, these performances are lower the MRLs recommended by the European Legislation in wine and grapes (e.g., Directives 90/642/CE).

Cyprodinil, fludioxonil, pyrimethanil, tebuconazole, azoxystrobin, fluazinam, kresoxim-methyl, mepanipyrim, and tetraconazole in grapes, must, and wine organic extracts are determined by GC/MS–SIM analysis with LODs of 0.05 mg/kg for cyprodinil, pyrimethanil, and kresoxim-methyl, and 0.10 mg/kg for the other analytes (Cabras et al.,

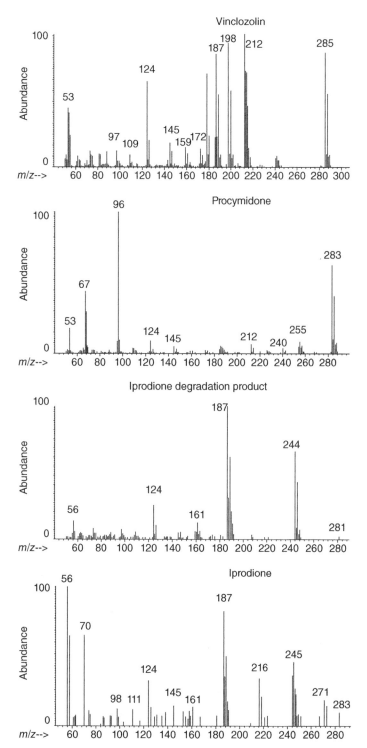

Figure 9.5. The EI fragmentation spectra of dicarboximide fungicides vinclozolin, iprodione, procymidone, and of (3,5-dichlorophenyl)hydantoin (the iprodione degradation product) recorded by stir bar sorptive extraction and thermal desorption–GC/MS analysis (SBSE–TD–GC/MS). (Reprinted from *Journal of Chromatography A*, 928, Sandra et al., Stir bar sorptive extraction applied to the determination of dicarboximide fungicides in wine. p. 121, Copyright © 2001, with permission from Elsevier.)

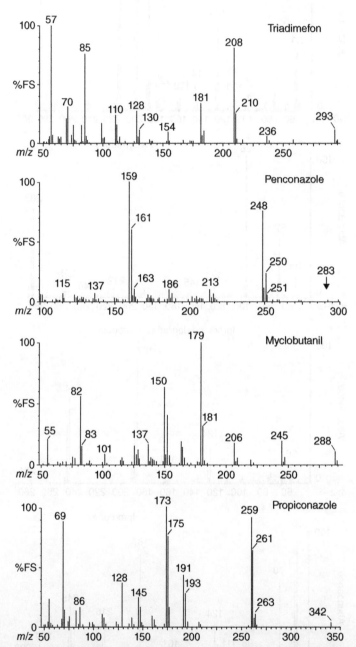

Figure 9.6. The GC/MS (EI 70 eV) mass spectra of triazoles. (Reprinted from *Journal of Chromatography A*, 967, Zambonin et al., Solid-phase microextraction and gas chromatography-mass spectrometry for the rapid screening of triazole residues in wines and strawberries. p. 258, Copyright © 2002, with permission from Elsevier.)

1997b; 1998). Analysis is performed using a fused silica 5% phenyl–95% methylpolysiloxane capillary column (30 m × 0.25 mm i.d.; 0.25-μm film thickness) and an oven temperature starting from 80 °C, raised to 300 °C at 10 °C/min, and held at 300 °C for 20 min. The signals at m/z 224 for cyprodinil, 248 for fludioxonil, 198 for pyrimethanil, 250 for tebuconazole, 344 for azoxystrobin, 371, 387, 417 for fluazinam, 116, 131, 206 for kresoxim-methyl, 222 for mepanipyrim, 336 for tetraconazole, at 109, 125 for the internal standard parathion methyl, and at 326 for the internal standard triphenylphosphate, are recorded. Figure 9.7 reports the mass spectra of fluazinam, mepanipyrim, tetraconazole, and pyrimethanil, which are not found in the commercially available libraries.

9.2.3   The GC/ITMS Analysis of Pesticides

Analysis of vinclozolin and iprodione in wine can be performed using a GC-ion trap (IT) system with the analytical conditions reported in Table 9.8. By coupling this method and the SPE sample preparation with the use of the porous carbon stationary phase, the analytes can be recovered with toluene, and the LOQs of 50 ng/L and 50 μg/L for vinclozolin and iprodione, respectively, are achieved (Matisová et al., 1996). An LOQ of 0.50 μg/L is achieved for analysis of metalaxyl in wine (Kakalíková et al., 1996).

A GC/ITMS analysis of propanil, acetochlor, myclobutanil, and fenoxycarb in grape juice and wine is performed by recording signals of the collision-produced ions formed by multiple mass spectrometry (MS/MS) reported in Table 9.9. Data in the table compare the GC/MS–SIM and GC/MS–MS methods used for analysis of these pesticides in grape juice and wine. The two methods have a similar precision and sensitivity in wine analysis, with IT providing a lower sensitivity in the grape juice analysis.

Analysis of captan and its metabolite THPI (Fig. 9.2) in grape, must, and wine extracts can be performed using GC/ITMS with liquid chemical ionization using methanol (Angioni et al., 2003). Figure 9.8 shows the chromatogram relative to a grape extract analysis performed with the analytical conditions reported in Table 9.10. In Figure 9.9 the mass spectrum of captan recorded with the GC/ITMS system operating in the EI mode is reported.

The GC/ITMS analysis of zoxamide in grape, must, wine, and spirits extracts can be performed operating in both EI and CI modes using the experimental conditions described in Table 9.11 (Angioni et al., 2005).

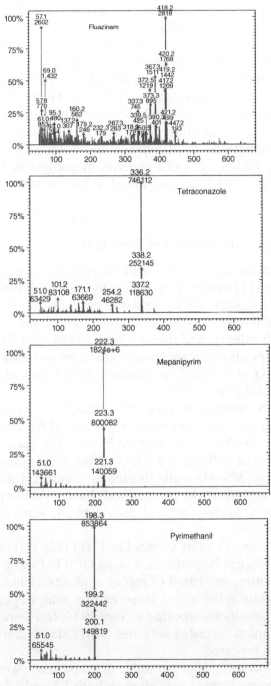

Figure 9.7. Mass spectra of fluazinam, mepanipyrim, tetraconazole, and pyrimethanil, pesticides not reported in the libraries commercially available. (Kindly provided by Prof. P. Cabras.)

TABLE 9.8. Analytical Conditions for Analysis of Vinclozolin and Iprodione in Wines[a]

Column	5% Diphenyl–95% dimethylpolysiloxane (30 m × 0.25 mm i.d.; 0.25-μm film thickness)
Injection mode	Splitless
Injected volume	4.0 μL
Carrier gas	He Flow rate 1 mL/min
Injector temperature	250 °C
Oven temperature	60 °C (isotherm for 1 min) → 250 °C (10 °C/min, isotherm for 20 min)
Detector	ITMS; 220 °C; EI (70 eV) Multiple ion detection (MID):
IT conditions	Vinclozolin signals: m/z 178, 180, 198, 200, 212, 215, 285, 287; iprodione signals: m/z 187, 189, 244, 247, 314, 316

[a]Matisová et al., 1996.

TABLE 9.9. Performances of GC/MS–SIM and GC/IT–MS/MS Methods for Analysis of Propanil, Acetochlor, Myclobutanil, and Fenoxycarb in Grape Juice and Wine[a]

		Precision (RDS, $n = 3$)		LOD (μg/L)	
	Product Ions (m/z)	White Wine	Grape Juice	White Wine	Grape Juice
GC/MS					
Propanil		5.3	5.9	1	0.1
Acetochlor		7.3	9.1	5	0.2
Myclobutanil		5.1	11.3	8	1.0
Fenoxycarb		4.1	14.4	4	0.3
GC/MS–MS					
Propanil	161 → 126,134	7.2	9.6	3	2
Acetochlor	223 → 146	3.1	13.0	15	5
Myclobutanil	179 → 125,152	2.2	17.7	2	10
Fenoxycarb	116 → 88	9.1	11.2	5	8

[a]Product ions are formed by collision-induced MS/MS. Analyses performed by immersion of a CAR/DVB 65-μm SPME fiber. Limit of detection (LOD) based on signal-to-noise ratio (S/N) of 3; RSD is calculated for three replicate analyses (Natangelo et al., 2002).

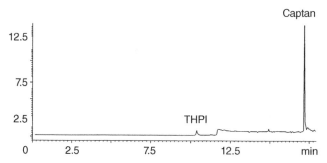

Figure 9.8. The GC/ITMS analysis of THPI (0.13 mg/kg) and captan (0.76 mg/kg) in a grape sample using the analytical conditions described in Table 9.10. (Reproduced from *Journal of Agricultural and Food Chemistry*, 2003, 51, p. 6763, Angioni et al., with permission of American Chemical Society.)

TABLE 9.10. The GC/ITMS Conditions Used for Captan and THPI Analysis Showed in Fig. 9.8[a]

Column	5% Diphenyl–95% dimethylpolysiloxane (30 m × 0.25 mm i.d.; 0.25-μm film thickness)
Injected volume	4.0 μL
Carrier gas	He Flow rate 1 mL/min
Injector temperature	From 60 to 150 °C at 30 °C/s
Oven temperature	60 °C (isotherm for 1 min) → 240 °C (3 °C/min)
Detector	ITMS, transfer line 200 °C IT Temperature 150 °C
IT conditions	Ionization mode: from 9 to 12-min liquid CI μSIS (methanol), from 12 to 18 EI SIS range mode (m/z 70–310) Quantitative ions: m/z 152 for THPI; m/z 79, 149, 264 for captan

[a]Angioni et al., 2003.

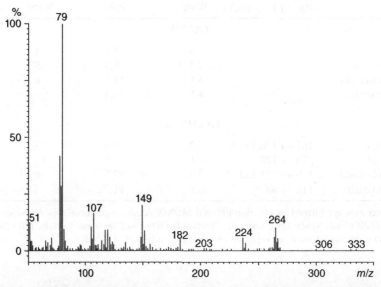

Figure 9.9. Mass spectrum of captan recorded by electron impact GC/ITMS analysis operating with the analytical conditions reported in Table 9.10. (Reproduced from *Journal of Agricultural and Food Chemistry*, 2003, 51, p. 6763, Angioni et al., with permission of American Chemical Society.)

The EI and CI fragmentation patterns of zoxamide proposed on the basis of MS/MS experiments are shown in Fig. 9.10. In EI mode, the molecular ion is not observed due to immediate cleavage of the H radical and the successive loss of HCl with consequent formation of

TABLE 9.11. GC/ITMS Conditions for Analysis of Zoxamide[a]

Column	5% Phenyl–95% methylpolysiloxane
	(30 m × 0.25 mm i.d.; 0.25-μm film thickness)
Injected volume	4.0 μL
Carrier gas	He Flow rate 1 mL/min
Injector temperature	From 60 °C (hold 1 min), to 150 °C at 30 °C/s (hold 20 min)
Injection mode	Splitless
Oven temperature	65 °C (isotherm for 1 min) → 280 °C (10 °C/min)
Detector	ITMS Operating in both EI and CI (acetonitrile)
	Ion trap, manifold, and transfer line temperatures 170, 100, and 200 °C, multiplier voltage 1400 V, emission current 80 μA (EI) and 30 μA (CI)
ITMS conditions	Prescan ionization time 1500 μs (EI) and 100 μs (CI)
	EI Selective ion storage range m/z 170–260
	SIS m/z 300 for CI
	MS/MS in waveform resonant mode, parent ion m/z 299

[a]Angioni et al., 2005.

the ions at m/z 335 and 299. The ions at m/z 271 and 214 are formed by the loss of CO of the ion at m/z 299 and 242, respectively. Conditions more commonly used in GC/IT–MS/MS analysis of vinclozolin, metalaxyl, captan, procymidone, folpet, and iprodione in wines, are reported in Table 9.12.

9.2.4 The LC/MS Analysis of Pesticides

Liquid chromatography is suitable for analysis of polar, low-volatile, and thermally labile pesticides, such as phenylureas and carbamates. In spite of the high sensitivity of postcolumn derivatization and fluorescence detection, or the robustness of UV detection, MS offers the advantages of high sensitivity and selectivity.

Multiresidue determination of pesticide residues in grape extracts can be performed using LC/MS/MS (Jansson et al., 2004; Banerjee et al., 2007; Venkateswarlu et al., 2007). Typical analytical conditions used are reported in Table 9.13; Table 9.14 reports the MS/MS parameters.

The LC/MS positive-ion mode analysis of grape carbamates reported in Fig. 9.11 (carbaryl, carbofuran, diethofencarb, ethiofencarb, fenobucarb, fenoxycarb, isoprocarb, methiocarb, metholcarb, oxamyl, pirimicarb, propoxur, and thiobencarb) was performed by matrix solid-phase dispersion (MSPD) extraction using either atmospheric pressure–chemical ionization (APCI) or electrospray ionization (ESI) (Fernández et al., 2000).

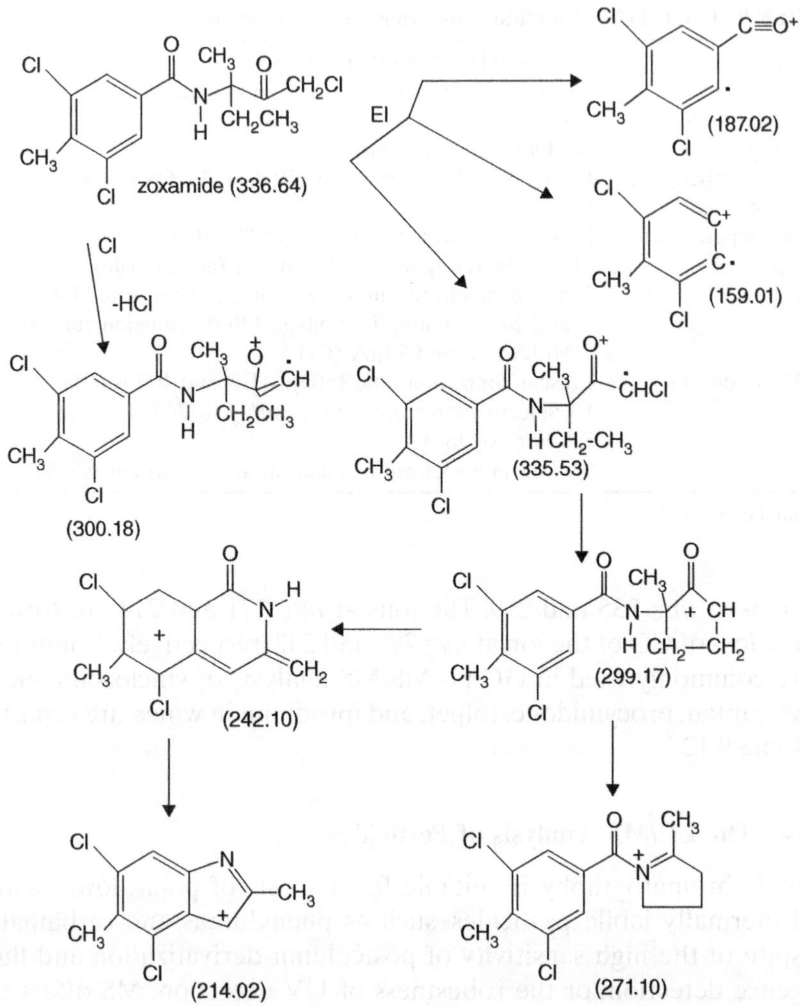

Figure 9.10. Zoxamide fragmentation patterns studied by GC/ITMS using both EI and CI. (Reprinted from *Journal of Chromatography A*, 1097, Angioni et al., Gas chromatographic ion trap mass spectrometry determination of zoxamide residues in grape, grape processing, and in the fermentation process. p. 166, Copyright © 2005, with permission from Elsevier.)

A C_8 LC column with elution using a methanol–water gradient was used. By replacing methanol with acetonitrile an improved chromatographic peak resolution was observed, as well as rapid contamination of the corona discharge needle in APCI. This effect is probably due to low ionizability of acetonitrile. The signals of the three main ions $[M+Na]^+$, $[M+H]^+$, and $[M+H-CH_3NCO]^+$ are observed in positive-ion mode with a cone voltage of 20V. The molecular mass is provided from both

TABLE 9.12. GC/IT–MS/MS Conditions for Analysis of Pesticides in Wine[a]

Pesticide	GC Retention Time (min)	MW	MS/MS Precursor Ion (m/z)		Quantitative Ion (m/z)	Excitation Storage Level (m/z)
Vinclozolin	13.9	286.1	212	→	172	93.4
Metalaxyl	14.3	279.3	206	→	162	90.7
Captan	17.3	300.6	264	→	236	116.3
Procymidone	17.3	284.1	283	→	255	124.7
Folpet	17.5	296.6	260	→	232	114.5
Iprodione	23.3	330.2	315	→	245	138.3

[a]GC column 5% diphenyl–95% dimethylpolysiloxane 30 m × 0.25 mm i.d.; 0.25-μm. Oven temperature from 80 to 300 °C at 5 °C/min; ionization type EI; isolation window 3.0 m/z; MS/MS waveform resonant mode; excitation amplitude 0.40 V; excitation time 20 mse.

TABLE 9.13. LC/MS Conditions Used for Multiresidue Analysis of Pesticides in Grape[a]

Column	C_{18} (100 × 3 mm.; 4 μm)
	Binary solvent: (A) methanol, (B) NH_4^+ formate 10 mM pH 4.0.
Elution mode	Gradient: from 0 to 90% A in 15 min, isocratic for 5 min, from 90 to 0% A in 3 min
Detector	MS Triple quadrupole with ES ion source operating in positive and negative-ion mode
MS conditions	Nebulizing gas N_2 90 L/h; drying gas N_2 400 °C at 600 L/h; capillary Voltage switched between +4.0 and −3.5 kV; source block temperature 120 °C; cone voltage between 10 and 70 eV, collision energy between 5 and 50 eV

[a]Jansson et al., 2004.

APCI and ESI. *N*-Methylcarbamate insecticides are labile compounds and can undergo collision-induced decomposition even when operating with a low-cone voltage (e.g., at 20 V, the base peak of the oxamyl APCI positive spectrum was the ion at *m/z* 163, formed by methylisocyanate loss). The authors observed that operating in positive mode, ESI produces both $[M+H]^+$ and $[M+Na]^+$ adducts, whereas APCI only yields the $[M+H]^+$ ion. Better sensitivity is achieved in positive mode; negative APCI shows formation of $[M–CONHCH_3]^-$ ion for most compounds, and of $[M–H]^-$ for diethofencarb and fenoxycarb. The softer ESI ionization induces lower fragmentation of oxamyl with respect to APCI, negative fragment ions of carbamates are formed with APCI, but not with ESI. Table 9.15 reports the principal species and fragment ions of carbamates formed in both positive- and negative-ion APCI, and by positive ESI.

TABLE 9.14. Parameters for LC/MS/MS Analysis of Pesticides[a,b]

Pesticide	[M+H]+	Q1	DP(V)	CE(V)	CXP(V)	Q2	CE(V)	CXP(V)
Acephate	184	143	48	14	5	125	29	4
Acetamiprid	223	126	60	27	6.6	56	35	3.5
Atrazine	216	174	65	28	8	104,96	30	2
Azinphos methyl	318	160	54	13	7	132	24	5
Azoxystrobin	404	372	53	22	4	311	32	2
Benalaxyl	326	208	65	24	11	148	23	7
Benfuracarb[c]	411	251.8						
Bitertanol	338	269	45	19	4	70	19	2
Buprofezin	306	201	32	20	9	116	24	7
Butachlor	312	238	31	18	12	162,91	35,40	8,5
Carbendazim	192	160	33	30	7	132	43	5.5
Carbaryl	202	145	53	13	6	127	40	6
Carbendazim[e]	192	160						
Carbofuran	222	165	55	20	8	123	28	6
Carbofuran-3-OH	238	163	32	17	8	163,107	22,46	8,4.4
Clofentezine[e]	303MW	138						
Chlorpyrifos[d]	350.6MW	197.8	43	29	15			
Clothianidin	250	169	50	20	5	132	29	6
Cymoxanil	199	111	48	31	4.3	128	22	6
Cyprodinil	226	93						
Demeton-S-methyl	231	89	34	18	4.8	155,61	25,47	8.3
Demeton-S-methyl sulfone	263	169	62	22	8	121	22	5
Diazinon	305	169	15	31	8	153,97	34,50	4
Dichlofluanid	333	224	33	24	11	123	39	6
Dichlorvos	221	109	65	27	2	127	28	5.5
Diethofencarb[c]	268	226						
Difenoconazole	406	337	74	25	4	251	34	13

308

Compound							
Dimethoate	230	199	50	18	1	125	29	4
Dimethomorph	388	301	55	30	3	165	49	8
Disodium methylarsonate (DMSA)	201	137	40	14	6	92	27	5
Diniconazole	326	159	74	51	7.2	70	53	2
Diuron[e]	233[MW]	72						
Emamectin benzoate	886.5	158	187	48	7	82.3	95	2
Ethion	385	199	25	16	1	171	25	1
Etrimfos	293	125	127	40	6	265,79	24,57	13,2
Fenamidone	312	236	53	21	5	92	35	3
Fenarimol	331	268	90	35	10	81	55	4
Fenobucarb[e]	208	116,88	12	25	4.6	152	14	8
Fenoxycarb[e]	302							
Fenpyroximate	422	366	63	27	2	135,138	50	6
Fenthion	279	247	10	16	2.3	169,105	27,35	8,2,4
Flufenoxuron[e]	489[MW]	158						
Flusilazole	316	165	13	37	8	247	28	2
Forchlorfenuron	248	129	56	25	5.6	155	25	5,6
Hexaconazole	314	70	52	38	2	159	38	6
Hexythiazox[e]	353[MW]	228	52	34	7.8	201	40	2
Imazalil	297	159	55	21	11	175	29	8
Imidacloprid	256	209	81	21	7	249,56	25,55	3,2
Indoxacarb	528	203	51	13	10	186,119	18,25	10,5
Iprovalicarb	321	203	38	19	11.0	189,145	34,49	9,6
Isoprothiolane	291	231	57	35	2	165	20	9
Isoproturon	207	72	46	37	6	205	15	10
Iprobenfos	289	91	58	10	9	206,116	10,21	9,5
Kresoxim methyl	314	267	62	19	6.0	285,99	13,42	4
Lufenuron[e]	511[MW]	158						
Malathion	331	127	15	17	10.0	99	42	4
Malaoxon	315	127						
Mandipropamid	412	328	68	18	6	356,125	15,48	7,5

TABLE 9.14. (*Continued*)

Pesticide	[M+H]⁺	Q1	DP(V)	CE(V)	CXP(V)	Q2	CE(V)	CXP(V)
Metalaxyl	280	192	58	26	8.0	220,160	20,27	8,9
Methamidophos	142	94	14	18	5.0	125	26	3,5
Methidathion	303	145	39	13	10.0	85	32	4
Methiocarb sulfoxide[e]	242	185,122						
Methiocarb sulphone[e]	258	122						
Methiocarb[e]	226	169						
Methomyl	163	106	34	17	2.0	88	17	2
Metribuzin	215	187	64	25	9	84	32	6
Mevinpho	225	193	43	15	9.0	127	22	6
Monocrotophos	224	127	52	16	3.0	98	20	3
Myclobutanil	289	70	67	50	2.0	125	29	5
Omethoate	214	125	45	35	9.0	109,183	42,20	4,10
Oxadixyl[e]	279	219						
Oxamyl[e]	219[MW]	237,72						
Oxydemeton methyl	247	169	48	20	8.8	229,109	17,19	12,4.5
Paraxon methyl	248	202	40	27	11.0	231,127	25,32	12,6
Penconazole	284	159	56	36	8.0	70	45	2
Phenthoate	321	163	18	20	8.0	275,247	11,17	4.4,13
Phosalone	368	182	68	30	9.0	138,111	48,60	6,4
Phosmet	318	160	109	10	9.0	133	50	5
Phosphamidon	300	174	68	21	6.0	127	30	5
Pirimicarb[e]	239	182						
Profenophos	373	303	75	28	6.0	311,207	19,38	4,10
Propargite	368	231	12	17	10.0	175	24	8.5

Propiconazole	342	159	30	33	8.3	69	40	2
Pyraclostrobin	388	194	20	18	10	163,296	40,18	7,8,3
Quinalphos	299	147	58	37	6.2	163,243	37,26	6,2,11
Simazine	202	132	60	27	5.8	124,96	27,34	6,3
Spinosyn A	732	142	90	38	7.6	99	101	4
Spinosyn D	746	142	93	35	7.0	99	100	5
Spiroxamine[e]	298	144						
Tebuconazole	308	70	61	55	4	125	59	8
Teflubenzuron[e]	381[MW]	158						
Temefos	467	419	92	30	6.0	341,125	40,49	3,5,6
Tetraconazole	372	70	66	68	5.0	169	40	6,5
Thiamethoxam	292	211	52	18	10.0	132	31	6
Thiacloprid	253	126	65	29	6	186	24	6
Thiodicarb	355	88	10	26	4.3	193,163	14,13	9,8
Thiometon	247	89	12	10	4.5	61	50	3,6
Triazophos	314	162	29	25	7.0	119	49	4,8
Triadimefon	294	197	58	21	8.0	115,69	18,33	5,8,2
Triadimenol	296	70	35	25	5.0	227	14	10
Trifloxystrobin	409	186	10	25	9.7	206,116	19	10,4,4
Triphenyl phosphate (I.S.)	327	215	80	38	10	152,77,51	55,65,125	7,7,5

[a]Banerjee et al., 2007.
[b]Protonated parent ion = [M+H]$^+$; MW = molecular weight; Q1 = quantifier ion; Q2 = second transition; DP = declustering potential; CE = collision energy; CXP = collision cell exit potential.
[c]Sagratini et al., 2007.
[d]Venkateswarlu et al., 2007.
[e]Jansson et al., 2004; Hernández et al., 2006.

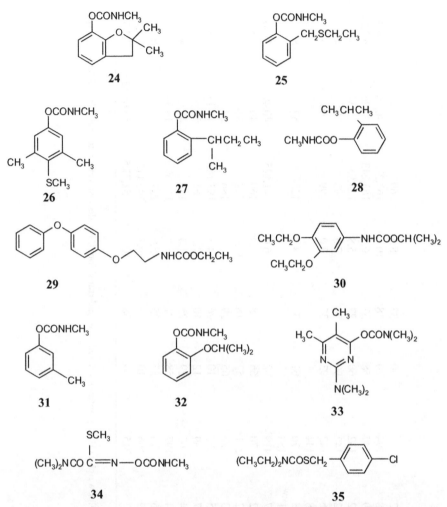

Figure 9.11. Carbamates determined by LC-atmospheric pressure chemical ionization (APCI) or electrospray (ES) in positive-ion mode (Fernández et al., 2000). (**24**) carbofuran, (**25**) ethiofencarb, (**26**) methiocarb, (**27**) fenobucarb, (**28**) isoprocarb, (**29**) fenoxycarb, (**30**) diethofencarb, (**31**) metholcarb, (**32**) propoxur, (**33**) pirimicarb, (**34**) oxamyl, (**35**) thiobencarb. Structure of carbaryl is reported in Fig. 9.1.

Carbamate and phenylurea pesticides can be determined in the same LC/ESI–MS run. Selected ions and corresponding fragmentator cone voltage (V_f) used in positive ESI/MS are reported in Table 9.16.

Vinclozolin, procymidone, iprodione, and 3,5-dichlorophenyl hydantoin (the iprodione degradation product) in wine were determined by SBSE and negative-ion liquid desorption-APCI (fragmentor voltage 70V, capillary voltage 4000V, mass range m/z 200–350) (Sandra et al.,

TABLE 9.15. Principal Species and Fragment Ions Observed in Positive- and Negative-Ion APCI and Positive ESI Analysis and Their Relative Abundances (R%)^a

Compound	MW	APCI Positive-Ion Mode m/z and Tentative Ions (R%)	APCI Negative-Ion Mode m/z and Tentative Ions (R%)	ESI Positive-Ion Mode m/z and Tentative Ions (R%)
Carbaryl	201	202 $[M+H]^+$ (100%) 234 $[M+H+CH_3OH]^+$ (13%)	143 $[M-H-CH_3NCO]^-$ (100%)	202 $[M+H]^+$ (95%) 145 $[M+H-CH_3NCO]^+$ (100%) 224 $[M+Na]^+$ (75%)
Carbofuran	221	222 $[M+H]^+$ (100%)	163 $[M-H-CH_3NCO]^-$ (100%)	222 $[M+H]^+$ (100%) 244 $[M+Na]^+$ (20%)
Diethofencarb	267	268 $[M+H]^+$ (100%) 182 $[M+H-(CH_3)_2CH_2NCO]^+$ (22%)	226 $[M-H]^-$ (100%)	268 $[M+H]^+$ (100%) 290 $[M+Na]^+$ (20%)
Ethiofencarb	225	226 $[M+H]^+$ (100%)	167 $[M-H-CH_3NCO]^-$ (100%)	226 $[M+H]^+$ (100%) 248 $[M+Na]^+$ (62%) 107 $[M-CH_3CH_2S-CH_3NCO]^+$ (50%)
Fenobucarb	207	208 $[M+H]^+$ (100%)	149 $[M-H-CH_3NCO]^-$ (100%)	164 $[M-CH_3CH_2S]^+$ (50%) 208 $[M+H]^+$ (100%) 226 $[M+NH_4]^+$ (25%)
Fenoxycarb	301	302 $[M+H]^+$ (100%) 230 $[M+H-(CH_3)_2NCO]^+$ (28%)	185 $[M-H-(CH_3)_2CH_3NCO_2]^-$ (100%)	302 $[M+H]^+$ (100%) 324 $[M+Na]^+$ (41%)
Isoprocarb	193	194 $[M+H]^+$ (100%)	135 $[M-H-CH_3NCO]^-$ (100%)	194 $[M+H]^+$ (100%) 216 $[M+Na]^+$ (15%)
Methiocarb	225	226 $[M+H]^+$ (100%)	167 $[M-H-CH_3NCO]^-$ (100%) 152 $[M-H-CH_3NCO-CH_3]^-$ (60%)	226 $[M+H]^+$ (100%) 248 $[M+Na]^+$ (22%)
Metholcarb	165	166 $[M+H]^+$ (100%)	107 $[M-H-CH_3NCO]^-$ (100%)	166 $[M+H]^+$ (100%) 188 $[M+Na]^+$ (19%)

TABLE 9.15. (*Continued*)

Compound	MW	APCI Positive-Ion Mode m/z and Tentative Ions (R%)	APCI Negative-Ion Mode m/z and Tentative Ions (R%)	ESI Positive-Ion Mode m/z and Tentative Ions (R%)
Oxamyl	219	163 [M+H–CH$_3$NCO]$^+$ (100%)	161 [M–H–CH$_3$NCO]$^-$ (100%) 147 [M–(CH$_3$)NCO]$^-$ (40%)	242 [M+Na]$^+$ (100%) 258 [M+K]$^+$ (40%) 237 [M+NH$_4$]$^+$ (27%) 251 [M+CH$_3$OH]$^+$ (17%)
Pirimicarb	238	239 [M+H]$^+$ (100%) 261 [M+Na]$^+$ (29%)		239 [M+H]$^+$ (100%)
Propoxur	209	210 [M+H]$^+$ (100%)	151 [M–H–CH$_3$NCO]$^-$ (100%)	210 [M+H]$^+$ (100%) 168 [M+H–CH$_3$CH=CH$_2$]$^+$ (33%) 153 [M+H–CH$_3$NCO]$^+$ (20%)
Thiobencarb	257	258 [M+H]$^+$ (100%)	132 [M–CH$_2$C$_6$H$_4$Cl]$^-$ (100%)	258 [M+H]$^+$ (100%)

[a]Fragmentor voltages: 20V positive mode, –40V negative mode (Fernández et al., 2000).

TABLE 9.16. Selected Ions and the Corresponding Fragmentator Cone Voltage (V_f) Used for ESI/MS Analysis of Wine Pesticides[a]

Carbamate Pesticide	MW	m/z and Ions Selected	V_f (V)	Phenylurea Pesticide	MW	m/z and Ions Selected	V_f (V)
Carbaryl	201	202 $[M+H]^+$	30	Monuron	198	199 $[M+H]^+$	60
		145 $[M+H-CH_3NCO]^+$	60			221 $[M+Na]^+$	70
Methiocarb	225	226 $[M+H]^+$	30			72 $[C_3H_6NO]^+$	100
		169 $[M+H-CH_3NCO]^+$	60	Fluometuron	232	233 $[M+H]^+$	60
Propham	179	120 $[C_6H_5NCO+H]^+$	90			72 $[C_3H_6NO]^+$	100
		138 $[M+H-C_3H_6]^+$	60	Siduron	232	233 $[M+H]^+$	60
Promecarb	207	208 $[M+H]^+$	30			255 $[M+Na]^+$	120
		151 $[M+H-CH_3NCO]^+$	60	Diuron	232	233 $[M+H]^+$	60
Chlorpropham	214	154 $[M-C_3H_7OH]^+$	90			72 $[C_3H_6NO]^+$	100
		172 $[M-C_3H_6]^+$	60	Linuron	248	249 $[M+H]^+$	60
Barban	258	258 $[M]^+$	30	Neburon	274	275 $[M+H]^+$	50
		178 $[M+H-81]^+$	60			297 $[M+Na]^+$	70

[a]Capillary voltage 4500 V, positive ion mode (Wu et al., 2002).

TABLE 9.17. Principal *m/z* Signals Recoded in the Negative LC/MS Mass Spectra of Procymidone, Iprodione, and Vinclozolin[a]

Pesticide	MW	*m/z* Signals
Procymidone	283	317.9, 316.1, 286.1, 284.0, 275.9, 274.1, 257.9, 255.8, 243.8, 242.0 (100)
Iprodione	329	246.8, 245.0, 242.9 (100)
Vinclozolin	285	316.1, 314.0 (100), 302.0, 299.9, 284.0, 281.9

[a]Sandra et al., 2001.

2001). After sampling, the stir bar was desorbed in acetonitrile and analysis of the extract was carried out with a C_{18} column and a binary solvent composed of water (solvent A) and 10% tetrahydrofuran (THF) in methanol (B), with a linear gradient elution program from 0 to 80% B in 20 min. Formation of the $(M+CH_3OH-H)^-$ ion was observed for vinclozolin (MW 285) and procymidone (MW 283). For iprodione (MW 329) formation of $[M-CONHCH(CH_3)_2]^-$ ion is due to the thermo instability of the compound. LC/MS mass spectra of the three compounds are characterized from the *m/z* signals reported in Table 9.17. Negative APCI resulted in a better and more robust method than positive ionization, and than ESI in both positive and negative ion mode (Sandra et al., 2001). Accurate mass spectra of procymidone and iprodione recorded by LC/ESI-TOF-MS, are reported in Figure 9.12.

9.3   ISOTHIOCYANATES IN WINE

Allyl isothiocyanate is a contaminant used to protect the wine from the *Candida Mycoderma* yeast attack and to sterilize the air in wine storage containers. Illegal additions of methyl isothiocyanate to the wines are made to prevent spontaneous fermentations and are used as a soil fumigant for nematodes, fungi, and other diseases in fruit and vegetables (Saito et al., 1994; Gandini and Riguzzi, 1997).

Determination of methyl isothiocyanate in wine can be performed by extraction with ethyl acetate and GC/MS analysis (Uchiyama et al., 1992). Spiked samples showed recoveries ranging between 83 and 90% in white wines, and 75 and 82% in red wines, with an LOD of 0.05 mg/L. Also, analysis by direct injection using 1,4-dioxan as an internal standard was proposed (Fostel and Podek, 1992).

Analysis of methyl isothiocyanate in wine also can be performed by headspace SPME–GC/MS using a CAR/DVB 65-μm fiber. A volume

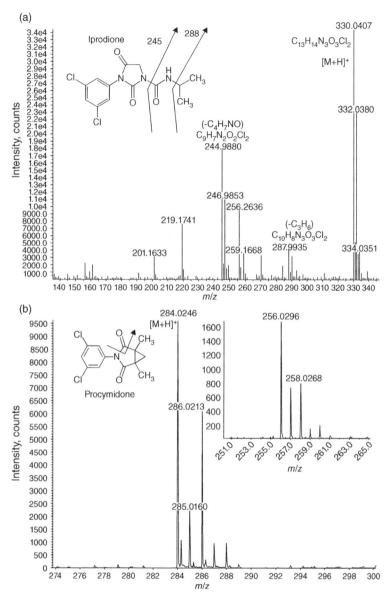

Figure 9.12. Accurate LC/ESI–TOF–MS positive-ion mass spectra of iprodione (a) and procymidone (b) (capillary voltage 4000 V, nebulizer pressure 40 psi, drying gas 9 L/min, gas temperature 300 °C, skimmer voltage 60 V, octapole DC 1:37.5 V, octapole RF 250 V, scan range m/z 50–1000, CID fragmentor voltages 190 and 230 V). (Reprinted from Rapid Communications in Mass Spectrometry 19, García-Reyes, et al., Searching for non-target chlorinated pesticides in food by liquid chromatography/time-of-flight mass spectrometry, pp. 2784, 2785, Copyright © 2005, with permission from John Wiley & Sons, Ltd.)

of 5 mL of the sample is placed in a 10-mL vial and added with 1.25 g NaCl performing extraction at room temperature with stirring for 30 min (Gandini and Riguzzi, 1997). A GC/MS analysis is made using a fused silica PEG capillary column (30 m × 0.25 mm i.d.; 0.25-μm film thickness) and an oven temperature program starting from 40 °C for 2 min, temperature is increased to 60 °C at 20 °C/min and held for 1 min, then to 75 °C at 1 °C/min and held for 2 min, finally to 220 °C at 8 °C/min and held for 5 min. The ions at m/z 73 (M^+), 72, and 45, are recorded (Saito et al., 1994) by operating in SIM-mode. Under similar conditions and by also recording the allyl isothiocyanate signals at m/z 99 (M^+) and 72, simultaneous detection of both compounds is achieved (Przyborski et al., 1995).

REFERENCES

Anastassiades, M., Lehotay, S.J., Štajnbaher, D., and Schenck, F.J. (2003). Fast and easy multiresidue method employing acetonitrile extraction/partitioning and "dispersive solid-phase extraction" for the determination of pesticide residues in produce, *J. AOAC Int.*, **86**(2), 412–431.

Angioni, A., Garau, A., Caboni, P., Russo, M., Farris, G.A., Zara, S., and Cabras, P. (2005). Gas chromatographic ion trap mass spectrometry determination of zoxamide residues in grape, grape processing, and in the fermentation process, *J. Chromatogr. A*, **1097**(1–2), 165–170.

Angioni, A., Garau, V.L., Aguilera Del Real, A., Melis, M., Minelli, E.V., Tuberoso, C., and Cabras, P. (2003). GC-ITMS determination and degradation of captan during winemaking, *J. Agric. Food Chem.*, **51**(23), 6761–6766.

Banerjee, K., Oulkar, D.P., Dasgupta, S., Patil, S.B., Patil, S.H., Savant, R., and Adsule, P.G. (2007). Validation and uncertainty analysis of a multiresidue method for pesticides in grapes using ethyl acetate extraction and liquid chromatography-tandem mass spectrometry, *J. Chromatogr. A*, **1173**, 98–109.

Briggs, S. (1992). *A basic guide to pesticides: Their characteristics and hazards.* Hemisphere Publishing Corporation, Washington, DC.

Cabras, P. and Caboni, P. (2008). Analysis of Pesticide Residues in Grape and Wine, In: *Hyphenated Techniques in Grape & Wine Chemistry*, Riccardo Flamini, (Ed.) John Wiley & Son, Ltd., pp. 227–248.

Cabras, P., Angioni, A., Garau, V.L., Melis, M., Pirisi, F.M., Farris, G.A., Sotgiu, C., and Minelli, E,V. (1997a). Persistence and metabolism of folpet in grapes and wine, *J. Agric. Food Chem.*, **45**(2), 476–479.

Cabras, P., Angioni, A., Garau, V.L., and Minelli, E.V. (1997b). Gas chromatographic determination of cyprodinil, fludioxonil, pyrimethanil, and tebuconazole in grapes, must, and wine, *J. AOAC Int.*, **80**(4), 867–870.

Cabras, P., Angioni, A., Garau, V.L., Pirisi, F.M., and Brandolini, V. (1998). Gas chromatographic determination of azoxystrobin, fluazinam, kresoxim-methyl, mepanipyrim and tetraconazole in grapes, must and wine, *J. AOAC Int.*, **81**(6), 1185–1189.

Cabras, P., Diana, P., Meloni, M., Pirisi, F.M., and Pirisi, R. (1983). Reversed-phase high-performance liquid chromatography of pesticides. Analysis of Vinclozolin, Iprodione, Procymidone, Dichlozolinate and their degradation product 3,5-dichloroaniline on white must and wine extracts, *J. Chromatogr.*, **256**, 176–181.

Cabras, P., Tuberoso, C., Melis, M., and Martini, M.G. (1992). Multiresidue method for pesticide determination in wine by high-performance liquid chromatography, *J. Agric. Food Chem.*, **40**(5), 817–819.

Fernández, M., Picó, Y., and Mañes, J. (2000). Determination of carbamate residues in fruits and vegetables by matrix solid-phase dispersion and liquid chromatography-mass spectrometry, *J. Chromatogr. A*, **871**, 43–46.

Flamini, R. and Panighel, A. (2006). Mass spectrometry in grape and wine chemistry. Part II: The Consumer Protection, *Mass Spectrom. Rev.*, **25**(5), 741–774.

Fostel, H. and Podek, F. (1992). Detection of methyl isothiocyanate (methyl mustard oil) in wine, *Ernaehrung*, **16**(3), 157–158.

Gandini, N. and Riguzzi, R. (1997). Headspace solid-phase microextraction analysis of methyl isothiocyanate in wine, *J. Agric. Food Chem.*, **45**, 3092–3094.

García-Reyes, J.F., Ferrer, I., Thurman, E.M., Molina-Díaz, A., and Fernández-Alba, A.R. (2005). Searching for non-target chlorinated pesticides in food by liquid chromatography/time-of-flight mass spectrometry, *Rapid Comm. Mass Spectrom.*, **19**, 2780–2788.

Hatzidimitriou, E., Darriet, P., Bertrand, A., and Dubourdieu, D. (1997). Folpet hydrolysis—incidence on the initiation of the alcoholic fermentation, *J. Int. Sci. Vigne Vin*, **31**(1), 51–55.

Hayasaka, Y., MacNamara, K., Baldock, G.A., Taylor, R.L., and Pollnitz, A.P. (2003). Application of stir bar sorptive extraction for wine analysis, *Anal. Bioanal. Chem.*, **375**, 948–955.

Hernández, F., Pozo, O.J., Sancho, J.V., Bijlsma, L., Barreda, M., and Pitarch, E. (2006). Multiresidue liquid chromatography tandem mass spectrometry determination of 52 non gas chromatography-amenable pesticides and metabolites in different food commodities, *Chromatogr. A*, **1109**, 242–252.

Holland, P.T., McNaughton, D.E., and Malcolm, C.P. (1994). Multiresidue analysis of pesticides in wines by solid-phase extraction, *J. AOAC Int.*, **77**(1), 79–86.

Jansson, C., Pihlstrom, T., Osterdahl, B.G., and Markides, K.E. (2004). A new multiresidue method for analysis of pesticides residues in fruit and vegetables using liquid chromatography with tandem mass spectrometric detection, *J. Chromatogr. A.*, **1023**(1), 93–104.

Kakalíková, L., Matisová, E., and Leško, J. (1996). Analysis of metalaxyl residues in wines by SPE in combination with HRCGC and GC/MS, *Z. Lebensmittel Untersuchung Forsch.*, **203**(1), 56–60.

Kaufmann, A. (1997). Fully automated determination of pesticides in wine, *J. AOAC Int.*, **80**(6), 1302–1307.

Matisová, E., Kakalíková, L., Leško, J., and de Zeeuw, J. (1996). Application of porous carbon for solid-phase extraction of dicarboxyimide fungicide residues from wines in combination with high-resolution capillary gas chromatography and gas chromatography-mass spectrometry, *J. Chromatogr. A*, **754**, 445–454.

Millán, S., Sampedro, M.C., Unceta, N., Goicolea, M.A., Rodríguez, E., and Barrio, R.J. (2003). Coupling solid-phase microextraction and high-performance liquid chromatography for direct sensitive determination of halogenated fungicides in wine, *J. Chromatogr. A*, **995**, 135–142.

Natangelo, M., Tavazzi, S., and Benfenati, E. (2002). Evaluation of solid phase microextraction-gas chromatography in the analysis of some pesticides with different mass spectrometric techniques: application to environmental waters and food samples, *Anal. Lett.*, **35**(2), 327–338.

Przyborski, H., Wacha, C., and Bandion, F. (1995). Detection and evaluation of methylisothiocyanate and allylisothiocyanate in wine, *Mitteilungen Klosterneuburg*, **45**(4), 123–126.

Sagratini, G., Manes, J., Giardiná, D., Damiani, P., and Picó, Y. (2007). Analysis of carbamate and phenylurea pesticide residues in fruit juices by solid-phase microextraction and liquid chromatography–mass spectrometry, *J. Chromatogr. A*, **1147**, 135–143.

Saito, I., Yamada, S., Oshima, H., Ikai, Y., Oka, H., and Hayakawa, J. (1994). Analysis of Methyl isothiocyanate in wine by gas chromatography with dual detection, *J. AOAC Int.*, **77**(5), 1296–1299.

Sandra, P., Tienpont, B., Vercammen, J., Tredoux, A., Sandra, T., and David, F. (2001). Stir bar sorptive extraction applied to the determination of dicarboximide fungicides in wine, *J. Chromatogr. A*, **928**, 117–126.

Uchiyama, S., Takeda, H., Kobayashi, A., Ito, S., Sakurai, H., Tada, Y., Aoki, G., Hosgai, T., Yamanaka, T., Maekawa, Y., Yoshikawa, R., Ito, K., and Saito, Y. (1992). Determination of methyl isothiocyanate in wine by GC and GC/MS, *J. Food Hygien. Soc. Jpn.*, **33**(6), 603–608.

Tomlin, C. (1994). *The Pesticide Manual*, 10th ed: British Crop Protection Council: Faenham, England.

Venkateswarlu, P., Mohan, K.R., Kumar, Ch.R., and Seshaiah, K. (2007). Monitoring of multi-class pesticide residues in fresh grape samples using liquid chromatography with electrospray tandem mass spectrometry, *Food Chem.*, **105**, 1760–1766.

Vitali, M., Guidotti, M., Giovinazzo, R., and Cedrone, O. (1998). Determination of pesticide residues in wine by SPME and GC/MS for consumer risk assessment, *Food Addit. Contam.*, **15**(3), 280–287.

Wong, J.W. and Halverson, C.A. (1999). Multiresidue analysis of pesticides in wines using C-18 solid-phase extraction and gas chromatography-mass spectrometry, *Am. J. Enol. Vitic.*, **50**(4), 435–442.

Wong, J.W., Webster, M.G., Halverson, C.A., Hengel, M.J., Ngim, K.K., and Ebeler, S.E. (2003). Multiresidue pesticide analysis in wines by solid-phase extraction and capillary gas chromatography-mass spectrometric detection with selective ion monitoring, *J. Agric. Food Chem.*, **51**, 1148–1161.

Wu, J., Tragas, C., Lord, H., and Pawliszyn, J. (2002). Analysis of polar pesticides in water and wine samples by automated in-tube solid-phase microextraction coupled with high-performance liquid chromatography-mass spectrometry, *J. Chromatogr. A*, **976**, 357–367.

Zambonin, C.G., Cilenti, A., and Palmisano, F. (2002). Solid–phase microextraction and gas chromatography-mass spectrometry for the rapid screening of triazole residues in wines and strawberries, *J. Chromatogr. A*, **967**, 255–260.

10

PEPTIDES AND PROTEINS OF GRAPE AND WINE

10.1 INTRODUCTION

Most potential plant pathogens are stopped by physical and chemical barriers and by defence responses. These mechanisms are induced by specific receptor-mediated recognition of the pathogen or plant cell wall-derived molecules, termed exogenous or endogenous elicitors, respectively (Ebel and Cosio, 1994). Pathogen recognition, signal-transduction pathways involving ion fluxes, protein kinase activation, and active oxygen species production, are activated (Blumwald et al., 1998; Nümberger and Scheel, 2001). These events are frequently associated with a localized cell death known as the hypersensitive reaction. Plants can also develop systemic acquired resistance that reduces subsequent infection of healthy tissues by a broad range of pathogens. These mechanisms are controlled by signaling molecules, including salicylic acid, jasmonate, and ethylene. The former compound is often implicated in the resistance to biotrophic pathogens (Dempsey et al., 1999), while ethylene and jasmonate are active against necrotrophic fungi (Berrocal-Lobo et al., 2002; Ton et al., 2002). Some pathogenic strains are able to bypass plant defences, necrotrophs (e.g., *Botrytis*

Mass Spectrometry in Grape and Wine Chemistry, by Riccardo Flamini
and Pietro Traldi
Copyright © 2010 John Wiley & Sons, Inc.

cinerea) are able to degrade host cell walls (Staples and Mayer, 1995), detoxify plant products (Gil-ad and Mayer, 1999) to block defence responses (Cessna et al., 2000).

The major pathogenic-related (PR) proteins in grape are chitinases-and thaumatin-like proteins. Both of these proteins persist through the vinification process and cause hazes and sediments in bottled wines. Traditional methods of wine protein analysis include dialysis, ultrafiltration, precipitation, exclusion chromatography, one- or two-dimensional (1D or 2D) electrophoresis, capillary electrophoresis (CE), isoelectric focusing, affinity chromatography, immunodetection, high-performance liquid chromatography (HPLC) and fast-protein liquid chromatography (Kwon, 2004). Liquid chromatography–electro-spray mass spectrometry (LC/ESI–MS), ESI–MS, and matrix-assisted laser desorption ionization (MALDI) coupled with time-of-flight (TOF) analyzer mass spectrometry, are successfully applied to the study of grape and wine proteins and in the differentiation of grape varieties. On-line coupling of CE to ESI interface and MS detector (CE–MS) offers separations with high resolution and provides important information on the structure of a number of proteins (Simó et al., 2004). Both ESI and MALDI coupled with quadrupolar, magnetic sector, or TOF analyzers and MS tandem systems (MS/MS) provide structural information on the amino acids forming the proteins (Moreno-Arribas et al., 2002).

Separation and identification of several thousand proteins are achieved by use of two-dimension electrophoresis (2-DE) coupled to MS. This procedure, which can be automated, involves excision of the protein spots from the 2-DE gel followed by individual enzymatic proteolysis with trypsin and MS analysis of the mixture (Ashcroft, 2003).

The 2-DE polyacrylamide gel electrophoresis (2D-PAGE) provides resolution based on both molecular size and m/z differences. The MS data and genome database searching allows identification of protein sequence (Opiteck and Scheffler, 2004). For ambiguous identification, ESI–MS/MS analyses allow generation of a sequence tag of peptide, and by further database searching with both the peptide molecular mass and sequence tag information, unambiguous protein identification can be achieved (Ashcroft, 2003). When these sequences are auto-mated, protein identification is possible by database searching using algorithms, such as KNEXUS, SEQUEST, THEGPM, PHOENIX, and MASCOT. Each protein sequence from the database (e.g., SWISS-PROT, TrEMBL, NCBI) is virtually digested according to the specific-ity of the used protease. The resulting peptides that match the measured mass of the peptide ion are now identified. In the next step, the experimentally derived MS/MS spectrum of the peptide ion is compared to the theoretical spectra obtained by virtual fragmentation of candidate peptide sequences. Finally, a score is calculated for each peptide sequence by matching the predicted fragment ions to the ions observed in the experimental spectrum. Because database searches can generate false positives and negatives, depending on the parameters used, a manual evaluation of the data is opportune to confirm protein identification (Glinski and Weckwerth, 2006).

The MALDI–TOF (see Section 1.5) is widely used in protein analysis (Weiss et al., 1998). An acidic solution containing an energy-absorbing molecule (matrix) is mixed with the analyte and highly focused laser pulses are directed at the mixture. Proteins are desorbed, ionized, and accelerated by a high electrical potential. The ions arrive at the detector in the order of their increasing m/z ratio. This technique is used regularly to perform generation of a mass map of proteins after enzymatic digestion, due to robustness, tolerance to salt- and detergent-related impurities, and ability to be automated (Ashcroft, 2003). An α-cyano-4-hydroxycinnamic acid (CHCA) matrix is commonly used for analysis of peptides and small proteins; sinapinic acid (SA) is used for the analysis of higher molecular weight (MW) proteins (10–100 kDa). Advantages of MALDI–TOF methods are their good mass accuracy (0.01%) and sensitivity (proteins in femtomole range can be detect), and that they require very little sample for analysis.

A review on applications of MS in the study of grape and wine proteins was recently reported (Flamini and De Rosso, 2006).

10.1.1   Grape Proteins

The grape berry contains a large number of proteins even if present in low amounts compared to other fruits. The most abundant are synthesized by the plant after veraison and accumulate during ripening in conjunction with sugars (Tattersall et al., 1997). By exploiting a proteomic approach, 66 different protein components were identified using MALDI–TOF/MS in the pulp of the *Gamay* grape (Sarry et al., 2004) and *Cabernet Sauvignon* grape skins (Deytieux et al., 2007), most involved in energy metabolism, biotic or abiotic stress, and primary metabolism. At the harvest, the dominant proteins in skins are involved in defence mechanisms (Deytieux et al., 2007).

In addition to metabolic enzymes, other major proteins present in the mature grape berry are pathogenesis-related (PR) proteins, including chitinases (PR3 family) and thaumatin-like proteins (TLP, PR5 family). They are produced during ripening of the berry and are involved in the defence mechanisms of the plant against fungal pathogen attacks being able to hydrolyze chitin, a structural component of the cell wall of the invading fungus. Their production can increase as a consequence of wounding or exposure to pathogens (Boller, 1987; Linthorst, 1991). In general, PR proteins are acid soluble and resistant to proteases; TLP and chitinases have antifungal properties (Punja and Zhang, 1993; Stintzi et al., 1993; Graham and Sticklen, 1994; Cheong et al., 1997).

All grape cultivars synthesize a set of PR proteins that was observed to be identical to those involved in the haze in wine forming, a number of isoforms exist within individual varieties with a MW that might differ slightly among varieties (Pocock et al., 1998; 2000; Pocock and Waters 1998; Waters and Williams, 1996; Waters et al., 1996; 1998; Robinson et al., 1997; Derckel et al., 1996; Jacobs et al., 1999; Busam et al., 1997). Grape chitinases identified by MS have a MWs of 25–26 kDa, the most important isoform seems to be a class IV chitinase (containing a chitin-binding domain), which is highly expressed during berry ripening (Pocock et al., 2000; Robinson et al., 1997). The main TLP isoform in grape is VVTL1, but another minor form can be present. The MW of these proteins, determined by ESI–MS, is 21.272 and 21.260 kDa, respectively (Pocock et al., 2000). Moreover, TLP with a MW 23.881 kDa and an isoelectric point (pI) 4.67 was identified in the *Cabernet Sauvignon* grape cluster (Vincent et al., 2006), and in the *Gamay* mesocarp (Sarry et al., 2004) with a MW 24.0 kDa (pI 5.1). Environmental and/or pathological factors prevailing during development and maturation of berries can determine the level and composition of both TLP and chitinases, but neither drought stress nor the physical damage deriving from mechanical harvesting seem to influence these proteins (Monteiro et al., 2003; Pocock et al., 1998; 2000).

The low-MW proteins profile of grape seeds shows signal clusters in the ranges between m/z 4,000 and 5,000, 5,500 and 6,500, and 12,500 and 15,000 (Pesavento et al., 2008).

10.1.2   Wine Proteins

Although proteins and peptides are minor constituents of wine, they make a significant contribution to the quality of product and play an important role in the wine quality as they affect taste, clarity, and stability. Chitinases and TLP persist through the vinification process and may cause hazes and sediments in bottled wines during the storage due to protein denaturing and aggregating with mechanisms not fully understood yet. Some yeast proteins reduce haze in white wine, while other grape proteins can induce it.

A number of other proteins contribute to the formation and stability of foam in sparkling wines. Peptides exhibit surfactant and sensory properties that can influence the organoleptic characteristics of product.

Wine proteins have MWs ranging from 9 to 62 kDa and pIs between 3 and 9 (Brissonnet and Maujean, 1993; Hsu and Heaterbell, 1987); also, the presence of high MW mannoproteins from yeasts was reported (Gonçalves et al., 2002). However, main proteins (grape chitinases and TLP) have MWs of 20–30 kDa and pIs 4–6 (Waters et al., 1996). In addition, grape invertase (MW 62–64 kDa) seems to be one of the most abundant proteins of wine (Dambrouck et al., 2005).

10.2 ANALYTICAL METHODS

10.2.1 The MS Analysis of Grape Peptides and Proteins

10.2.1.1 Extraction. If grape proteins are studied in relation to wine characteristics and technology, it may be convenient to start from the "free run juice" obtained by recovering the liquid from squeezing of the grape berries that is representative of the must used for winemaking. For preparation, the suspension obtained from berry squeezing is filtered through a cloth and the suspended particles are separated by centrifugation. The soluble proteins can be concentrated by ultrafiltration on membranes with a 10-kDa cutoff (Waters et al., 1998). Proteins can be collected from the solution by the classical methods for protein precipitation (saturation with 80% ammonium sulfate) or by addition of organic solvents to a final concentration of 80%. In this case, only the proteins remaining soluble under the conditions arising from the preparation of the must will be present in the extract. If all proteins have to be extracted, for example, for the study of grape biology, it is necessary to adopt a multistep method in order to minimize protein losses occurring with rupture of the cells and to protein interaction with phenolic compounds (Curioni et al., 2008).

To detect the maximum quantity of the protein components belonging to the pulp, it is necessary to introduce into the extraction buffer some protective agents, such as reducing substances, ascorbic acid, and PVPP, in order to minimize protein modifications and losses (Tattersall et al., 1997). Recently, a method to achieve high recovery of pulp proteins by using a solution containing 12.5% of trichloroacetic acid (TCA) in cold acetone and 2-mercaptoethanol as the reducing agent was developed (Sarry et al., 2004). The procedure is summarized in

TABLE 10.1. Method for Proteins Extraction from Grape Pulp[a]

Washing of fresh berries in tap water
Crushing in 1:10 (w/v) of cold trichloroacetic acid/acetone 12.5:87.5 (v/v) solution
    containing 2-mercaptoethanol 28 mM
Filtration on 40-μm Miracloth mesh
Incubation at −20°C for 60 min
Centrifugation at 10,000 g for 15 min
Washing twice with 85% ethanol

[a]Sarry et al., 2004.

TABLE 10.2. Protocol for Proteins Extraction from Grape Skins[a]

Grounding of 20 berry skins to a fine powder with liquid nitrogen
Extraction with 3 volumes (v/w) of buffer pH 7.5 (Tris-HCl 0.1 M,
    ethylenediaminetetraacetic acid (EDTA) 5 mM, phenylmethanesulfonylfluoride
    1 mM, 2-mercaptoethanol 2%, KCl 0.1 M, sucrose 0.7 M, PVPP 1%) under stirring
    at 4°C for 1 h
Addition of equal volume of phenol–Tris-HCl pH 7.5 and agitation at 4°C for 1 h
Centrifugation 9000 g at 4°C for 30 min and collection of the lower (phenolic)
    solution
Reextraction of aqueous phase for 30 min with −2 mL extraction buffer +2 mL
    phenol solution
Washing of phenolic phase three times with equal volume of extraction buffer
Addition to phenolic phase of 5 volumes of ammonium acetate 0.1 M in methanol
Incubation overnight at −20°C
Centrifugation 9000 g at 4°C for 30 min
Washing and centrifugation of pellet: once with ammonium acetate 0.1 M in
    methanol, twice with cold methanol, twice with ice-cold 80% acetone
Drying under nitrogen stream

[a]Deytieux et al., 2007.

Table 10.1. A solid-phase extraction (SPE) using a C_{18} cartridge can be used for purification of extract from phenolic compounds. This procedure may lead to losses of proteins, but in general does not affect the protein composition (Waters et al., 1992).

The procedure summarized in Table 10.2 was efficient for extraction of proteins from skins in terms of the number of different proteins detected (Deytieux et al., 2007).

Other than large quantities of lipids and polyphenols, grape seeds contain a significant amount of proteins and peptides (Yokotsuka and Fukui, 2002). In this case, it is particularly necessary to minimize the effects of the high-polyphenol content, such as to perform extraction using a high pH buffer (>10.0) containing PEG, followed by

TABLE 10.3. Methods for Extraction of Peptides and Proteins from Grape Seeds

Method proposed by Famiani et al. (2000)

Grounding of seeds with liquid nitrogen

50 mg of powder extracted with 400 μL of ice-cold containing 2-amino-2-methyl-1-propanol (AMPS) 0.5 M (pH 10.8), SDS 1%, PEG-6000 1%, and dithiothreitol (DTT) 50 mM

Centrifugation at 12,000 g for 5 min and collection of the supernatant

Precipitation of proteins with 3 volumes of 80% acetone. Placing in liquid nitrogen for 10 min

Centrifugation at 12,000 g for 5 min and collection of the protein pellet

Method proposed by Pesavento et al. (2008)

10 g of grape seeds previously washed in water and powdered with liquid nitrogen

5 g of seeds powder defatted by 100 mL of *n*-hexane with stirring for 30 min

Organic solvent removed and the residue left to dry at room temperature

500 mg of residue suspended in 5-mL water and dialyzed against double distilled water at 4 °C for 48 h

Centrifugation at 3000 rpm for 15 min

10-mg sample extracted by 1-mL water containing trifluoroacetic acid 0.1% for 5 min with ultrasonic waves

The supernatant is collected and analyzed

precipitation with cold acetone (Famiani et al., 2000). Recently, a method for extraction of the grape seed peptides (2000–20000 kDa) finalized to MALDI analysis was proposed (Pesavento et al., 2008). Different methods of extraction were studied under the same conditions by suspending the seeds powder in the three solvents (H_2O + 0.1% trifluoroacetic acid, acetonitrile, methanol/acetonitrile 1:1 v/v). A H_2O + 0.1% trifluoroacetic acid solution gave the best results in terms of signal intensity. The two methods are summarized in Table 10.3.

10.2.1.2 Analysis. Mass spectrometry techniques mainly used for analysis of proteins are LC/MS and MALDI–TOF; direct ESI/MS is used in the variety characterization as a complement to DNA methods.

The LC/MS of grape juice proteins (MW 13–33 kDa) can be performed by direct injection of the concentrated juice using a C_8 reverse-phase column (e.g., 250 × 1 mm) equilibrated with a mixture of 0.05% trifluoroacetic acid (TFA) in water (solvent A) and 0.05% (v/v) TFA in 90% aqueous acetonitrile (solvent B) 3:1 v/v (Hayasaka et al., 2001). Elution is performed with a linear gradient program from 25 to 90% solvent B for 60 min at a flow rate of 15 μL/min, then isocratic for 30 min. The column is directly connected to an ESI source or connected to a UV–Vis detector operating at a 220 nm wavelength coupled on-line with the mass spectrometer. Analysis is performed in positive-ion mode with ESI and orifice potentials at 5.5 kV and 30 V, respectively (curtain gas N_2 and nebulizer air at 8 and 10 units, respectively). The mass spectrum of the protein peak, consisting of the multiple charge ions, is processed to determine the most probable MW of the proteins. The identity of individual proteins is determined on the basis of both LC elution order and MW. Distribution and intensities of the multiply charged molecular ions produced by ESI are directly related to the number of basic amino acids and the structural conformation of the protein.

Protein trap-ESI/MS was used in the study of juice proteins by loading of the concentrated juice onto a protein trap (3×8 mm) cartridge directly connected to the mass spectrometer (Hayasaka et al., 2001). The cartridge was equilibrated with a mixture of formic acid/ H_2O 2:98 (v/v) (solvent A) and of 2% (v/v) formic acid in 80% (v/v) aqueous acetonitrile (solvent B) 70:30 (v/v). The proteins were eluted in one or two broad fractions by using the gradient program from 30 to 60% of B in 10 min, isocratic for 10 min, then B increased to 80% in 10 min, finally isocratic for 5 min (the cartridge was washed sequentially with 3 mL of 50% and 3 mL of 80% aqueous acetonitrile containing 2% formic acid before being reused). This approach enhanced the detectability of TLP with respect to LC/MS 10-fold, even with complicated mass spectra due to coelution of proteins. The minimum amount of total proteins required for this method was 150 ng. Analysis of PR proteins with MW in the ranges 21,239–21,272 and 25,330–25,631 Da were useful for differentiation of grape varieties. The ESI/MS patterns of PR-proteins in the juice of three different white grape varieties are shown in Fig. 10.1 (Hayasaka et al., 2003).

A MALDI–TOF study of 2000–20,000-kDa peptides in grape seeds finalized for variety characterization, was recently reported (Pesavento et al., 2008). The MALDI analysis was performed using a 2,5-dihydroxybenzoic acid (DHB) matrix prepared by dissolving 10 mg of DHB in 1 mL of H_2O (0.1% TFA)/acetonitrile 1:1 (v/v) solution. Compared with α-cyano-4-hydroxycinnamic acid (CHCA) and sinapinic acid (SA) matrices, the DHB shown lead to spectra of the highest quality with detection of a high number of proteins and a significantly lower signal-to-noise (S/N) ratio. The worst results were shown by SA, while CHCA seems more effective to promote ionization of low molecular weight peptides. Washing seeds powered by hexane followed by sample dialysis proved to reduce the chemical noise in the low m/z region of the mass spectra. Figure 10.2 shows the MALDI mass spectra of proteins in defatted and dialyzed powder sample of grape seeds from three different grape varieties.

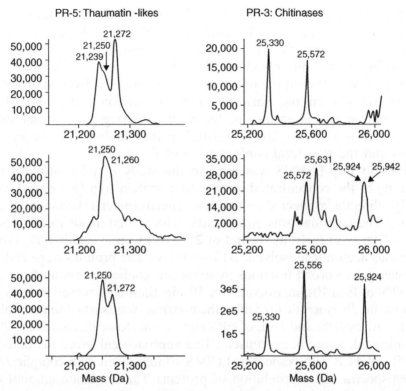

Figure 10.1. The ESI/MS of the pathogenesis-related protein of juice from three different white grape varieties (Hayasaka et al., 2003). (This figure was first published by the *Australian and New Zealand Wine Industry Journal*, May–June 18(3), 2003, reproduced with kind permission of the publisher, Winetitles www.winebiz.com.au).

The seeds peptide MALDI profile proved to be suitable in characterizing grape varieties. This result was confirmed excluding, or by evaluating as being of minor influence, the other factors that might contribute to the protein profile: the harvest year, zone of production and vineyard treatments. To achieve it, seeds of *Raboso Piave* grape samples collected from different vineyards and harvested in different years were studied. The 3D plot of Fig. 10.3 shows that the profile is maintained: in particular the species at *m/z* 6113, which is characteristic of this variety, is detected in the spectra of all samples.

10.2.2   The MS Analysis of Wine Peptides and Proteins

10.2.2.1   Extraction. Since the protein level in wine is normally < 100 mg/L, and interfering substances (e.g., salts, acids, and polyphenols)

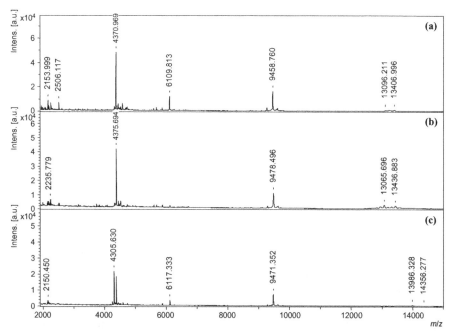

Figure 10.2. The MALDI mass spectra of proteins extracted from *Raboso Piave* (a), *Prosecco* (b), and *Malvasia Nera di Brindisi* (c) grape seeds (measurements in the positive-ion linear mode of ions formed by a pulsed nitrogen laser at λ = 337 nm with a repetition rate of 50 ps, ion source voltage 1: 25, 2: 23.35 kV, ion source lens voltage 10.5 kV, sample mixed with DBH solution 1:1 v/v, 1 µL of mixture deposited on the stainless steel sample holder) (Pesavento et al., 2008).

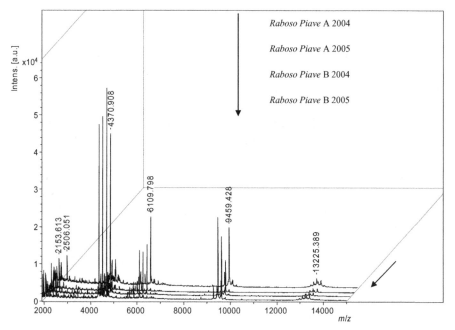

Figure 10.3. Peptide MALDI profiles of *Raboso Piave* seed extracts of grapes harvested from two different vineyards in two different years (Pesavento et al., 2008).

TABLE 10.4. A Method for Quantitative Extraction of Proteins from Wine[a]

1-mL wine added to 10-µL SDS 10%
Incubation at 100 °C for 5 min
Addition of 250-µL KCl 1 M
Incubation at room temperature for 2 h
Centrifugation 14,000 g at 4 °C for 15 min
Washing pellet with KCl 1 M, then centrifugation 14,000 g at 4 °C for 15 min (three times)
Resuspension of protein pellet in a minimum volume of water or buffer

[a]Vincenzi et al., 2005.

are present, it can be useful to perform concentration and purification of the sample prior to analysis. It can be done by precipitation of proteins with ammonium sulfate, solvents (e.g., ethanol or acetone), and acids (e.g., trichloroacetic, sulfosalicylic, or phosphotungstic acid) (Moreno-Arribas et al., 2002). Purification either before or after concentration is performed by dialysis using low cut-off membranes. In any case, the drastic conditions occurring in precipitation with acids could promote denaturation of proteins.

A practical method for concentration and purification of wine is to perform an ultrafiltration followed by dialysis of the sample by low cut-off membranes (i.e., 3.5 kDa). If quantitative recovery is not required, it is possible to remove most polyphenols by passing the sample through a C$_{18}$ cartridge (Curioni et al., 2008). A method for quantitative recovery of proteins from wine and to remove phenols is summarized in Table 10.4 (Vincenzi et al., 2005).

Analysis of peptides is usually performed after separation from the high MW proteins and polysaccharides by ultrafiltration on appropriate cut-off membranes and/or by gel filtration (e.g., on Sephadex LH-20 or G-10 gels) of the concentrated sample (Desportes et al., 2000; Moreno-Arribas et al., 1996; 1998). One or more peptide fractions are achieved, and interfering substances (e.g., salts, amino acids, phenols, organic acids, and sugars) are removed.

A method of sample preparation proposed for MS/MS analysis of proteins in wine is by enzymatic hydrolysis of gel pieces from sodium dodecyl sulfate-polyacrylamide gel electrophoresis (SDS-PAGE) using trypsin (Kwon, 2004). Hydrolysis occurs at the carboxyl side of lysine and arginine residues. Protocols of protein precipitation, enzymatic digestion, and sample preparation are reported in Table 10.5.

10.2.2.2 Analysis. The LC–ESI/MS analysis of di- and tripeptides in wine, after ultrafiltration of a sample with a MW 1000 cut-off

TABLE 10.5. The SDS–PAGE, Enzymatic Digestion, and Sample Preparation Protocols for MS/MS Analysis of Wine Proteins[a]

SDS–PAGE

20-mL wine centrifuged at 10,000 g for 30 min and filtered at 0.22 μm

Salting-out precipitation of proteins by 80% (v/v) saturated ammonium sulfate

Formation of protein pellets by centrifugation at 14,000 g for 30 min,

Formation of soluble proteins by centrifugation at 4000 g for 30 min using a 5-kDa regenerated cellulose membrane centrifugal filter tube

Protein pellets dissolved in 50 μL of SDS buffer containing 12.5% (v/v) of 0.5 M Tris-HCl, 10% (v/v) glycerol, SDS 2% (w/v), 2-mercaptoethanol 5% (v/v), and bromophenol blue 0.001% (w/v), and boiled for 5 min

Cooling and centrifugation at 10,000 g for 1 min

The supernatant loaded onto a homemade 12% (w/v) polyacrylamide gel (80 × 70 × 1.5 mm)

Application of constant voltage 150 V to the gel for 50 min at room temperature

Gel stained by colloidal Coomassie Blue, 1-cm^3 pieces protein bands excised from the gel slab

Pieces destained by NH$_4$HCO$_3$ 25 mM in 50% of methanol/50% water (v/v) solution (three times for 10 min), washed with 10% acetic acid/50% methanol/40% water (v/v/v) solution (three times 1 h each), and swollen in water (two times 20 min each)

Enzymatic Digestion and Sample Preparation

Gel pieces dehydrated with acetonitrile and dried

Hydrated with modified porcine trypsin 10 ng/L in 50 mM NH$_4$HCO$_3$ solution

In-gel digestion at 37 °C overnight

First extraction of tryptic peptides with acetonitrile/H$_2$O/TFA 50:45:5 (v/v/v)

Second extraction of peptides with acetonitrile/H$_2$O/TFA 75:24.9:0.1 (v/v/v)

Extracts combined, dried and cleaned with C$_{18}$ pipette tip

[a]Kwon, 2004.

membrane, can be performed by operating in positive-ion mode using the Phe-Arg dipeptide as an internal standard. With this approach, nine dipeptides were identified in Champagne wine: Ile-Arg was found to be the most abundant (2.2–7.0 mg/L), followed by Arg-Ile, Ile-Val, and Tyr-Lys. The other dipeptides were in concentrations <1 mg/L (De Person et al., 2004). Liquid chromatography is performed by using an alkyl-amide reversed-phase column (150 × 4.6 mm, 5 μm) at 30 °C, and with a binary solvent composed of 2 mM aqueous nonafluoropentanoic acid with pH 2.8 (solvent A) and acetonitrile (solvent B) and elution gradient program from 0 to 10% of B in 4 min, 10–30% of B in 13 min, isocratic 30% B for 13 min, at a flow rate of 1 mL/min. The MS data of some peptides investigated in wine are listed in Table 10.6, where the major collisionally activated dissociation (CAD) fragments

TABLE 10.6. Peptides Identified in Champagne Wine: m/z Values of Protonated Peptides and CAD Fragments[a]

m/z

A-B

Peptide	[M+H]+	[M+H-NH3]+	[A+H]+	[B+H]+ (y1)	[A+H-H2O]+ (b1)	[B+H-H2O]+	[A Im]+ (a1)	[B Im]+	Others
Ile-Val	231						86	72	69
Val-Ile	231			132			72	86	55, 185
Arg-Ile	288	271	175		157			86	70**, 112, 229
Ile-Arg	288			175			86		70**, 116*, 158
Lys-Phe	294	277	147		129		101		84, 259
Phe-Lys	294	277		147		129	120	101	84, 259
Lys-Tyr	310	293	147	182	129		101	136	247, 275
Tyr-Gln	310	293	182	147			136		107, 119, 130, 276
Tyr-Lys	310	293		147		129	136		91, 130
Phe-Arg	322	305		175			120		70**, 116*

A-B-C

Peptide	[M+H]+	[M+H-NH3]+	[M+H-H2O]+	[B+H]+	[C+H]+ (y1)	[A+H-H2O]+ (b1)	[A Im]+ (a1)	Others
Phe-Arg-Arg	478		460	175	175		120	116*, 287, 322, 418
Lys-Met-Asn	392	375			133	129		357, 264, 260, 84

[a]The fragments exceeding 2% of the total fragment ion abundance are reported. [Im]+: [H2N=CH-R]+ with R amino acid residue. In parentheses: typical sequence a, b, y fragment ions. Underlined: fragments formed from [A Im]+ or [B Im]+ by ammonia loss. (*) Fragments formed by guanidino group loss and cyclization of Arg residue according to Dookeran et al. (1996). (**) Ions formed from the corresponding (*) fragments by carboxylic group loss. The MS conditions: triple quadrupole operating in positive-ion mode, selective reaction monitoring (SRM) with collision energy from 5 to 30 eV, curtain and collision gas N_2, ion spray temperature and voltage 400°C and 5.2 kV, declustering potential 20V, focusing potential 200V, entrance potential 10V, dwell time 250 ms (De Person et al., 2004).

are reported. The MS/MS experiments on the $[M+H]^+$ precursor ion were performed with a collision energy from 10 to 50 eV. Dipeptides showed $[M+H–NH_3]^+$, y_1, and a_1 ions as the principal fragments. By increasing the collision energy, the major fragment ion abundance showed a maximum range from 20 eV for most dipeptides, and 30 eV for the tripeptide Phe-Arg-Arg. Two MS/MS transitions and LC retention times of peptides are reported in Table 10.7.

Nano-ESI utilizes a very low solvent flow rate that is carried on by the charge applied to the capillary (see Section 1.1.4). Compared to the standard ESI, the S/N ratio is enhanced. A small aliquot of sample is introduced for ~30 min. This allows to perform several MS/MS sequence tag analyses on a single sample (Ashcroft, 2003).

By nano-LC/MS, 80 peptides corresponding to 20 proteins reported in Table 10.8 (5 derived from grape, 12 from yeast, 2 from bacteria, and 1 from fungi) were identified in a *Sauvignon Blanc* wine (Kwon, 2004). After sample preparation, as described in Table 10.5, 2 μL of peptide solution in acetonitrile/H_2O/acetic acid 2:97.9:0.1 (v/v/v) (solvent A) was analyzed by a capillary C_{18} column (50 mm × 75 μm i.d., 5-μm particle size, 300-Å pore diameter) and peptides were eluted with a gradient from 5 to 80% of solvent B (acetonitrile/H_2O/acetic acid 90:9.9:0.1 v/v/v) for 10 min at a flow rate of 0.3 μL/min. The MS/MS spectra were acquired in a data-dependent mode that determines the masses of the parent ions, and the fragments used for the protein

TABLE 10.7. The MS/MS Transitions and LC Retention Times (RT) of Peptides Studied in *Champagne* Wine[a]

RT (min)	Peptide	Main MS/MS Transition	Confirmation MS/MS Transition Q1 > Q2
5.6	Tyr-Gln	310 > 147	310 > 129
11.9	Ile-Val	231 > 86	231 > 69
12.2	Lys-Met-Asn	392 > 129	392 > 264
12.9	Val-Ile	231 > 72	231 > 132
12.9	Tyr-Lys	310 > 129	310 > 147
13.7	Ile-Arg	288 > 175	288 > 86
13.9	Lys-Tyr	310 > 129	310 > 147
14.3	Phe-Lys	294 > 129	294 > 84
14.5	Phe-Arg (I.S.)	322 > 175	322 > 120
14.9	Arg-Ile	288 > 175	288 > 86
15.1	Lys-Phe	294 > 129	294 > 84
15.7	Phe-Arg-Arg	478 > 175	478 > 322

[a]De Person et al., 2004.

TABLE 10.8. Proteins Identified by Nano-LC/MS in a Sauvignon Blanc Wine[a]

Identified Protein	Mass (kDa)	gi Number	Identified Peptide	Species
Laccase 2	63.4	15022489	(K)SPANFNLVNPPR (R)YDSSSTVDPTSVGVTPR	B. fuckeliana
Succinyl-CoA-synthetase	41.2	26990878	(K)ATIDPLVGAQPFQGR (K)ELYLGAVVDR (R)LEGNNAELGAK (K)QLFAEYGLPVSK	P. putida KT2440
Translation elongation factors	77.1	23470603	(K)IATDPFVGTLTFVR (K)LAQEDPSFR	P. syringae pv. syringae B728a
YJU1	21.8	4814	(K)DGSSYIFSSK (K)EGSESDAATGFSIK (K)FDDDKYAWNEDGSFK (K)LGSGSGSFEATITDDGK (R)SGSDLQYLSVYSDNGTLK	Saccharomyces cerevisiae
Endo-β-1,3-glucanase	34.1	6321721	(K)AALQTYLPK (K)ESTVAGFLVGSEALYR (K)HWGVFTSSDNLK (K)(KESTVAGFLVGSEALYR (R)NDLTASQLSDK (NDLTASOLSDKINDVR) (K)STSDYETELQALK (R)SWADISDSDGK	S. cerevisiae
GP38	37.3	297485	(R)GVLSVTSDK (K)NAVGAGYLSPIK (K)RGVLSVTSDK (K)SALESIFP (K)WFFDASKPTLISSDIIR	S. cerevisiae
Target of SBF	47.9	6319638	(K)AAVIFNSSDK (R)EGIPAYHGFGGADK (K)USHIHDGODGGTQDYFERPTDGTLK	S. cerevisiae

Protein	MW	GI number	Peptides	Organism
ECM33 protein precursor	48.3	1351738	(K)KVNVFNINNNR (K)VGQSLSIVSNDELSK (K)VNVFNINNNR	*S. cerevisiae*
Putative glycosidase	49.9	6320795	(K)NSGGTVLSSTR (K)YQYPQTPSK	*S. cerevisiae*
Acid phosphatase	52.7	6319568	(K)QSETQDLK (K)YDTTYLDDIAK (R)YSYGQDLVSFYQDGPGYDMIR	*S. cerevisiae*
Putative glycosidase	52.7	6321628	(R)GEFHGVDTPTDK (K)TTWYLDGESVR (K)VIVTDYSTGK	*S. cerevisiae*
β-1,3-Glucanosyltransferase	59.5	6323967	(K)IPVGYSSNDDEDTR (R)KIPVGYSSNDDEDTR (K)KLNTNVIR (K)LNTNVIR (K)TLDDFNNYSSEINK (K)YGLVSIDGNDVK	*S. cerevisiae*
Invertase 4 precursor	60.5	124705	(K)FSLNTEYQANPETELINLK (K)GLEDPEEYLR (K)IEIYSSDDLK (R)KFSLNTEYQANPETELINLK	*S. cerevisiae*
Endo-β-1,3-glucanase	63.5	6320467	(R)QFIEAQLATYSSK (K)SPVVGIQIVNEPLGGK (K)TWITEDDFEQIK	*S. cerevisiae*
Daughter cell specific secreted protein	12.1	6324395	(R)DVANPSEKDEYFAQSR (K)DWVNSLVR (K)IGSSVGFNTIVSESSSNLAQGILK (K)NEESSEDYNFAYAMK (R)SETFVEEEWQTK	*S. cerevisiae*

TABLE 10.8. (*Continued*)

Identified Protein	Mass (kDa)	gi Number	Identified Peptide	Species
Basic extracellular β-1,3-glucanase precursor	14.6	4151201	(K)HWGLFLPNK (K)TYNSNLIQHVK	*V. vinifera*
Putative thaumatin-like protein	20.1	7406714	(R)CPDAYSYPK (R)TNCNFDASGNGK (K)TRCPDAYSYPK	*V. vinifera*
WTL1	23.9	2213852	(K)CTYTVWAAASPGGGR (R)LDSGQSWTITVNPGTTNAR (R)RLDSGQSWTITVNPGTTNAR	*V. vinifera*
Class IV endochitinase	27.5	2306813	(R)AAFLSALNSYSGFGNDGSTDANK (R)AAFLSALNSYSGFGNDGSTDANKR	*V. vinifera*
Vacuolar invertase 1	71.5	1839578	(R)DPTTMWVGADGNWR (K)GWASLQSIPR (R)ILYGWISEGDIESDDLK (K)KGWASLQSIPR (K)TFFCTDLSR (R)VLVDHSIVEGFSQGGR (R)ILYGWISEGDIESDDLKK (R)SSLAVDDVDQR (R)TAFHFQPEK (K)YENNPVMVPPAGIGSDDFR (R)VYPTEAIYGAAR (R)SCITTRVYPTEAIYGAAR	*V. vinifera*

[a]Reprinted from *Journal of Agricultural and Food Chemistry* 52, Sung Wong Know, Profiling of soluble proteins in wine by nano-high-performance liquid chromatography/tandem mass spectrometry. p. 7260, Copyright © 2004, with permission from American Chemical Society.

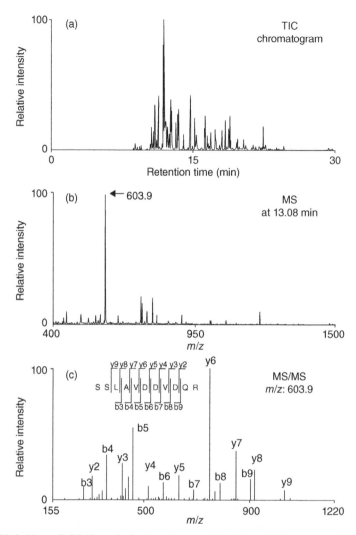

Figure 10.4. Nano-LC/MS analysis of a wine peptide: (a) total ion current (TIC) chromatogram of the tryptic digest (MW range 60–75 kDa in SDS–PAGE); (b) *m/z* 400–1500 MS spectrum of the signal at the retention time of 13.08 min; (c) MS/MS spectrum of the ion at *m/z* 603.9 identified the peptide SSLAVDDVDQR. (Reprinted from *Journal of Agricultural and Food Chemistry* 52, Sung Wong Kwon, Profiling of soluble proteins in wine by nano-high-performance liquid chromatography/tandem mass spectrometry, p. 7262, Copyright © 2004, with permission from American Chemical Society.)

identification. Figure 10.4 shows the chromatogram, mass, and MS/MS spectra of an identified peptide. In this study, the three strongest parent ions in the full MS spectrum were selected and fragmented. The m/z 700–1300 spectra were recorded and each MS/MS spectrum was checked against the NCBI nonredundant protein sequence database using the Knexus program. A manual confirmation of protein identification was performed using as criteria: (1) the major isotope-resolved peaks should match fragment masses of the identified peptide; (2) y, b, and a ions and their water or amine loss peaks (Table 10.6) are considered; (3) to emphasize the isotope-resolved peaks; (4) seven major isotope-resolved peaks are matched to theoretical masses of the peptide fragments; (5) all redundant proteins are removed by confirming the unique peptides; (6) to confirm the unique peptides, all amino acid sequences of the identified proteins are listed and each peptide is examined.

Two different methods of sample preparation can be performed for MALDI–TOF analysis of proteins in wine: (1) the wine sample is mixed with an SA saturated acetonitrile/water/TFA solution and 2 µL of solution is applied to the sample holder and dried; (2) 50 mL of wine are lyophilized, the residue is dissolved in a water/urea solution, proteins are precipitated with ethanol, and again dissolved in urea. After a second precipitation, the residue is dissolved in an H_2O/TFA solution and mixed with SA. Better resolution of the peak in the m/z 15,000–18,000 range was found using the latter procedure (Szilàgyi et al., 1996). For analysis of lower MW proteins (0–15 kDa), CHCA is usually used as the matrix (Weiss et al., 1998). In general, of the proteins in wine that have masses between 7 and 86 kDa, 21.3 kDa are the major proteins, and other significant masses of 7.2, 9.1, 13.1, and 22.2 kDa were found. Several equally spaced peaks observed suggest the presence of a glycoprotein with a difference between the neighboring peaks of 162 Da that correspond to a hexose residue. At least 22 components differing in the number of sugar residues were observed for this glycoprotein. Two further glycoproteins in the m/z 8,800–9,500 and 10,500–12,200 ranges containing 5 and 11 sugar units, respectively, were observed. Formation of multiply charged ions and dimers can be influenced from the matrix and laser energy. Since desorption–ionization depends on the size and nature of individual proteins, it is not possible to make a direct comparison of relative intensities between different proteins, and an accurate protein quantification is possible only with the use of internal standards very similar to each analyte.

Also, surface-enhanced-laser–desorption/ionization time-of-flight mass spectrometry (SELDI–TOF–MS) was applied to analysis of peptides and proteins in wine (Weiss et al., 1998). This affinity MS (AMS) technique utilizes functional groups on inert platforms to capture molecules from the sample. The use of agarose beads containing an iminodiacetate-chelated copper ion as a functional group (IDA–Cu), which interacts with specific amino acid residues of wine proteins, induces formation of interactions with histidine, lysine, tryptophan, cysteine, aspartic acid, and glutamic acid. Wine proteins and peptides determined by SELDI–TOF–MS show peaks quite similar to MALDI–TOF. By coupling the two techniques, MALDI–TOF shows the greatest number of peaks, while SELDI–TOF provides an increased sensitivity, as well as selectivity, for some protein fractions.

REFERENCES

Ashcroft, A.E. (2003). Protein and peptide identification: the role of mass spectrometry in proteomics, *Nat. Prod. Rep.*, **20**, 202–215.

Berrocal-Lobo, M., Molina, A.A., and Solano, R. (2002). Constitutive expression of ETHYLENE-RESPONSE-FACTOR 1 in Arabidopsis confers resistance to several necrotrophic fungi, *Plant J.*, **29**, 23–32.

Blumwald, E., Aharon, G.S., and Lam, B.C.H. (1998). Early signal transduction pathways in plant-pathogen interactions, *Trends Plant Sci.*, **3**, 342–346.

Boller, T. (1987). Hydrolytic enzymes in plant disease resistance, *Plant-Microbe Interactions: Molecular and Genetic Perspectives*, **2**, 385–413.

Brissonnet, F. and Maujean, A. (1993). Characterization of foaming proteins in a champagne base wine, *Am. J. Enol. Vitic.*, **44**, 297–301.

Busam, G., Kassemeyer, H.H., and Matern, U. (1997). Differential expression of chitinases in Vitis vinifera 1. Responding to systemic acquired resistance activators or fungal challenge, *Plant Physiol.*, **115**, 1029–1038.

Cessna, S.G., Sears, V.E., Dickman, M.B., and Low, P.S. (2000). Oxalic acid, a pathogenicity factor for Sclerotinia sclerotiorum, supresses the oxidative burst of the host plant, *Plant Cell*, **12**, 2191–2199.

Cheong, N.E., Choi, Y.O., Kim, W.Y., Kim, S.C., Bae, I.S., Cho, M.J., Hwang, I., Kim, J.W., and Lee, S.Y. (1997). Purification and characterization of an antifungal PR-5 protein from pumpkin leaves, *Mol. Cells*, **7**, 214–219.

Curioni, A., Vincenzi, S., and Flamini, R. (2008). Proteins and Peptides in Grape and Wine, In: Hyphenated Techniques in Grape & Wine Chemistry, Flamini, R. (ed.) John Wiley & Sons, Ltd., pp. 249–288.

Dambrouck, T., Marchal, R., Cilindre, C., Parmentier, M., and Jeandet, P. (2005). Determination of the grape invertase content (using PTA–ELISA) following various fining treatments versus changes in the total protein content of wine. Relationships with wine foamability, *J. Agric. Food Chem.*, **53**, 8782–8789.

De Person, M., Sevestre, A., Chaimbault, P., Perrot, L., Duchiron, F., and Elfakir, C. (2004). Characterization of low-molecular weight peptides in champagne wine by liquid chromatography/tandem mass spectrometry, *Anal. Chim. Acta*, **520**, 149–158.

Dempsey, D.A., Shah, J., and Klessing, D.F. (1999). Salicylic acid and disease resistance in plants, *Crit. Rev. Plant Sci.*, **18**, 547–575.

Derckel, J.P., Legendre, L., Audran, J,C., Haye, B., and Lambert, B. (1996). Chitinases of the grapevine (Vitis vinifera L.): five isoforms induced in leaves by salicylic acid are constitutively expressed in other tissues, *Plant Sci.*, **119**, 31–37.

Desportes, C., Charpentier, M., Duteurtre, B., Maujean, A., and Duchiron, F. (2000). Liquid chromatographic fractionation of small peptides from wine, *J. Chrom. A*, **893**, 281–291.

Deytieux, C., Geny, L., Lapaillerie, D., Claverol, S., Bonneau, M., and Donèche, B. (2007). Proteome analysis of grape skins during ripening, *J. Experim. Botany*, **58**, 1851–1862.

Dookeran, N.N., Yalcin, T., and Harrison, A.G. (1996). Fragmentation reactions of protonated α-amino acids, *J. Mass Spectrom.*, **31**, 500–508.

Ebel, J. and Cosio, E.G. (1994). Ellicitors of plant defense responses, *Int. Rev. Cyto.*, **148**, 1–36.

Famiani, F., Walker, R.P., Tecsi, L., Chen, Z.H., Proietti, P., and Leegood, R.C. (2000). An immunoistochemical study of the compartmentation of metabolism during the development of grape (Vitis vinifera L.) berries, *J. Experim. Botany*, **51**, 675–683.

Flamini, R. and De Rosso, M. (2006). Mass Spectrometry in the Analysis of Grape and Wine Proteins, *Exp. Rev. Proteom.*, **3**, 321–331.

Gil-ad, N. and Mayer, M. (1999). Evidence for rapid breakdown of hydrogen peroxide by Botrytis cinerea, *FEMS (Fed. Eur. Microbiol. Soc.)*, Lett. **176**, 455–461.

Graham, L.S. and Sticklen, M.B. (1994). Plant chitinases, *Can. J. Bot.*, **72**, 1057–1083.

Glinski, M. and Weckwerth, W. (2006). The role of mass spectrometry in plant systems biology, *Mass Spectrom. Rev.*, **25**, 173–214.

Gonçalves, F., Heyraud, A. de Pinho, M.N., and Rinaudo, M. (2002). Characterization of white wine mannoproteins, *J. Agric. Food Chem.*, **50**, 6097–6101.

Hayasaka, Y., Adams, K.S., Pocock, K.F., Baldock, G.A., Waters, E.J., and Hoj, P.B. (2001). Use of electrospray mass spectrometry for mass determination of grape (Vitis vinifera) juice pathogenesis-related proteins: a potential tool for varietal differentiation, *J. Agric. Food Chem.*, **49**, 1830–1839.

Hayasaka, Y., Baldock, G.A., Pocock, K.F., Waters, E.J. Pretorius, I., and Hoj, P.B. (2003). Varietal differentiation of grape juices by protein fingerprinting, *Aust. New Zealand Wine Ind. J.*, **18**(3), 27–31.

Hsu, J.C. and Heatherbell, D.A. (1987). Isolation and characterization of soluble proteins in grapes, grape juice and wine, *Am. J. Enol. Vitic.*, **38**, 6–10.

Jacobs, A.K., Dry, I.B., and Robinson, S.P. (1999). Induction of different pathogenesis-related cDNAs in grapevine infected with powdery mildew and treated with ethephon, *Plant Pathol.*, **48**, 325–336.

Kwon, S.W. (2004). Profiling of soluble proteins in wine by nano-high-performance liquid chromatography/tandem mass spectrometry, *J. Agric. Food Chem.*, **52**, 7258–7263.

Linthorst, H.J.M. (1991). Pathogenesis-related proteins of plants, *Crit. Rev. Plant Sci.*, **10**, 123–150.

Monteiro, S., Piçarra-Pereira, M.A., Teixeira, A.R., Loureiro, V.B., and Ferreira, R.B. (2003). Environmental conditions during vegetative growth determine the major proteins that accumulate in mature grapes, *J. Agric. Food Chem.*, **51**, 4046–4053.

Moreno-Arribas, M.V., Pueyo, E., and Polo, M.C. (2002). Analytical methods for the characterization of proteins and peptides in wines, *Anal. Chim. Acta*, **458**, 63–75.

Moreno-Arribas, V., Pueyo, E., and Polo, M.C. (1996). Peptides in must and wines. Changes during the manufacture of Cavas (Sparkling wines), *J. Agric. Food Chem.*, **44**, 3783–3788.

Moreno-Arribas, V., Pueyo, E., Polo, M.C., and Martin-Alvarez, P.J. (1998). Changes in the amino acid composition of the different nitrogenous fractions during the aging of wine with yeasts, *J. Agric. Food Chem.*, **46**, 4042–4051.

Nümberger, T. and Scheel, D. (2001). Signal trasmission in the plant immune response, *Trends Plant Sci.*, **6**, 372–379.

Opiteck, G.J. and Scheffler, J.E. (2004). Target class strategies in mass spectrometry-based proteomics, *Exp. Rev. Proteom.*, **1**(1), 57–66.

Pesavento, I.C., Bertazzo, A., Flamini, R., Dalla Vedova, A., De Rosso, M., Seraglia, R., and Traldi, P. (2008). Differentiation of Vitis vinifera Varieties by MALDI–MS analysis of the Grape Seeds Proteins, *J. Mass Spectrom.*, **43**(2), 234–241.

Pocock, K.F., Hayasaka, Y., McCarthy, M.G., and Waters, E.J. (2000). Thaumatin-like proteins and chitinases, the haze forming proteins of wine, accumulate during ripening of grape (Vitis vinifera) berries and drought stress does not affect the final levels per berry at maturity, *J. Agric. Food Chem.*, **48**, 1637–1643.

Pocock, K.F., Hayasaka, Y., Peng, Z., Williams, P.J., and Waters, E.J. (1998). The effect of mechanical harvesting and long-distance transport on the concentration of haze-forming proteins in grape juice, *Aust. J. Grape Wine Res.*, **4**, 23–29.

Pocock, K.F. and Waters, E.J. (1998). The effect of mechanical harvesting and transport of grapes, and juice oxidation, on the protein stability of wines, *Aust. J. Grape Wine Res.*, **4**, 136–139.

Punja, Z.K. and Zhang, Y.Y. (1993). Plant chitinases and their roles in resistance to fungal disease, *J. Nematol.*, **25**, 526–540.

Robinson, S.P., Jacobs, A.K., and Dry, I.B. (1997). A class IV chitinase is highly expressed in grape berries during ripening, *Plant Physiol.*, **114**, 771–778.

Sarry, J.E., Sommerer, N., Sauvage, F.X., Bergoin, A., Rossignol, M., Albagnac, G., and Romieu, C. (2004). Grape berry biochemistry revisited upon proteomic analysis of the mesocarp, *Proteomics*, **4**, 201–215.

Simó, C., Elvira, C., González, N., San Román, J., Barbas, C., and Cifuentes, A. (2004). Capillary electrophoresis-mass spectrometry of basic proteins using a new physically adsorbed polymer coating. Some applications in food analysis, *Electrophoresis*, **25**, 2056–2064.

Staples, R.C. and Mayer, A.M. (1995). Putative virulence factor of Botrytis cinerea acting as a wound pathogen, *FEMS (Fed. Eur. Microbiol. Soc.) Lett.*, **134**, 1–7.

Stintzi, A., Heitz, T., Prasad, V., Wiedemannmerdinoglu, S., Kauffman, S., Geoffroy, P., Legrand, M., and Friting, B. (1993). Plant pathogenesis-related proteins and their role in defense against pathogens, *Biochemie*, **75**, 687–706.

Szilágyi, Z., Vas, G., Mády, G., and Vékey, K. (1996). Investigation of macromolecules in wines by matrix-assisted laser desorption/ionization time-of-flight mass spectrometry, *Rapid Commun. Mass Spectrom.*, **10**, 1141–1143.

Tattersall, D.B., van Heeswijck, R., and Høj, P.B. (1997). Identification and characterization of a fruit-specific, thaumatin-like protein that accumulates at very high levels in conjunction with the onset of sugar accumulation and berry softening in grapes, *Plant Physiol.*, **114**, 759–769.

Ton, J., Van Pelt, J.A., Van Loon, L.C., and Pieterse, C.M.J. (2002). Differential effectiveness of salicylate-dependent and jasmonate/ethylene-dependent induced resistance in Arabidopsis, *Mol. Plant-Microbe Interact.*, **15**, 27–34.

Vincent, D., Wheatley, M.D., and Cramer, G.R. (2006). Optimization of protein extraction and solubilization for mature grape berry cluster, *Electrophoresis*, **27**, 1853–1865.

Vincenzi, S., Mosconi, S., Zoccatelli, G., Dalla Pellegrina, C., Veneri, G., Chignola, R., Peruffo, A.D.B., Curioni, A., and Rizzi, C. (2005). Development of a new procedure for protein recovery and quantification in wine, *Am. J. Enol. Vitic.*, **56**, 182–187.

Waters, E.J., Hayasaka, Y., Tattersall, D.B., Adams, K.S., and Williams, P.J. (1998). Sequence analysis of grape (Vitis vinifera) berry chitinases that cause haze formation in wines, *J. Agric. Food Chem.*, **46**, 4950–4957.

Waters, E.J., Shirley, N.J., and Williams, P.J. (1996). Nuisance proteins of wine are grape pathogenesis related proteins, *J. Agric. Food Chem.*, **44**, 3–5.

Waters, E.J. and Williams, P.J. (1996). Protein instability in wines. *Proceedings of the 11th International Oenological Symposium*, June 3–5, 1996, Sopron, Hungary; Internationale Vereinigung für Oenologie, Betriebsfürung und Winemarketing: Breisach, Germany.

Waters, E.J., Wallace, W., and Williams, P.J. (1992). Identification of heat-unstable wine proteins and their resistance to peptidases, *J. Agric. Food Chem.*, **40**, 1514–1519.

Weiss, K.C., Yip, T.T., Hutchens, T.W., and Bisson, L.F. (1998). Rapid and sensitive fingerprinting of wine proteins by matrix-assisted laser desorption/ionisation time-offlight (MALDI–TOF) mass spectrometry, *Am. J. Enol. Vitic.*, **49**(3), 231–239.

Yokotsuka, K. and Fukui, M. (2002). Changes in nitrogen compounds in berries of six grape cultivars during ripening over two years, *Am. J. Enol. Vitic.*, **53**, 69–77.

INDEX

Mass Spectrometry in Grape and Wine Chemistry, by Riccardo Flamini
and Pietro Traldi
Copyright © 2010 John Wiley & Sons, Inc.

WILEY-INTERSCIENCE SERIES IN MASS SPECTROMETRY

Series Editors

Dominic M. Desiderio
Departments of Neurology and Biochemistry
University of Tennessee Health Science Center

Nico M. M. Nibbering
Vrije Universiteit Amsterdam, The Netherlands